Advances in Computational Intelligence for Climate Change Security and Sustainability

Imdad Ali Shah
School of Computing Science, Taylor's University, Malaysia

N.Z. Jhanjhi
School of Computing Science, Taylor's University, Malaysia

IGI Global
Scientific Publishing
Publishing Tomorrow's Research Today

Vice President of Editorial	Melissa Wagner
Director of Acquisitions	Mikaela Felty
Director of Book Development	Jocelynn Hessler
Production Manager	Mike Brehm
Cover Design	Jose Rosado

Published in the United States of America by
IGI Global Scientific Publishing
701 East Chocolate Avenue
Hershey, PA, 17033, USA
Tel: 717-533-8845
Fax: 717-533-7115
Website: https://www.igi-global.com E-mail: cust@igi-global.com

Library of Congress Cataloging-in-Publication Data

LCCN: 2024054524 (CIP Data Pending)
ISBN13: 979-8-3693-9132-7
Isbn13Softcover: 979-8-3693-9133-4
EISBN13: 979-8-3693-9134-1

British Cataloguing in Publication Data
A Cataloguing in Publication record for this book is available from the British Library.

Table of Contents

Detailed Table of Contents

Small and Medium Enterprises (SMEs) are the backbone of Pakistan's economy. Yet these businesses face issues because of climate change and restricted flexibility. However, the incorporation of advanced computational intelligence (ACI) techniques like data analytics, machine learning, and predictive modeling improves the environmental sustainability, assesses the environmental risks, maximizes the utilization of resources, and applies the flexible strategies suitable to their working environments. The study presents a new model that integrates digital transformation with environmental sustainability, offering a structure for the SMEs to shift toward climate-proof models. Findings emphasize how information-driven insights can aid decision-making and decrease the carbon footprint. This research highlights the essential role of ACI in authorizing the SMEs to thrive during the weather uncertainty, and the creativity of the ecosystems. Keywords: Advanced computational intelligence, Sustainability Performance, SMEs, Pakistan.

Climate change poses an unprecedented threat to ecosystems, human livelihoods, and global sustainability. Addressing this challenge requires innovative approaches that go beyond traditional methods. This chapter also focuses on the role of advanced Computational Intelligence encompassing Machine Learning, deep learning, evolutionary algorithms, and fuzzy systems in facilitating effective climate change adaptation strategies. By integrating large-scale climate data, environmental models, and socioeconomic parameters, ACI enables predictive analytics, risk assessment, and decision-support systems tailored to dynamic climate conditions. Furthermore, this chapter also highlights recent advancements from the past five years, showcasing how computational intelligence is being leveraged across disciplines to build resilience and support sustainable adaptation pathways. Challenges such as data uncertainty, model interpretability, and ethical considerations are discussed, along with future research directions.

This chapter focuses on ecological control methods as critical instruments to enforce legal compliance with environmental standards, and for sustainable management of our natural resources. Primary mechanisms of compliance such as regulatory sanctions, market instruments, technological monitoring and voluntary compliance. Various methodological approaches are also discussed and these include scientific and analytical methods. Six types of control are addressed - inspection, investigation, administrative supervision, monitoring, environmental assessment and audit, the last of which is presented as the most comprehensive method. Similarly there is attention in the chapter to the preparatory phase of control which includes information collection and analysis, comparison of data to find anomalies which directs the planning of verification actions. General controls aim to guarantee the legality, effectiveness, efficiency and transparency of environmental action, while detecting any malfunctions or abuses that may ultimately be necessary to protect the environment.

Subsequently, AI has been instrumental toward suppressing the climate change issue considering it sets creative techniques across the various quandaries associated with the global changes. The following areas of use have been identified: energy control, reduction of greenhouse gas emissions, climatic simulation, and prediction of storms. The strength of AI that is its suitability to work on vast, intertwined datasets can be used to explain patterns and findings which would help researchers and policymakers in eradicating climate change. Looking at the impact of climate change, AI is employed to improve energy, and decrease its usage as well as adopt to green energy. AI systems, by nature, are highly dependent on data, underline the motivating computational requirements and energy utilization, in general, and are linked with the carbon footprint. In addition, lack of data, data quality issues, and relevance and originality of professional practice could also avert the application of intelligent climate solutions based on artificial intelligence, particularly in the developing world.

Heavy metal contamination, particularly from nickel (Ni) and cobalt (Co), is a significant issue in agricultural regions, especially those influenced by industrial activities. This study investigates the impact of Ni and Co stress on maize (Zea mays) by evaluating the plant's morphological, physiological, and biochemical responses to varying metal concentrations. We measured leaf number, leaf length, root length, fresh and dry weights of aerial and root parts, as well as chlorophyll content, proline, total sugars, polyphenols, and flavonoids. The results reveal that higher concentrations of Ni and Co caused notable reductions in growth and alterations in metabolic processes, with increased levels of stress-related markers. This research emphasizes the detrimental effects of heavy metals on maize, highlighting the need for sustainable agricultural practices to mitigate pollution. By understanding the impact of heavy metals on crop health, this study contributes to the broader goal of enhancing food security and improving resilience in regions facing soil contamination challenges.

This chapter explores the salinity stress response of Lavandula dentata under monoculture and co-cultivation with halophytes (Atriplex prostrata, Plantago macrorhiza) using an integrative approach combining explainable artificial intelligence, deep learning, and causal inference. Dimensionality reduction (PCA, t-SNE, UMAP) and clustering algorithms (K-Means, DBSCAN) identified stress- and culture-dependent phenotypic patterns, while supervised learning models (Random Forest, MLP) predicted cultivation conditions with high accuracy, highlighting the key role of root volume, proline, and water content. SHAP values offered model interpretability, and causal inference quantified the direct effects of co-cultivation on biomass. Results revealed that co-cultivation mitigates certain stress traits but may reduce biomass, suggesting trade-offs in physiological adaptation. This chapter demonstrates how data-driven frameworks can support agroecological strategies for salt-affected environments under climate change.

Salt stress is one of the major challenges for sustainable agriculture, particularly in areas vulnerable to the effects of climate change. This chapter explores the impact of cultivation in association with halophytes on the salt tolerance of Rosmarinus officinalis, a glycophyte plant. Experience has shown that salt stress negatively affects the growth of rosemary. However, association with halophytes significantly improved morphological and physiological parameters by regulating cell hydration,

Preface

In an era marked by escalating climate challenges and environmental uncertainty, the role of advanced computational intelligence (ACI) has emerged as a cornerstone for sustainable adaptation and mitigation strategies. The intersection of artificial intelligence, environmental science, and data-driven decision-making now offers transformative tools to address climate change, resource management, and global sustainability.

This edited volume, *Advances in Computational Intelligence for Climate Change Security and Sustainability*, brings together pioneering research, case studies, and technological insights from scholars and institutions across the globe. The chapters presented herein explore how ACI, including machine learning, deep learning, geospatial modeling, and explainable AI, is being harnessed to understand, predict, and respond to climate-related challenges in various sectors—from agriculture and industry to energy systems and environmental policy.

The book's contributors include interdisciplinary experts from Pakistan, Malaysia, Morocco, India, China, and the United States, reflecting a global commitment to climate resilience through innovation. Their collective work emphasizes practical, scalable solutions for both developing and developed nations and aligns with the United Nations' Sustainable Development Goals (SDGs), particularly those targeting climate action, industry innovation, sustainable agriculture, and clean energy.

This book is intended for researchers, practitioners, policymakers, and students who seek an in-depth understanding of how computational intelligence can be applied in real-world environmental contexts. By bridging scientific inquiry with actionable strategies, we aim to inspire collaborative, data-driven efforts toward a more secure and sustainable planet.

CHAPTER OVERVIEW

- **Chapter 1** explores a new paradigm for climate resilience in Pakistani SMEs through the integration of ACI, proposing a model that enhances sustainability, resource optimization, and risk mitigation.
- **Chapter 2** provides a foundational review of recent advancements in ACI for climate adaptation, highlighting tools such as deep learning and fuzzy systems for climate modeling and decision support.
- **Chapter 3** introduces geospatial modeling techniques for flood risk assessment, particularly within the Indus River Basin, offering climate-informed planning tools for infrastructure and land use.
- **Chapter 4** addresses environmental control and enforcement mechanisms, including scientific and regulatory methods essential for maintaining ecological integrity.
- **Chapter 5** details the diverse roles of AI in climate change mitigation, from greenhouse gas monitoring to energy optimization and predictive simulation models.
- **Chapter 6** examines the biochemical and morphological impact of heavy metals on maize crops, providing insights into sustainable agricultural practices under pollution stress.
- **Chapter 7** employs explainable AI and deep learning to analyze salinity stress responses in lavender under co-cultivation, offering strategies for sustainable agroecological systems.
- **Chapter 8** investigates halophyte intercropping as a biological solution to soil salinity, showing enhanced plant tolerance and resilience under climate stress.
- **Chapter 9** concludes with a review of Moroccan climatic impacts on solar mirror durability, offering vital insights for renewable energy infrastructure in arid regions.

This volume showcases how the thoughtful application of computational intelligence can enable societies to not only adapt to climate change but also to thrive in spite of it. We thank the contributors for their significant efforts and invite readers to engage with the ideas presented here in the spirit of innovation, resilience, and sustainability.

Chapter 1
Advanced Computational Intelligence and Climate Change Adaptation for Sustainability in SMEs of Pakistan:
A New Paradigm for Business Resilience

Muhammad Naeem
https://orcid.org/0000-0001-6678-3536
Sindh Institute of Management and Technology, Karachi, Pakistan

Salman Memon
Emaan Institute of Management and Sciences, Karachi, Pakistan

Anita Larik
Sindh Institute of Management and Technology, Karachi, Pakistan

Syed Rizwan Mehdi
Sindh Institute of Management and Technology, Karachi, Pakistan

Hasan Ahmed Faridi
Sindh Institute of Management and Technology, Karachi, Pakistan

Khalida Khan
Mehran University Engineering and Technology, Jamshoro, Pakistan

Sana Zafar
Iqra University, Pakistan

Manoj Kumar
Indus Univeristy, Pakistan

ABSTRACT

Small and Medium Enterprises (SMEs) are the backbone of Pakistan's economy.

DOI: 10.4018/979-8-3693-9132-7.ch001

Yet these businesses face issues because of climate change and restricted flexibility. However, the incorporation of advanced computational intelligence (ACI) techniques like data analytics, machine learning, and predictive modeling improves the environmental sustainability, assesses the environmental risks, maximizes the utilization of resources, and applies the flexible strategies suitable to their working environments. The study presents a new model that integrates digital transformation with environmental sustainability, offering a structure for the SMEs to shift toward climate-proof models. Findings emphasize how information-driven insights can aid decision-making and decrease the carbon footprint. This research highlights the essential role of ACI in authorizing the SMEs to thrive during the weather uncertainty, and the creativity of the ecosystems. *Keywords: Advanced computational intelligence, Sustainability Performance, SMEs, Pakistan.*

1. INTRODUCTION

ACI or Artificial Intelligence (AI) is rising too rapidly throughout a huge number of industries globally, evolving business operations, improving decision-making, and enhancing sustainability practices (Naeem et al., 2025a; Fateh et al., 2025). AI has been delivering solutions to every issue being asked to resolve and has emerged as an essential tool for firms striving to enhance efficiency, reduce costs, and achieve sustainability goals (Kulkov et al., 2024; Shah et al., 2025). In terms of SMEs, AI delivers unlimited benefits to effectively allocate resources, enhance productivity, and maximize sustainable performance. Compared to large firms that have financial resources allocated to sustainability teams, but in terms of SMEs face significant challenges in combining sustainable actions because of finite fiscal assets, poor technology, and insufficient expertise (Naeem et al., 2025b; Jhanjhi et al., 2025a). Although AI tools like automation, predictive analytics, and resourceful monitoring can remove this research gap by offering cost-effective results for improving the sustainability of environmental, social, and economic. The rising worldwide trend on sustainability outcomes, assisted by global enterprises like the Sustainable Development Goals (SDGs), the Paris Climate Accord and has emphasized the necessity for SMEs to adopt AI tools and methods for long-term success (Ameh, 2024; Shah et al., 2024a).

The SMEs of Pakistan perform an important role in the economic improvement of nations, contributing to the GDP growth rate, employment creation, and industrial evolution in Pakistan (Anser et al., 2024; Shah et al., 2024b, c; Farooq et al., 2023; Farooq & Ahmad, 2023). However, the SMEs of Pakistan tackle a huge amount of challenges, involving energy mismanagement, wastage of resources, high manufacturing costs, and continuous barriers from the government. Sustainability is a primary focus all over the world, many SMEs strive to balance profits with social responsibilities (Prasanna et al., 2019; Naeem et al., 2024). The utilization of AI in SME practices gives a huge opportunity to reduce the complexities through the implementation of novel AI-powered energy management, optimizing supply chain methods via AI, and improving the decision-making process through real-time data analytics (Roberts, 2025). Regardless of the advantages of AI, the implementation of AI for the sustainability solutions of SMEs in Pakistan is in its initial stages. Parameters like the shortage of surrounding awareness, major initial expenses, and issues due to the impact of AI on jobs have delayed its adoption in the SMEs of Pakistan. Through studying the literature, it was assessed that AI can assist in achieving sustainable performance in SMEs, which is important for the implementation of targeted strategies that increase AI adoption and awareness of its benefits (Naeem et al., 2025c).

Sustainable performance using AI in the SME sector provides a complete system that involves economic advancements with social and environmental responsibilities (Ali et al., 2023). This includes the utilization of resources in an efficient manner for production processes, mitigating waste, emphasizing fair labor principles, and using operations with ethical practices in the firms. Regardless of huge enterprises, which have the most accessible technological and financial resources, SMEs usually have to face a tough time in the implementation of sustainability in their operational models (Gregurec et al., 2021). AI assists with the help of a combination of novel tools like data-driven tactics, automation in sustainability methods, and improving efficiency to allocate resources wisely (Zong & Guan, 2024). Predictive maintenance through the integration of AI can assist SMEs in reducing the failure of equipment and downtime, which helps to decrease operational costs and enhance efficiency concerning cost and energy (Waqar et al., 2024). Likewise, AI can assist in forecasting demand, The AI contains the techniques that can assist SMEs in buying the optimal amount of inventory, ultimately, which will mitigate the excess manufacturing and minimize negative influence over the environment. The capability of AI techniques to implement sustainability into SME operations makes it important for strategic long-term success in Pakistan.

AI techniques like machine learning, processing of programming languages, and big data analytics play an essential role in the struggle to increase sustainability inside SMEs (Iyelolu et al., 2024). The AI methods increase the firm's efficiency in analyzing large amounts of data, predicting trends, and making timely and efficient decisions that increase operational efficiency. Related to the SMEs of Pakistan, AI adoption is continuously gaining momentum in SMEs, particularly within the areas of manufacturing, retail, and agriculture sectors. Automation with AI methods is helping SMEs to minimize operational costs, optimize their operations, and increase customer involvement. Similarly, AI-assisted smart grids and energy-producing systems enable firms to investigate energy usage and implement energy conservation methods. Regardless of the achievements of AI, there lies a significant gap in the adoption of AI to promote sustainability. A large number of SMEs are unaware of AI's functions, however, the rest of the SMEs face hurdles associated with the limitations of infrastructure and a lack of knowledge. Considering these hurdles, SMEs have to adopt a comprehensive technique that includes capacity-building approaches and financial leverages to encourage AI policy support adoption in the SME sector of Pakistan.

This chapter highlights the adoption of AI to increase sustainability inside the SMEs of Pakistan by investigating its benefits, applications, and challenges. Additionally, it aims to identify how AI gives solutions to achieve the sustainability goals of SMEs while enhancing efficiency and competitiveness. Moreover, this study tends to provide fact-based investigations on the advantages and constraints of the adoption of AI in the SMEs of Pakistan, investigating the empirical data and case studies. Via understanding of the key factors and challenges of AI execution, this chapter participates in identifying the evolving factors of AI that help sustainability and its implementation in the SMEs of Pakistan. It also provides practical implications for SME owners, managers, policymakers, and academicians. With the rising value of sustainability in firm practices, this study provides valuable approaches that tell how AI can be implemented as a strategic tool for long-term earnings and success.

This chapter is organized as follows: Section 1 provides an introduction to the study, and Section 2 highlights the theoretical framework and literature review. Section 3 discusses the methodology of this proposed study, however, the results are provided in Section 4. Ultimately, section 5 comprises the conclusion of this study.

2. THEORETICAL FRAMEWORK & LITERATURE REVIEW

AI is revolutionizing the current era of businesses, in which it has changed the production models and improved the sustainability practices for small as well as large businesses (Farayola et al., 2023). The influence of AI in various types of businesses is increasing day by day. AI assists businesses in providing innovative solutions to reduce costs, enhance the efficiency of processes, and promote environmentally friendly approaches. AI provides multiple opportunities but also brings challenges to the SMEs of Pakistan. This chapter explored the theoretical perspective concerning the implementation of AI in businesses having different sustainability models, with the challenges faced by SMEs in Pakistan, and the impact of AI on sustainability.

2.1 AI in Business: A Theoretical Perspective

The use of AI in SMEs is described in various studies that explain the role of AI in improving sustainability, innovation, and efficiency. The resource-based view (RBV) theory states that businesses having unique, valuable, and non-replaceable assets get a competitive advantage (Ali et al., 2022; Farooq et al., 2024; Naeem et al., 2024). Considering AI as a technological asset, it allows firms to improve production, analyze market trends, and enhance the decision process. Using AI for analytics and automation techniques, SMEs can optimize asset allocation and improve their competitive advantages.

The TBL theory illustrates the role of AI in sustainability aspects. AI assists firms in balancing environmental, economic, and social performance (Nogueira et al., 2025). AI techniques can help in all three aspects by minimizing influence on the environment, reducing costs, and improving social well-being via workforce supervision and ethical decision-making processes. These theories provide a theoretical exposure that helps businesses in the adoption of AI in SMEs.

2.2 Sustainability Models and Their Relevance to AI

The worth of sustainability in SME firms is often represented via published models that give structured approaches to environmental, social, and economic concerns with corporate strategies (Evans et al., 2017). One of the frequently used models in this perspective is the Brundtland model of sustainable development, which explains sustainability as evolution that fulfills current needs without neglecting the needs of future generations, with the ability to meet their desires (Barbosa et al., 2014). AI aids the above model by exploring sustainable production, reducing waste, and predictive maintenance by guaranteeing business feasibility for the long term (Rane et al., 2024).

Additionally, another widely used and important model is the Natural Capitalism Model, which suggests the usage of sustainability of natural assets while enhancing the profitability of the firms. The resource optimization models driven by AI, like AI-powered energy management tools and smart grid systems, align with the AI model by mitigating environmental damage and preserving economic efficiency (Challoumis, 2024). Similarly, the Circular Economy Model helps in recycling, reusing, and reducing waste materials, which is a closed-loop approach. The study shows that AI promotes the functionality of this model via intelligent sorting of waste, optimization of supply chain tactics, and automation of sustainability reporting (Seyyedi et al., 2024).

From the SMEs' point of view, the models of sustainability must be customized to allocate the resources efficiently, which reduces limitations and market constraints. AI methods assist SMEs in incorporating approaches that can be cost-effective for sustainability without bearing the burden of finance associated with classic models. The AI and sustainability integrated models ensure that SMEs can increase their performance while maintaining the environmental and ethical standards of the world.

2.3 Challenges Faced by SMEs in AI Adoption for Sustainability

AI assists in numerous ways for the SME sector, however, several challenges in Pakistan restrict its widespread adoption (Naeem et al., 2025a). One of the primary hurdles is the higher development cost for the implementation of AI, which includes AI software, initial investment in integration with existing methods, and AI-skilled personnel. Unlike large enterprises, which have higher reserved financial amounts, SMEs have to allocate budgets for AI initial costs to promote sustainable growth. The limited availability of fiscal assistance and government restrictions further extend this challenge.

Another limitation is the lack of technical knowledge within the SME personnel. AI demands special expertise in machine learning, data science, and automation, which can assist many small firms (Ikpe, 2024). The shortage of employees which knows AI and the unwillingness to invest in professional training in AI restrict SMEs from adopting AI capabilities. In addition to this, the limitations related to the privacy of data and cybersecurity impose challenges, as SMEs have a shortage of the necessary tools to protect data from internet threats.

Government regulatory and compliance hurdles are the current challenges, as many SMEs work in the markets with increasing demand for AI governance (Joswig & Kurz, 2025). The lack of appropriate guidelines on ethical AI, transparency, and accountability creates uncertainty, which discourages SMEs from adopting AI-driven sustainability actions. Additionally, cultural and organizational frictions limit the changes and adoption of AI, as many classic SMEs favor classic business models that restrict the usage of digital transformation.

2.4 AI and Sustainability

Due to the challenges in adopting AI for SMEs, AI has demonstrated high potential in increasing sustainability among various SME sectors. One of the most common and notable applications of AI is associated with energy management and optimization of resources. The AI-driven smart grids aid SMEs in checking and optimizing energy consumption, implementing conservation strategies, and determining inefficiencies. Implementing AI automation methods in business can decrease operational expenses a decrease their carbon footprint.

AI's role in the supply chain process is to promote sustainability and help SMEs optimize transportation methods, logistics, and inventory management. AI-powered predictive analytics helps enterprises predict demand forecasting, controlling the overproduction and wastage of excessive inventory. Moreover, the AI-driven blockchain technique in the process of the supply chain increases transparency, ethical sourcing, and responsible trade.

In manufacturing firms, predictive maintenance with AI-powered technologies assists SMEs in predicting and enhancing the lifespan of the machines by decreasing frequent replacements and reducing industrial wastage. The automation techniques with AI sensors indicate the equipment failure before its occurrence, which prevents downtime and makes it efficient. This method not only promotes production efficiency but also assists in increasing environmental sustainability via decreasing energy consumption and material wastage.

With the aspect of social sustainability, AI assists in the fields of human resource management and workforce development. Integrating HR analytics with AI helps SMEs to check employee well-being, optimize work-life balance, and improve labor schedules. Moreover, decision-making with AI tools decreases bias in hiring processes, enhancing diversity and implementation in SMEs.

The sustainability procedures with AI-powered techniques can help in checking and facilitating SMEs to comply with environmental regulations. Automatic reporting tools help to investigate business activities, make decisions about reducing carbon footprints, create sustainability reporting, and help comply with enterprise standards. This decreases the administrative loads on SMEs and increases transparency in sustainability implementation.

H1a: Advanced computational intelligence has a significant effect on economic sustainability.

H1b: Advanced computational intelligence has a significant effect on environmental sustainability.

H1c: Advanced computational intelligence has a significant effect on social sustainability.

2.5 Conceptual Framework

The conceptual model of this research consists of the independent variable and dependent variable stated below in Figure 1.

Figure 1. Independent variable and dependent variable

3. METHODOLOGY

This research employs a quantitative research method to highlight the impact of AI on sustainable performance for the SMEs of Pakistan. The quantitative research method applies to this study, and it allows for the achievement of the objectives, statistical testing of associations, and results findings (Anser et al., 2024). This study follows a research methodology of a deductive approach, which links with existing theories and models on AI, sustainability models, and SME performances. Deductive reasoning starts with the established theories, like the RBV and the TBL models, and applies them to the specific configurations of the Pakistan SMEs. By analyzing the predefined hypotheses, this study highlights a systematic evaluation of AI's effect on sustainability performance.

3.1 Data Collection and Sampling

This research focuses on primary data collection, using a structured questionnaire as the important data-gathering method. The questionnaire is designed to analyze the key points on AI adoption, sustainability performance, and challenges in SMEs. It contains closed-ended questions utilizing a five-point Likert scale to indicate perceptions of respondents and experiences quantitatively. The structured format determines the consistency and facilitates the statistical analysis.

The targeted population involves SME owners, managers, and decision-makers in Pakistan, especially those working in industries where AI adoption plays an important role in sustainability, such as retail, manufacturing, and services. This research adopts a purposive sampling method, selecting SMEs that have either implemented AI practices or have the potential to implement them. Purposive sampling is applicable because it highlights the inclusion of relevant participants, which can provide meaningful participation in the research topic.

3.2 Data Analysis Technique

This study implements PLS-SEM utilizing Smart PLS as the primary data tool. PLS-SEM is appropriate for this study because of its ability to handle difficult models with multiple constructs and relationships. It is specifically useful for those studies that have latent variables, such as AI adoption and sustainability performance, which cannot be directly observed but are investigated via multiple indicators.

PLS-SEM is utilized to test the proposed hypotheses, decide the validity of the measurement model (including reliability and construct validity), and determine the structural model for path associations. The method tells the researchers to determine direct, indirect, and mediating effects inside the conceptual model, giving deeper insights into how AI impacts sustainability performance in SMEs.

4. RESULTS

4.1 Demographic Result

Table 1. Demographic Statistics

Demographics	Frequency	Percentage
Gender		
Male	221	53.77
Female	190	46.23
Age		
18-25 Years	98	23.84
26-35 Years	145	35.27
36-45 Years	82	19.95
46 and above	86	20.92
Qualification		
Matriculation	45	10.94
Intermediate	121	29.44
Bachelors	107	26.03
Masters	95	23.11
PhD	43	10.46
Total	411	100.0

The demographic analysis gives comprehensive details of the respondents' characteristics, providing insights into the composition of the sample. In the case of gender distribution, the major portion of the respondents are male (53.77%), however, the females make up 46.23% of the sample size. The above suggests some relatively balanced respondents, who increase the reliability of the investigations by implementing different perspectives from both genders.

Concerning the age distributions, the major proportion of respondents (35.27%) lie within the 26-35 years of age bracket, demonstrating that a significant portion of the sample contains young professionals and mid-career persons. The second largest group of respondents includes ages 18-25 years (23.84%), followed by ages 46 and above (20.92%) and 36-45 years (19.95%). This distribution of respondents indicates that this study contains insights from many generational cohorts, which may impact the perspectives on AI adoption and its sustainability in SMEs.

The data on educational qualification reveals that the largest group of respondents has finished an intermediate level of education (29.44%), followed by the degree holders, including a bachelor's degree (26.03%) and a master's degree (23.11%). A minor proportion of the respondents have a PhD degree (10.46%) or education at the matriculation level (10.94%). This differentiation in educational levels indicates that this study considers a mix of academic qualifications, which will impact the understanding of the respondents and the implementation of AI-powered sustainability actions in SMEs.

4.2 Measurement Model

Table 2. Discriminant Validity

	Factor loading	SD	T-Statistics	CR	AVE	VIF
Artificial Intelligence				0.821	0.593	
AI1	0.792	0.042	27.101			1.291
AI2	0.738	0.037	26.192			1.372
AI3	0.829	0.010	15.135			1.237
AI4	0.799	0.021	11.129			1.283
AI5	0.708	0.031	17.136			1.197
Economic Sustainability				0.834	0.676	
Eco_Sus1	0.873	0.029	31.147			1.184
Eco_Sus2	0.851	0.028	19.101			1.551
Eco_Sus3	0.784	0.031	30.251			2.392

continued on following page

Table 2. Continued

	Factor loading	SD	T-Statistics	CR	AVE	VIF
Eco_Sus4	0.813	0.020	28.137			2.212
Environmental Sustainability				0.809	0.572	
Env_Sus1	0.842	0.017	19.514			1.737
Env_Sus2	0.703	0.020	28.139			1.109
Env_Sus3	0.736	0.028	21.199			1.270
Social Sustainability				0.793	0.549	
Soc_Sus1	0.899	0.027	18.415			1.201
Soc_Sus2	0.801	0.029	19.103			1.299
Soc_Sus3	0.735	0.038	15.321			1.198
Soc_Sus4	0.721	0.043	18.210			1.236

The discriminant validity findings indicate that the measurement model is well-framed and that the constructs are different from each other. The factor loadings of all items exceeded the recommended boundary of 0.7, illustrating the strong item reliability. AI3 and AI4 contain the highest factor loadings within the AI construct, indicating that these factors influence the AI in this study. Likewise, within Economic Sustainability parameters, Eco_Sus1 and Eco_Sus2 give strong loadings, confirming their impact in explaining economic sustainability. The Environmental Sustainability construct also illustrates adequate factor loadings, with Env_Sus1 having the largest contribution. For Social Sustainability, Soc_Sus1 shows the largest factor loading, focusing on its significance in determining social sustainability.

The Composite Reliabilities (CR) for all constructs exceed the acceptable threshold of 0.7, confirming the reliability of the internal consistency. Economic Sustainability has the largest CR (0.834), demonstrating that its indicators continuously measure the intended construct. Similarly, Environmental and Social Sustainability constructs revealed reliability with CR values of 0.809 and 0.793, respectively. The Average Variance Extracted (AVE) values exceeded 0.5 for all constructs, which meets the criteria for convergent validity. This confirms that a significant portion of the variance in the observed variables is explained by the latent constructs they represent.

The Variance Inflation Factor (VIF) values for all items remain below the critical threshold of 5, indicating that multicollinearity is not a concern in this model. However, Eco_Sus3 and Eco_Sus4 have slightly higher VIF values (2.392 and 2.212, respectively), suggesting a moderate level of collinearity within Economic Sustainability. Despite this, the values remain within acceptable limits. The T-statistic values are all significant, demonstrating that each indicator contributes significantly to its respective construct. The highest T-statistic is observed for AI1, reinforcing its robustness in measuring AI.

Table 3. Heterotrait-Monotrait ratio (HTMT) matrix

	AI	Eco_Sus	Env_Sus	Soc_Sus
AI				
Eco_Sus	**0.732**			
Env_Sus	0.410	**0.719**		
Soc_Sus	0.314	0.409	**0.736**	

Table 4. Farnell–Larcker criterion matrix

	AI	Eco_Sus	Env_Sus	Soc_Sus
AI	**0.732**			
Eco_Sus	0.394	**0.699**		
Env_Sus	0.320	0.432	**0.629**	
Soc_Sus	0.309	0.330	0.299	**0.654**

The Heterotrait-Monotrait Ratio (HTMT) matrix in Table III assesses discriminant validity by measuring the degree of correlation between latent constructs. The HTMT values should ideally be below 0.85 (or 0.90 in some cases) to confirm discriminant validity (Naeem et al., 2024b). In this study, the highest HTMT value is 0.736 (between Environmental Sustainability and Social Sustainability), which is well within the acceptable threshold. This indicates that all constructs exhibit sufficient discriminant validity, meaning they are conceptually distinct from each other.

The Fornell–Fornell–Larcker criterion present in Table IV gives another test for the validity of the discriminant. Regarding this criterion, the square root of the AVE of each construct (diagonal values) should be higher than its correlations for the other constructs (off-diagonal values). In this scenario, each diagonal value is higher than the respective off-diagonal correlations, validating that each construct is different from the others and describes more variance with its indicators than with other constructs.

4.3 Hypotheses Measurement

Table 5. Direct Effect

Path	Path Coefficient	T-Statistics	P-value	Hypothesis
AI → Eco_SP	0.293	2.73	0.000***	H1: Supported
AI → Env_SP	0.219	3.19	0.000***	H2: Supported
AI → Soc_SP	0.138	2.97	0.003**	H3: Supported

The results demonstrate that AI significantly impacts all three variables of sustainability performance- economic, environmental, and social inside the SMEs in Pakistan. The direct impact of AI on sustainability performance is showing a significant effect, with all p-values below 0.05, evidencing the robustness of the findings.

The positive and significant association between AI and economic sustainability performance (Eco_SP) ($\beta = 0.293$, $p = 0.000$) indicates that AI adoption increases the financial and operational efficiency of SMEs. These results validate the RBV theory, which emphasizes that firms using advanced technologies to get a competitive advantage by optimizing the resources and decreasing inefficiencies (Barney et al., 2011). The Previous researchers also validated with integration to AI, which helped to make optimized production, cost reduction, and profitability enhancements in terms of the SMEs (Chatterjee et al., 2023).

Likewise, the influence of AI on sustainability performance via the environment (Env_SP) ($\beta = 0.219$, $p = 0.000$) dictates the role of AI in reducing energy consumption, mitigating waste, and generating eco-friendly businesses. This investigation is validating the TBL model, which primarily focuses on incorporating environmental sustainability into SMEs. Previous researchers have demonstrated that AI-focused automation methods like intelligent monitoring and predictive analytics assist SMEs in achieving environmental goals, including carbon footprint reduction and sustainable resource management (Naeem et al., 2025b).

The relationship between AI and social sustainability performance (Soc_SP) ($\beta = 0.138$, $p = 0.003$) shows that AI helped to achieve stakeholder involvement, ethical labor practices, and CSR initiatives. The results validated the results of Pai and Chandra (2024), which focused on AI's role in enhancing inclusivity, workplace improvements, and increasing consumer trust through ethical AI solutions.

Ultimately, the findings indicate that AI adoption in SMEs in Pakistan provides multidimensional benefits in terms of sustainability. These findings provide empirical assistance for relating AI-powered solutions with the strategies of SMEs to achieve growth in sustainable businesses. This research contributed to the ongoing advancements in highlighting the role of AI in sustainability by validating previous research in the context of Pakistani SMEs, providing practical implications for SME leaders and policymakers.

5. CONCLUSION

The outcomes of this study indicate that AI significantly increases the environmental, economic, and social sustainability performance of SMEs in Pakistan. The results assist in evidencing the arguments that the adoption of AI provides resource optimization, financial efficiency, and ethical business actions, which help to achieve a business with sustainable business framework. This study validates the established concepts involving the RBV, TBL, and Stakeholder Theory, suggesting the role of AI in the transformation of traditional SMEs into responsible enterprises. These results help the literature on the integration of AI and sustainability by giving empirical evidence of an emerging economy. The research enlightens the role of AI as a critical driver of innovation and sustainable performance, emphasizing that SMEs should incorporate AI-powered solutions to remain resilient in a continuously evolving business environment. The study suggests that through AI's transformative potential, policymakers and industry leaders should develop long-term strategies that facilitate AI adoption while adopting ethical and sustainable business actions. This study provides an insight and foundation for further exploring the potential of AI's long-term impact on sustainability, specifically in developing countries where SMEs play an important role in economic growth.

The implications of this research suggest that business leaders, policymakers, and researchers understand the role of AI in sustainable businesses. For SME owners and managers, the outcomes suggest that the adoption of AI can influence financial stability, increase environmental efficiency, and enhance social responsibility, which ultimately leads to competitive advantages in the evolving markets. Policymakers should utilize the above insights to design influential policies, like incentives for AI-powered sustainability programs, training employees for the adoption of AI in SMEs, and ethical guidelines for AI utilization. This study also suggests to investors and financial institutions the potential benefits of investing in AI-powered projects to align with sustainability goals. From a theoretical point of view, this research contributes to the prior literature on AI and sustainability, reinforcing the necessity for interdisciplinary research that integrates technology, business strategies, and sustainability practices. By emphasizing the impact of AI on SMEs, businesses and governments can develop an environment that implements innovation, sustainable growth, and long-term economic benefits.

Despite its valuable impacts, this study has certain limitations that should be addressed in future research. First, the study solely focuses on SMEs in Pakistan, which may restrict the generalizability of the outcomes to other regions with varying economic and technological landscapes. Second, the research implements a cross-sectional design, recording data at a single point in time, which may not reflect the overall strategic impact of AI on sustainability performance. Third, this study focuses on the self-reported data from SME managers, which may introduce bias related to thinking and reporting accuracy. Moreover, the research does not include the industry-specific differentiation in AI adoption, which may impact sustainability outcomes differently. Future studies should consider a longitudinal approach, comparative analyses across firms and countries, and mixed-method research structures to get a more comprehensive understanding of AI's role in the sustainable performance of SMEs.

REFERENCE

Ali, S., Murtaza, G., Hedvicakova, M., Jiang, J., & Naeem, M. (2022). Intellectual capital and financial performance: A comparative study. *Frontiers in Psychology*, *13*, 967820. DOI: 10.3389/fpsyg.2022.967820 PMID: 36081720

Ameh, B. (2024). Technology-integrated sustainable supply chains: Balancing domestic policy goals, global stability, and economic growth. *Int J Sci Res Arch*, *13*(2), 1811–1828. DOI: 10.30574/ijsra.2024.13.2.2369

Anser, M. K., Naeem, M., Ali, S., Huizhen, W., & Farooq, S. (2024). From knowledge to profit: Business reputation as a mediator in the impact of green intellectual capital on business performance. *Journal of Intellectual Capital*, *25*(5/6), 1133–1153. DOI: 10.1108/JIC-01-2024-0025

Barbosa, G. S., Drach, P. R., & Corbella, O. D. (2014). A conceptual review of the terms sustainable development and sustainability. *Journal of Social Sciences*, *3*(2), 1.

Barney, J. B., Ketchen, D. J.Jr, & Wright, M. (2011). The future of resource-based theory: Revitalization or decline? *Journal of Management*, *37*(5), 1299–1315. DOI: 10.1177/0149206310391805

Challoumis, C. (2024, October). Building a sustainable economy-how ai can optimize resource allocation. In *XVI International Scientific Conference* (pp. 190-224).

Chatterjee, S., Chaudhuri, R., Kamble, S., Gupta, S., & Sivarajah, U. (2023). Adoption of artificial intelligence and cutting-edge technologies for production system sustainability: A moderator-mediation analysis. *Information Systems Frontiers*, *25*(5), 1779–1794. DOI: 10.1007/s10796-022-10317-x

Evans, S., Vladimirova, D., Holgado, M., Van Fossen, K., Yang, M., Silva, E. A., & Barlow, C. Y. (2017). Business model innovation for sustainability: Towards a unified perspective for creation of sustainable business models. *Business Strategy and the Environment*, *26*(5), 597–608. DOI: 10.1002/bse.1939

Farayola, O. A., Abdul, A. A., Irabor, B. O., & Okeleke, E. C.Oluwatoyin Ajoke FarayolaAdekunle Abiola AbdulBlessing Otohan IraborEvelyn Chinedu Okeleke. (2023). Innovative business models driven by ai technologies: A review. *Computer Science & IT Research Journal*, *4*(2), 85–110. DOI: 10.51594/csitrj.v4i2.608

Farooq, M., & Ahmad, N. (2023). Nexus between board characteristics, firm performance and intellectual capital: An emerging market evidence. *Corporate Governance (Bradford)*, *23*(6), 1269–1297. DOI: 10.1108/CG-08-2022-0355

Farooq, M., Noor, A., & Naeem, M. (2023). Does family ownership moderate the relationship between board characteristics and corporate social responsibility? Evidence from an emerging market. *Asian Journal of Business Ethics, 12*(1), 71–99. DOI: 10.1007/s13520-022-00164-z

Fateh, S., Shah, I. A., Sial, Q., & Jhanjhi, N. Z. (2025). Deep Learning Applications for Healthcare Risk Assessment. In *Generative AI Techniques for Sustainability in Healthcare Security* (pp. 59–76). IGI Global Scientific Publishing.

Gregurec, I., Tomičić Furjan, M., & Tomičić-Pupek, K. (2021). The impact of COVID-19 on sustainable business models in SMEs. *Sustainability (Basel), 13*(3), 1098. DOI: 10.3390/su13031098

Ikpe, E. O. (2024). Adoption and implementation of artificial intelligence in small businesses in selected developing countries. *Journal of Health. Applied Sciences and Management, 8*(1), 26–34.

Iyelolu, T. V., Agu, E. E., Idemudia, C., & Ijomah, T. I. (2024). Driving SME innovation with AI solutions: overcoming adoption barriers and future growth opportunities. *International Journal of Science and Technology Research Archive, 7*(1), 036-054.

Jhanjhi, N. Z., Shah, I. A., & Brohi, S. N. (2025). Hardware Vulnerabilities: Taxonomy and Business Security Models. In *Generative AI for Web Engineering Models* (pp. 129-146). IGI Global.

Joswig, T., & Kurz, W. (2025). Regulatory and Compliance Requirements for SMEs Operating AI Systems through Data Centers in the EU, with a Focus on Data Protection Challenges in Germany. *Journal of Next-Generation Research 5.0.*

Kulkov, I., Kulkova, J., Rohrbeck, R., Menvielle, L., Kaartemo, V., & Makkonen, H. (2024). Artificial intelligence-driven sustainable development: Examining organizational, technical, and processing approaches to achieving global goals. *Sustainable Development (Bradford), 32*(3), 2253–2267. DOI: 10.1002/sd.2773

Naeem, M., Ali, S., Rehman, A., & Farooq, S. (2025). Ethical Concerns in Artificial Intelligence (AI): The Role of Governance Mechanism in Finance. In *Generative AI for Web Engineering Models* (pp. 193-202). IGI Global.

Naeem, M., Khan, A., Rehman, A., Farooq, S., Mehboob, A., Abdali, A. S., & Ahmad, B. (2025). Does Artificial Intelligence With Blockchain Reduce the Costs of the Financial Sector? In *Generative AI for Web Engineering Models* (pp. 147-160). IGI Global.

Naeem, M., Mehboob, A., Shaikh, M. R., & Rehman, A. (2024). Does Green Innovation Linked between Environmental Governance and Firm Performance? A System of Generalised Method of Moments in Pakistan.

Naeem, M., Siraj, M., Ali, S., Rehman, A., & Farooq, S. (2025). The Role of Artificial Intelligence in Risk Management: Practices of the Banking Sector. In *Generative AI for Web Engineering Models* (pp. 83–106). IGI Global.

Nogueira, E., Gomes, S., & Lopes, J. M. (2025). Unveiling triple bottom line's influence on business performance. *Discover Sustainability*, *6*(1), 43. DOI: 10.1007/s43621-025-00804-x

Pai, V., & Chandra, S. (2022). Exploring factors influencing organizational adoption of artificial intelligence (AI) in corporate social responsibility (CSR) initiatives. *Pacific Asia Journal of the Association for Information Systems*, *14*(5), 4. DOI: 10.17705/1pais.14504

Prasanna, R. P. I. R., Jayasundara, J. M. S. B., Naradda Gamage, S. K., Ekanayake, E. M. S., Rajapakshe, P. S. K., & Abeyrathne, G. A. K. N. J. (2019). Sustainability of SMEs in the competition: A systemic review on technological challenges and SME performance. *Journal of Open Innovation*, *5*(4), 100. DOI: 10.3390/joitmc5040100

Rane, N. L., Choudhary, S. P., & Rane, J. (2024). Artificial Intelligence and machine learning in renewable and sustainable energy strategies: A critical review and future perspectives. *SSRN*, *2*(3), 80–102. DOI: 10.2139/ssrn.4838761

Roberts, O. (2025). A Qualitative Exploration on Artificial Intelligence and Renewable Energy Integration in Supply Chains.

Seyyedi, S. R., Kowsari, E., Gheibi, M., Chinnappan, A., & Ramakrishna, S. (2024). A comprehensive review integration of digitalization and circular economy in waste management by adopting artificial intelligence approaches: Towards a simulation model. *Journal of Cleaner Production*, *460*, 142584. DOI: 10.1016/j.jclepro.2024.142584

Shah, I. A., Jhanjhi, N. Z., & Ashraf, H. (2024). Logistics With the Internet of Things: Challenges, Perspectives, and Applications. In *Navigating Cyber Threats and Cybersecurity in the Logistics Industry* (pp. 172-195). IGI Global.

Shah, I. A., Jhanjhi, N. Z., & Brohi, S. (2024). Cybersecurity issues and challenges in civil aviation security. *Cybersecurity in the Transportation Industry*, 1-23.

Shah, I. A., Jhanjhi, N. Z., & Brohi, S. N. (2025). Machine Learning Models for Detecting Software Vulnerabilities. In *Generative AI for Web Engineering Models* (pp. 1–40). IGI Global.

Shah, I. A., Laraib, A., & Hussain, F. (2024). Logistics Industry in the Context of the Blockchain Technology. In *Navigating Cyber Threats and Cybersecurity in the Logistics Industry* (pp. 214–235). IGI Global. DOI: 10.4018/979-8-3693-3816-2.ch008

Waqar, M., Bhatti, I., & Khan, A. H. (2024). AI-powered automation: Revolutionizing industrial processes and enhancing operational efficiency. *Revista de Inteligencia Artificial en Medicina*, *15*(1), 1151–1175.

Zong, Z., & Guan, Y. (2024). AI-driven intelligent data analytics and predictive analysis in Industry 4.0: Transforming knowledge, innovation, and efficiency. *Journal of the Knowledge Economy*, 1–40. DOI: 10.1007/s13132-024-02001-z

Chapter 2
Advanced Computational Intelligence for Climate Change Adaptation

Imdad Ali Shah
https://orcid.org/0000-0003-2015-1028
FEST Department, Iqra University, Karachi, Pakistan

N. Z. Jhanjhi
https://orcid.org/0000-0001-8116-4733
School of Computing Science and Engineering, Taylor's University, Malaysia

ABSTRACT

Climate change poses an unprecedented threat to ecosystems, human livelihoods, and global sustainability. Addressing this challenge requires innovative approaches that go beyond traditional methods. This chapter also focuses on the role of advanced Computational Intelligence encompassing Machine Learning, deep learning, evolutionary algorithms, and fuzzy systems in facilitating effective climate change adaptation strategies. By integrating large-scale climate data, environmental models, and socioeconomic parameters, ACI enables predictive analytics, risk assessment, and decision-support systems tailored to dynamic climate conditions. Furthermore, this chapter also highlights recent advancements from the past five years, showcasing how computational intelligence is being leveraged across disciplines to build resilience and support sustainable adaptation pathways. Challenges such as data uncertainty, model interpretability, and ethical considerations are discussed, along with future research directions.

DOI: 10.4018/979-8-3693-9132-7.ch002

INTRODUCTION

AI is an ambiguous phrase that has recently become widely used to describe a group of methods intended to use machines to approximate some human or animal cognition feature. AI comes in a variety of forms. It looks at climate change, which is a huge issue in and of itself and one field that is ready for greater artificial intelligence. Without action, the destabilising climate would bury towns, cause deadly natural disasters, and endanger vital food and water supplies, according to a federal circuit court(Bolton & Zanna, 2019; Callaghan et al., 2021). Carbon emissions "rose 1.7% last year and hit a new record" due to increased worldwide energy consumption, despite the strong and consistent scientific evidence that we need to limit our global carbon emissions. The United States only achieved short-term worldwide carbon reductions comparable to the 8% experts estimate is required annually to meet climate-change targets during a global pandemic and economic shutdown(Challinor et al., 2018). Water-related management concerns have drawn the greatest amount of research on the application of AI (Chernozhukov et al., 2018). According to the survey's findings, the strongest consensus was found on the potential of AI and digitisation to improve governance procedures and policy coherence in the context of climate change. The study's findings indicate that, with the right precautions, artificial intelligence (AI) can be a useful tool in the worldwide endeavour to better comprehend and manage the myriad issues related to climate change. Figure 1Overview of Issues and challenges in computational Intelligence.

Figure 1. Overview of Issues and challenges in computational Intelligence

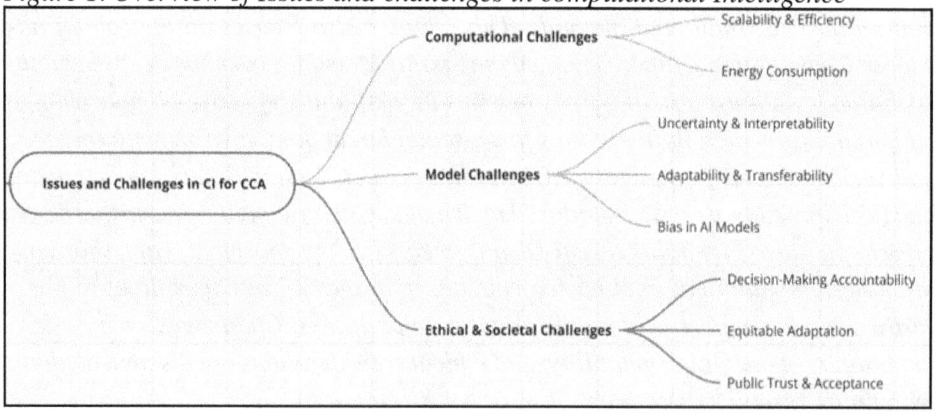

It is widely acknowledged that disruptive technological advancements create new possibilities for automating repetitive and typically time-consuming tasks, provide opportunities to recognise patterns in substantial amounts of unstructured data, and facilitate the self-learning of innovative algorithms, despite a comprehensive definition of artificial intelligence remaining elusive(Deines et al., 2021; Shah, Jhanjhi, & Ujjan, 2024). AI will undoubtedly change production methods in every industry and promote cutting-edge, creative, and sustainable answers to urgent sustainability issues. under the purview of studies with a sustainability focus(Harrison et al., 2016; Shah & Sial, 2024).The goal of the subfield of "natural computing" is to optimise algorithms by imitating environmental processes in computers and using biophysical materials as computational media. Expert systems, which rely on specialised knowledge and inference from databases and inference engines, are also emerging topics connected to AI's ability to address complex natural world problems, such as impact assessments of climate change, and support decision-making(He et al., 2021; Helmrich & Chester, 2022). Identify the salient features that characterise the interplay between human psychology and the global climate system. These characteristics include our incapacity to understand the effects of climate change because of their extreme scope and duration, our (single) dependence on risk-and uncertainty-capturing predictive models, the spatial and temporal remoteness of climate change impacts, and the inherent common-pool resource constraints of anthropogenic GHG emissions that impede collective mitigation efforts(Holman et al., 2019; Huntingford et al., 2019).

These characteristics help to explain why climate change has a variety of effects on livelihoods, from escalating socioeconomic inequality and human diseases to increasing temperatures and the frequency of extreme weather occurrences.

LITERATURE REVIEW

This work's primary research issue concerns investigating climate change mitigation techniques based on artificial intelligence. Integrated AI systems stand to become vital in high-level decision-making concerning scalability and other intangible factors. For this reason, the study is aimed at API's potential in solving difficult problems like vast environmental data analysis, resource allocation, and decision support (Ahmed et al., 2021). The findings of these research questions will help design mechanisms that aid in the management of global warming through the use of AI in ethically sustainable ways. Algorithms like neural networks and machine learning are among the AI techniques proposed for enhancing climate modeling and prediction accuracy (Ahmad et al., 2021). One advantage of AI climate models is their ability to simulate the world's complex climate systems more accurately. With AI-scaled models that can simulate weather extremities, ocean circulation, and carbon cycle patterns, scientists will be able to draw insightful conclusions and project changes to the Earth's climate system (Ahamadalipour et al., 2019). However, AI-enhanced clim is presumed to be the most effective in providing answers to questions related to deep learning supervised intermodal machine learning in geo-economics and geo-politics. Figure 2: Overview of Literature Review.

Figure 2. Overview of Literature Review

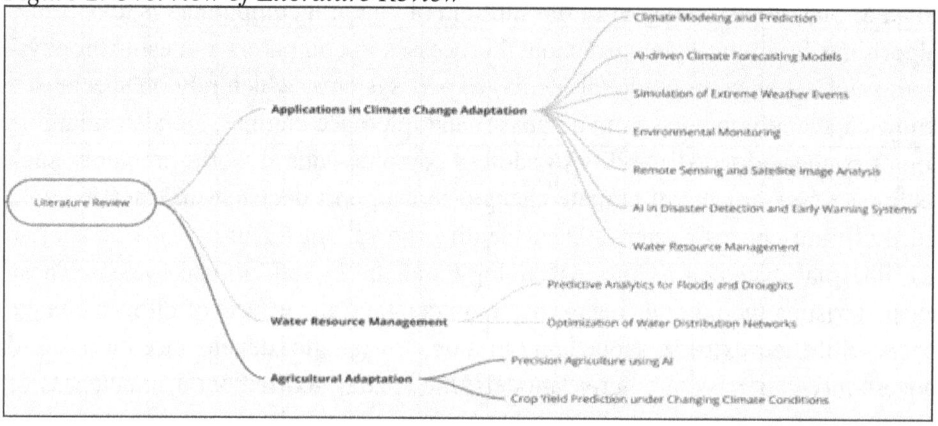

Existing climate models can be modified by machine learning algorithms to provide the most accurate climate projections for varying levels of greenhouse gases. Because machine learning can evaluate climatic risks and vulnerabilities, it is crucial to incorporate it in climate change models. In order to determine which areas and regions are most vulnerable to the various climate impacts and to prioritise adaptation efforts, artificial intelligence (AI) algorithms can now analyse global climate data and combine it with socioeconomic factors, infrastructure vulnerabilities, and information on ecosystem dynamics(Alemany et al., 2019). AI approaches, such as probabilistic forecasting and ensemble modelling, have a lot to offer in terms of their usefulness in supplying data on the likelihood and severity of extreme weather events, which can support risk management and other similar tactics. Understanding, predicting, and adapting to the effects of climate change requires the use of AI techniques in climate modelling(Shah, Jhanjhi, & Brohi, 2024). Researchers can create more accurate and comprehensive models of the Earth's climate by combining machine learning and neural networks, which will improve decision-making and the efficacy of climate action plans. However, there are still a number of significant issues that require further study and collaboration with the scientific communities in order to be properly addressed, including data availability, model validation, and computational capability(Ampatzidis et al., 2019).

OVERVIEW OF CLIMATE CHANGE AND ARTIFICIAL INTELLIGENCE

One of the more challenging issues confronting civilisation is climate change. However, more immediate and manageable issues that arise within an election cycle tend to dominate the political landscape because the worst of their effects are expected to occur later. As a defence mechanism, individuals are more inclined to downplay the most serious threats in society, as behavioural specialists often point out. A person's psychology can significantly benefit from denial, but their grandchildren will not reap similar advantages(Balogun et al., 2019). This section provides a brief overview of climate change, including examples of AI's general applications for reducing GHG emissions, with a focus on its technological and data-related challenges. Figure 3 Overview of climate change & AI.

Figure 3. Overview of climate change & AI

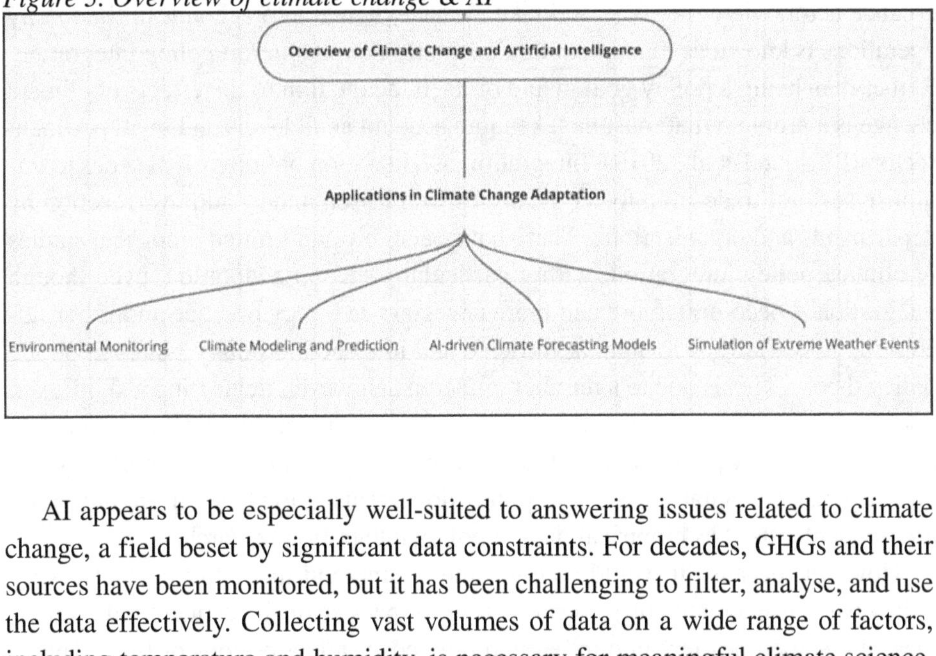

AI appears to be especially well-suited to answering issues related to climate change, a field beset by significant data constraints. For decades, GHGs and their sources have been monitored, but it has been challenging to filter, analyse, and use the data effectively. Collecting vast volumes of data on a wide range of factors, including temperature and humidity, is necessary for meaningful climate science, but handling such enormous data sets can be difficult(Bhatia, 2017). However, some contend that "immediate generation and analysis of large data sets are being made possible by rapid and continuous increases and improvements in the sophistication, affordability, compactness, and use of technology. Putting money into strategies to overcome these obstacles can pay off handsomely. Artificial intelligence is already being used to gather temperature, humidity, and carbon dioxide data to monitor the health of our forests and to analyse photos of shallow-water reefs to identify coral by colour to monitor the effects of climate change(Biesbroek et al., 2020). AI can also help democratise electricity by making it more accessible and enabling off-grid zero-carbon electrification through microgrid development, as almost a billion people worldwide lack access to electricity.

The potential of AI to mitigate climate change is among its most evident uses. According to studies conducted worldwide, nations, regions, and cities are increasingly creating policies, strategies, and actions specifically designed to adapt to the effects of climate change, both now and in the future. Increasing knowledge of climate hazards, creating innovative financial plans to boost resilience, altering current laws, and putting "hard" and "soft" adaptation strategies into practice are a few examples(Booth, 2018). As a result, it has become more evident that governments' capacity to incorporate or "mainstream" a focus on climate change across pertinent sectors, domains, and levels affects the effectiveness of adaptation efforts. The process by which climate change issues become an essential component of the structural aspects of sectoral public bureaucracies and impact how public governance actors view the issue and take climate change into account in their daily operations is known as mainstreaming adaptation or adaptation policy integration. Rather than being a policy goal in and of itself, adaptation to the effects of climate change is a problem that must be taken into account at all levels and in all pertinent sectors(Buckland et al., 2019). Integration of adaptation policies often seeks to minimise sectoral trade-offs, foster synergies, and lessen under- and overreaction by departments and organisations. There have been few and limited empirical studies of climate policy integration, with a particular focus on adaptation, even though integration is becoming more and more necessary in policy practice and is considered the holy grail for managing intricate and intersecting policy issues in public policy theory. This is due to a number of factors. However, neglecting a significant portion of relevant sectors and departments can become problematic as it does not provide a complete picture of how the government is planning for and implementing climate change adaptation. Second, policy documents or evidence from interviews are frequently used as empirical data in policy integration research.

However, because it would take too much time and money to evaluate every policy paper ever created by governments or to speak with all pertinent individuals, these are sometimes few in number and only cover a few periods of time. This is related to the difficulty of transparently and methodically coding the results, especially when the quantity of documents grows, which lowers (inter)coding reliability. Third, recognising climate change adaptation in policy texts is difficult. Despite the existence of formal definitions of adaptation, such as those offered by the IPC, there are numerous alternative interpretations that vary throughout nations and geographical areas, most notably between the interpretations that are centred on social vulnerability and climate change. As a result, most academics have adopted the practice of only mentioning adaptation when it is specifically mentioned in the text or mentioned by interviewees.

To monitor how policy integration changes and is incorporated into current bureaucracies, it is crucial to have a contextual understanding of adaptability. The quick development of tools and techniques in computational social sciences, especially around "text-as-data" methodologies, has been utilised in recent climate research and policy studies. Here, our goal is to investigate how computational social science techniques—more especially, machine learning techniques—can be used to assist researchers in locating, charting, and evaluating climate change initiatives within the government(Crane-Droesch, 2018).We make the assumption that knowledge about which departments and organisations have adapted to climate change can be used as a gauge for "subsystem involvement," which is a crucial element of policy integration analysis. Furthermore, these techniques might enable us to evaluate the two other essential components of policy integration analysis—the policy aim and adaptation tools both inside and between these departments and organisations. The format of this article is as follows: The following part will provide a quick overview of machine learning and document classification techniques, emphasising their key tenets, advantages, and disadvantages for the study of adaptation policy.

AI APPLICATIONS FOR RENEWABLE ENERGY EFFICIENCY

Renewable energy systems can surpass the constraints of grid management, energy generation optimisation, and the integration of renewable resources thanks to artificial intelligence (AI) algorithms. Numerous uses of AI in optimising renewable energy systems, particularly for solar and wind power, are revealed in the paragraph that follows. Conversely, demand response AI systems improve grid resilience and flexibility by allowing users to adjust their energy use to shifting energy costs and grid conditions. AI greatly improves grid management and optimisation of renewable energy generation. However, AI plays a major role in integrating renewable energy into the electrical system. The best locations and sizes for renewable energy projects can be determined by using machine learning to examine the temporal and spatial patterns of renewable energy generation(Sial et al., 2025).

Additionally, utilities and energy planners are unable to account for the regularity and unpredictability of renewable energy output and create suitable energy storage and backup plans because of the AI-powered forecast model. Renewable energy sources will be more widely used than fossil fuels based on renewable energy if AI-driven optimised procedures are applied to them. This will speed up the shift to a carbon-free society.

AI-ENABLED CLIMATE ADAPTATION AND RESILIENCE

The use of AI in response to the increasing frequency and intensity of climate-related disasters has improved resilience and adaptation initiatives in the regions most likely to experience the anticipated impacts. In order to provide resilience and adaptability in the face of the adverse effects of climate change, the next section will address the state-of-the-art applications of AI in early warning systems, catastrophe risk management, and climate-related infrastructure design(Doorn, 2021). Algorithms that apply machine learning can collect data on energy usage patterns, improve energy production and utilisation, and identify areas for energy conservation.

CONCLUSION

In conclusion, addressing this problem requires innovative solutions that extend beyond the traditional scope. ACI brings predictive analytics, risk assessment, and decision support systems for anticipating climate dynamics by integrating large-scale climate data with environmental models and socioeconomic factors. This also addresses uncertainties in data, model interpretation, ethics, and others while proposing future directions for research. There is an acute need for effective technological interventions and data-driven action to strengthen sustainability and resilience frameworks in the face of climate change. Advanced Computational Intelligence (ACI) is proving instrumental in untangling the multifaceted climate systems, predicting radical weather, and developing responsive strategies to various environmental changes that need to be pre-emptively tackled by adopting agile measures. Using machine learning, neural networks, fuzzy systems, and evolutionary computation, ACI improves forecasting models and makes analytics, as well as strategic decisions under uncertainty, much sharper and precise.

REFERENCES

Alemany, S., Beltran, J., Perez, A., & Ganzfried, S. (2019). Predicting hurricane trajectories using a recurrent neural network. Proceedings of the AAAI conference on artificial intelligence, Ampatzidis, Y., Partel, V., Meyering, B., & Albrecht, U. (2019). Citrus rootstock evaluation utilizing UAV-based remote sensing and artificial intelligence. *Computers and Electronics in Agriculture*, *164*, 104900.

Balogun, A.-L., Marks, D., Sharma, R., Shekhar, H., Balmes, C., Maheng, D., Arshad, A., & Salehi, P. (2019). Assessing the potentials of digitalization as a tool for climate change adaptation and sustainable development in urban centres. *Sustainable Cities and Society*, *58*, 1–12.

Bhatia, S. (2017). Artificial intelligence for better climate governance. *Journal of Artificial Intelligence Research & Advances*, *4*(3), 37–43.

Biesbroek, R., Badloe, S., & Athanasiadis, I. N. (2020). Machine learning for research on climate change adaptation policy integration: An exploratory UK case study. *Regional Environmental Change*, *20*(3), 85. DOI: 10.1007/s10113-020-01677-8

Bolton, T., & Zanna, L. (2019). Applications of deep learning to ocean data inference and subgrid parameterization. *Journal of Advances in Modeling Earth Systems*, *11*(1), 376–399. DOI: 10.1029/2018MS001472

Booth, M. (2018). Climate change and the neglected tropical diseases. *Advances in Parasitology*, *100*, 39–126. DOI: 10.1016/bs.apar.2018.02.001 PMID: 29753342

Buckland, C., Bailey, R., & Thomas, D. (2019). Using artificial neural networks to predict future dryland responses to human and climate disturbances. *Scientific Reports*, *9*(1), 3855. DOI: 10.1038/s41598-019-40429-5 PMID: 30846833

Callaghan, M., Schleussner, C.-F., Nath, S., Lejeune, Q., Knutson, T. R., Reichstein, M., Hansen, G., Theokritoff, E., Andrijevic, M., Brecha, R. J., Hegarty, M., Jones, C., Lee, K., Lucas, A., van Maanen, N., Menke, I., Pfleiderer, P., Yesil, B., & Minx, J. C. (2021). Machine-learning-based evidence and attribution mapping of 100,000 climate impact studies. *Nature Climate Change*, *11*(11), 966–972. DOI: 10.1038/s41558-021-01168-6

Challinor, A. J., Adger, W. N., Benton, T. G., Conway, D., Joshi, M., & Frame, D. (2018). Transmission of climate risks across sectors and borders. *Philosophical Transactions. Series A, Mathematical, Physical, and Engineering Sciences*, *376*(2121), 20170301. DOI: 10.1098/rsta.2017.0301 PMID: 29712795

Chernozhukov, V., Demirer, M., Duflo, E., & Fernandez-Val, I. (2018). *Generic machine learning inference on heterogeneous treatment effects in randomized experiments, with an application to immunization in India.*

Crane-Droesch, A. (2018). Machine learning methods for crop yield prediction and climate change impact assessment in agriculture. *Environmental Research Letters, 13*(11), 114003. DOI: 10.1088/1748-9326/aae159

Deines, J. M., Patel, R., Liang, S.-Z., Dado, W., & Lobell, D. B. (2021). A million kernels of truth: Insights into scalable satellite maize yield mapping and yield gap analysis from an extensive ground dataset in the US Corn Belt. *Remote Sensing of Environment, 253*, 112174. DOI: 10.1016/j.rse.2020.112174

Doorn, N. (2021). Artificial intelligence in the water domain: Opportunities for responsible use. *The Science of the Total Environment, 755*, 142561. DOI: 10.1016/j.scitotenv.2020.142561 PMID: 33039891

Harrison, P. A., Dunford, R. W., Holman, I. P., & Rounsevell, M. D. (2016). Climate change impact modelling needs to include cross-sectoral interactions. *Nature Climate Change, 6*(9), 885–890. DOI: 10.1038/nclimate3039

He, W., Hong, D., Scarpa, G., Uezato, T., & Yokoya, N. (2021). Multisource Remote Sensing Image Fusion. *Deep Learning for the Earth Sciences: A Comprehensive Approach to Remote Sensing, Climate Science, and Geosciences*, 136-149.

Helmrich, A. M., & Chester, M. V. (2022). Reconciling complexity and deep uncertainty in infrastructure design for climate adaptation. *Sustainable and Resilient Infrastructure, 7*(2), 83–99. DOI: 10.1080/23789689.2019.1708179

Holman, I. P., Brown, C., Carter, T. R., Harrison, P. A., & Rounsevell, M. (2019). Improving the representation of adaptation in climate change impact models. *Regional Environmental Change, 19*(3), 711–721. DOI: 10.1007/s10113-018-1328-4 PMID: 30956567

Huntingford, C., Jeffers, E. S., Bonsall, M. B., Christensen, H. M., Lees, T., & Yang, H. (2019). Machine learning and artificial intelligence to aid climate change research and preparedness. *Environmental Research Letters, 14*(12), 124007. DOI: 10.1088/1748-9326/ab4e55

Shah, I. A., Jhanjhi, N., & Brohi, S. N. (2024). Use of Emerging Technologies in Healthcare 4.0. *Advances in Medical Technologies and Clinical Practice, 4*(0), 280–296. DOI: 10.4018/979-8-3693-2333-5.ch015

Shah, I. A., Jhanjhi, N., & Ujjan, R. M. A. (2024). Industry 4.0: use of digitalization in healthcare. In *Advances in Computational Intelligence for the Healthcare Industry 4.0* (pp. 174-193). IGI Global.

Shah, I. A., & Sial, Q. (2024). *Advances in Computational Intelligence for the Healthcare Industry 4.0*. IGI Global. DOI: 10.4018/979-8-3693-2333-5

Sial, Q., Shah, I. A., & Jhanjhi, N. (2025). Generative AI Applications for Enhancing Medical Training. In *Generative AI Techniques for Sustainability in Healthcare Security* (pp. 161–174). IGI Global Scientific Publishing.

Chapter 3
Advances in Geospatial Modeling for Flood Risk Assessment in Changing Climate

Hafsa Muzammal
Agricultural Remote Sensing Lab, University of Agriculture, Faisalabad, Pakistan

Muhammad Zaman
https://orcid.org/0000-0003-1105-2343
Department of Irrigation and Drainage, University of Agriculture, Faisalabad, Pakistan

Muhammad Safdar
https://orcid.org/0009-0006-1779-6967
Agricultural Remote Sensing Lab, University of Agriculture, Faisalabad, Pakistan

Muhammad Adnan Shahid
Agricultural Remote Sensing Lab, University of Agriculture, Faisalabad, Pakistan

Zuhaib Nishtar
College of Electrical Engineering and New Energy, China Three Gorges University, Yichang City, China

Muhammad Bilal
https://orcid.org/0009-0003-8414-2873
Department of Irrigation and Drainage, University of Agriculture, Faisalabad, Pakistan

Muntaha Munir
https://orcid.org/0009-0005-3404-3636
Institute of Botany, University of the Punjab, Lahore, Pakistan

Mehar Muhammad Haseeb
Department of Space Science, University of the Punjab, Lahore, Pakistan

Aamir Raza
Precision Agriculture Center, Department of Soil, Water, and Climate, University of Minnesota, USA

Syed Intsar Hussain Shah
Agricultural Remote Sensing Lab, University of Agriculture, Faisalabad, Pakistan

DOI: 10.4018/979-8-3693-9132-7.ch003

Usman Zafar
Department of Irrigation and Drainage, University of Agriculture, Faisalabad, Pakistan

Hafiz Muhammad Bilawal Akram
https://orcid.org/0000-0001-6720-7266
Department of Agronomy, University of Agriculture, Faisalabad, Pakistan

Nalain E. Muhammad
Department of Irrigation and Drainage, University of Agriculture, Faisalabad, Pakistan

ABSTRACT

The chapter discusses advances in geospatial modeling for flood risk assessment in the context of climate change. It integrates Global Climate Models (GCMs) with geospatial techniques to explore the impact of changing climatic trends on flood frequency and intensity. By compiling historical climate data from 2000 to 2050, the study creates a baseline understanding of regional temperature fluctuations and precipitation changes. GCM results are downscaled to local levels, ensuring compatibility with hydrological models. The research provides a comprehensive framework for addressing the evolving flood landscape, accounting for risks related to GCM estimates and model assumptions. It highlights proactive adaptation strategies such as sustainable land use, flood-resistant infrastructure, and informed policymaking, especially for the Indus River Basin. This interdisciplinary approach offers an advanced analytical framework for understanding and mitigating flood-related vulnerabilities amid climatic uncertainty.

1. INTRODUCTION

Pakistan is especially affected by the negative impact of climate change because of its geography and agricultural sector. Pakistan has suffered significant changes in precipitation and temperature trends in the last few decades. The Meteorological Department of Pakistan PMD (2020) estimates that over the past century, average temperatures have risen by about 0.6°C. Increased and variable rainfall events caused by changes in precipitation patterns have raised the risk of floods. Flooding is a significant natural disaster in Pakistan, causing significant damage to crop and infrastructure. The frequency and severity of flooding have increased due to climate change factors like intense rainfall and glacier melting. In 2020, over 5 million people were affected by flooding. The agricultural sector, which supports a significant portion of the population and contributes to the national economy, suffers greatly from floods. The Food and Agriculture Organization (FAO) reports that floods in

Pakistan have caused billions of dollars in agricultural losses in recent years *(Muzammal et al. 2023)*. Usually, flooding causes a human disaster that destroys millions of people from their homes and increases poverty and food shortages. Because of the overcrowded conditions in temporary shelters, displaced people suffer a risk of waterborne diseases, malnutrition, and other health risks. The United Nations estimates that within the past ten years, flooding has caused many people in Pakistan to leave their homes. Because floods seriously damage infrastructure—roads, bridges, schools, and hospitals—essential services are limited, and recovery efforts cannot advance. Both emergency medical care and long-term infrastructure repair costs are major financial effects of floods in Pakistan. The World Bank determined that Pakistan's 2010 floods cost the economy $10 billion, or around 5% of the GDP.

1.1 Background and Motivation

Because of its complex interaction with rivers, mountains, and monsoons, Pakistan is quite sensitive to climate change. Pakistan's ecological equilibrium has been upset since 1901 by 0.8°C increase in average temperature. The Pakistan Meteorological Department claims that during the past century average temperatures have risen by about 0.6°C. Rising public health issues and major stress on ecosystems are resulting from the more frequent and strong heat waves. For example, approximately 1,900 deaths were recorded during a terrible heat wave Karachi had in 2010. Changing weather patterns have also made Pakistan more vulnerable to floods and other natural disasters *(Safdar et al., 2023)*. Crucially important to Pakistan's water security, the Indus River system mostly relies on the Himalayas as a freshwater reservoir. Long-term water security is seriously threatened by the doubling of glacier melt in the Hindu Kush area since the 1970s, according to 2019 research written for *The Cryosphere*. Rising temperatures are hastening the melting of glaciers, therefore affecting Pakistan's general water supply, hydropower generation, and agriculture. Furthermore, variations in the timing, intensity, and distribution of precipitation have produced some areas enduring protracted droughts while others endure more frequent and intense rainfall episodes. The Intergovernmental Panel on Climate Change (IPCC) forecasts that Pakistan's precipitation patterns would become more erratic due to climate change, therefore aggravating the likelihood of unanticipated floods. Based on the disastrous floods of 2010, which caused an estimated $10 billion in economic damage *(IPCC, 2014)*, the World Bank projects that some areas of Pakistan may see a 30% rise in severe precipitation episodes. Rising sea levels brought on by melting glaciers and warmer oceans put Pakistan's coastal regions, particularly in great danger. Little rises in sea levels can flood agricultural land and destroy coastal towns. Though ocean currents have little direct bearing on Pakistan, it suffers indirect effects such saltwater intrusion into coastal regions. Rising sea levels globally endanger infrastructure, ecosystems, and human communities; coastal flooding predicted in the next decades is projected to become more severe *(Hirabayashi et al., 2013).*

Particularly vulnerable to climate change are Pakistan's varied ecosystems, which include its mountain ranges and Indus River marshes. Rising temperatures, altered precipitation patterns, and extreme weather events are anticipated to disturb these ecosystems, therefore endangering wildlife and causing habitat degradation, changes in species ranges, and a loss of biodiversity. These developments gravely threaten Pakistan's environmental stability, claims Raza et al. (2024). Apart from compromising the food supply and agricultural sector of the nation, climate change also threatens threatened species such the snow leopard and the Indus River dolphin. Severe storms and climate change endanger water supplies, agriculture, wildlife, and their interactions. FAO (2021) projects that climate change might cause Pakistan's wheat yield to plummet by as much as 30% by 2025, therefore compromising food security and livelihoods. Moreover, as the IPCC and Bates et al. (2008) point out, floods are increasingly probable given the increasing frequency of severe storms and high rainfall. Catastrophic events gravely impact ecosystems, infrastructure, and human populations all throughout the planet (Hirabayashi et al., 2013). Successful overcoming these challenges requires a sharp awareness of probable hazards and inventive, flexible response possibilities. Apart from considerable property damage, floods cause significant population movement (Munich, 2020). Analyzing how climate change influences flood frequency might give crucial data for developing targeted adaptation and mitigating plans, hence accelerating the recovery of societies after such natural disasters (Dutta et al., 2003; Chang et al., 2016). Using GIS and GCM models will help us tackle these problems much more readily. Global circulation models (GCMs) might be able to predict significant variations in temperature and precipitation among other weather patterns (Hawkins and Sutton, 2009). Combining GCMs with GIS models yielded comprehensive assessments for locations at more risk of flooding, claims Arnobjerg-Nielsen et al. (2013). These concepts assist in elucidating the intricate relationships of water systems, climate, and geography when combined with hydrological data (Prudhomme et al., 2002). This strategy ensures that policies, urban design, and emergency response systems are developed with dependable knowledge on places prone to floods guiding hand. Geospatial modeling—which offers thorough geographic data—may let authorities maximize their time and resources, claims Srinivas and Nakagawa (2015). This study investigates how this phenomenon affects flood frequency utilizing spatial modeling and GCMs in order to better grasp and minimize the harmful consequences of climate change. Clarifying the intricate interactions between flood patterns and climate change is the key aim in order to raise society's resistance to floods. This strategy seeks to encourage environmentally friendly adaptation strategies and facilitate the development of suitable policies to handle environmental hazards.

1.2 Research Objectives

The possible influence of climate change on flood frequency is investigated in this study using geospatial modeling and Global Climate Models (GCMs..While historical data illuminates flooding patterns, GCM findings are aggregated with atmospheric variables to assess expected damage..An accurate simulation calls for exact modeling, so accuracy depends critically on the validation and calibration of these models. Vulnerability studies help to identify possible hazards to human populations, infrastructure, and ecosystems. On flood frequency, scenario analysis is used to ascertain the predicted effects very differently. Adaptation strategies seek to lessen the consequences of changing flood patterns by means of infrastructural enhancements, land-use planning, and programs of community resilience. It depends on good involvement of stakeholders and the sharing of research findings to raise awareness and provide adaptable solutions.

1.3 Scope and Limitations

Geospatial modeling and Global Climate Models (GCMs) are applied to investigate how climate change influences flood length with an eye on hazard identification and mitigating arising from shifting precipitation patterns and extreme occurrences. This work aims to foresee probable variations in flood frequency resulting from changing climatic conditions by clarifying the complex relationships among hydrological processes, land surface features, and climatic factors. Combining data from GCMs with GIS models that mimic hydrological processes at various spatial and temporal dimensions allows scientists to more precisely and reliably assess the likely consequences of climate change on flood occurrence. Finding basic patterns and causes of flood events depends on first looking at prior flood trends (Fowler et al., 2007). Geospatial modeling is crucial in exactly reproducing flood patterns and identifying susceptible locations by means of several spatial datasets—such as land cover, terrain elevation, and hydrological networks.

Including GCM data into future climate scenarios and their likely effects on flood frequency is another important element of the research. GCMs predict changes in important climatic variables, including temperature, precipitation, and humidity, under different greenhouse gas emission scenarios, thereby clearly showing broad climate patterns. Risk assessments combining ecological services, infrastructural vulnerabilities, and socioeconomic data examine the vulnerability of ecosystems, infrastructure, and human populations to changing flood frequencies. But GCMs frequently miss fine-scale features such topography variations and isolated storms because of their weak resolution. Downscaling techniques can help to overcome this limitation by raising the geographic resolution of GCM data; yet they also introduce uncertainties and prejudices. Geospatial modeling addresses quality, amount, and availability of input data as well as technological limits, socioeconomic concerns, institutional capability, and governance systems.

2 FLOOD FREQUENCY ANALYSIS

2.1 Overview of Flood Frequency Analysis

Flood frequency analysis is a crucial hydrological tool used to assess the likelihood of flood events in a watershed. It aids in infrastructure planning, risk analysis, and flood avoidance. Historical flood data, including frequency statistics and peak flows, are gathered to shape risk intervals. The Generalized Extreme Value (GEV) distribution is recommended for flood frequency studies, as it directly affects frequency estimate correctness. Statistical estimates of the distribution's location, size, and shape parameters are made using a probability distribution function. Flood percentages, including 10-year, 50-year, and 100-year floods, are calculated to show the degree of floods occurring with specific frequency.

While consistent historical flood data is assumed by flood frequency analysis, this may not always be accurate, particularly in areas with urbanisation, climatic fluctuation, or notable land use change. Historical data may have to be changed to fit non-stationarity and preserve frequency estimate accuracy *(*Ullah and Zhang, 2020*)*. Flood quantiles and the degree of forecast uncertainty are computed using flood frequency analysis. It facilitates the decision-making process about infrastructure design and flood risk control. Factors include inadequate historical data, measuring streamflow mistakes, and inherent unpredictability of major events generate uncertainty. Frequent usage of Monte Carlo simulations helps one to evaluate frequency estimate dependability and uncertainty. Forecasted flood percentages allow engineers to build hydraulic infrastructure; urban designers may create evacuation plans, set construction rules, and define floodplains to help to minimise effects on infrastructure and communities *(*Dutta et al., 2003; Chang et al., 2016*)*. **Figure 1** Flood risk assessment uses hydraulic modelling, loss analysis, transformation, baseflow evaluation, and routing procedures. It uses topographical data and rainfall records, addresses elevation data deficiencies with a Digital Elevation Model, and integrates stream networks with land use and precipitation data. A flood inundation map is created, indicating blue areas at risk, and a risk curve helps identify comprehensive flood risk and vulnerable areas. This method enhances community resilience and preparation for flood occurrences.

Figure 1. Overview of flood frequency analysis (Source: Omar M. Nofal & John W. van de Lindt, 2020)

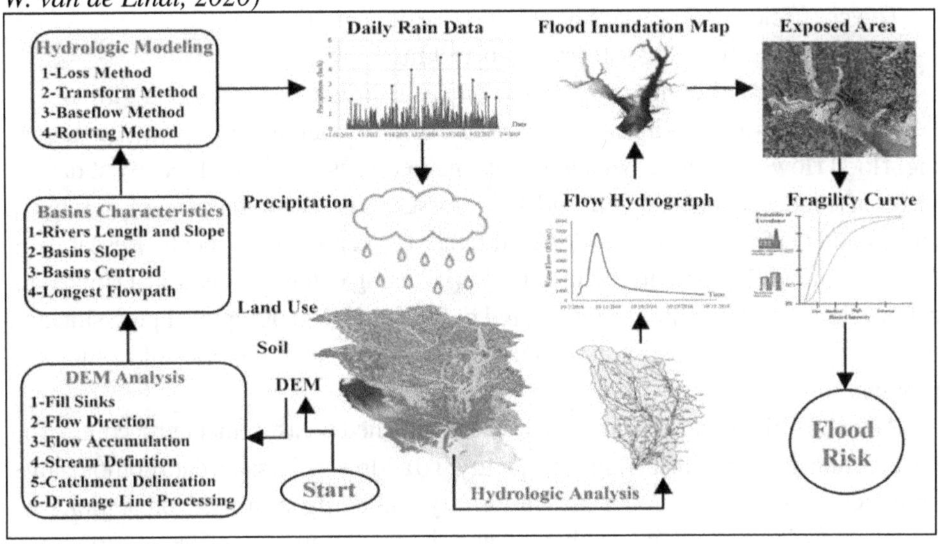

2.2 Analysis of Historical Flood Records

Pakistan's history is characterized by floods, primarily due to major rivers like the Indus and its tributaries. Analyzing this data is crucial to understanding their recurring occurrences, unique characteristics, and potential future trends. The Indus River System Authority (IRSA) maintains extensive historical flood data, including the highest water flows ever recorded. Hydrologists use statistical distributions like the Generalized Extreme Value (GEV) distribution to estimate the likelihood of exceeding a flood peak, but the accuracy depends on the historical record's quality and extent. Empirical formulas for flood estimation are also used.

a. **Dicken's formula** estimates flood discharge ((Q)) based on the area ((M)) of the basin.

[Q = C *M^3/4]

Here:

(Q): Discharge in cubic meters per second (cumecs).

(C): Coefficient (varies by region).
(M): Area of the basin in square kilometers.

b. **Ryve's formula** also estimates flood discharge ((Q)) using the basin area ((M)). It is given by: [Q = C *M^2/3]
The coefficient ((C)) varies based on specific conditions:
Coastal areas: (C) ranges from 6.8 to 15 km.
Inland areas: Different values based on basin size and location (e.g., hills).

Hydrologists and engineers use historical disaster mathematical statistics to predict flood flows and flood behavior. This method uses previous flood event data to forecast flood frequency, depth, and direct losses. It is used to develop vulnerability curves and perform quantitative or qualitative flood risk assessments. Researchers have combined expertise to evaluate flood risk using extensive historical flood data series. However, this approach is limited by inadequate geographical precision and temporal delay, making it suitable only for large geographic regions like cities and provinces.

Pakistan's historical flood records lack established measurement procedures and observation networks, affecting data quality. To address this issue, data homogeneity approaches can be used to ensure uniformity and comparability. Non-stationarity is crucial as land use, climate change, and deforestation can influence flood patterns over time. Identifying patterns in historical flood features can encourage the implementation of non-stationary models. Historical data provides crucial insights on flood duration, especially during major water flow events, essential for disaster management and recovery. Year-round analysis of past floods helps understand seasonal tendencies, enabling authorities to enhance resource allocation, strengthen infrastructure, and activate early warning systems.

2.3 Estimation of Design Floods

Flood management and infrastructure construction depend on accurate projections of flood designs; these projections are usually taken from historical flood data. Though, given the warming of our world, this approach is losing appeal. Streamflow analysis and precipitation evaluation help one to forecast when floods will strike by looking at historical rainfall patterns. Climate models might estimate future precipitation using scenarios for greenhouse gas emissions, therefore providing a high-level view of the frequency and degree of floods. Streamflow-based approaches use hydrological models to project the whole water cycle, including future precipitation and temperature fluctuations that would define flood severity. Still, exact calculations of design floods are not feasible given the erratic character of climate change. Among land use changes that could affect flood risk include urbanization, deforestation, and infrastructural development. Usually, scientists use estimations from many climate models to better understand prospective flood hazards. Forecasts of design floods increase flooding probability and intensity, therefore endangering infrastructure, people, and property in sensitive areas. Floodplain neighborhoods increase their likelihood of house loss and property damage. Floods might affect important infrastructure such transportation systems and electrical grids, therefore affecting businesses and requiring financial support. Communities could include designing flood forecasts into their development plans and growth strategies to help to better control environmental changes. Preventive strategies including early warning systems and floodplain management rules help to quickly handle floods and, ideally, save lives.

2.4 Uncertainty Assessment in Flood Frequency Analysis

Hydrological data and evaluations of model uncertainty are the backbone of flood frequency analysis. Using methods for data quality control, it locates and corrects issues in datasets. In contrast to sensitivity analysis, which identifies the key assumptions that affect the results, error propagation assesses the overall degree of uncertainty in data on flood frequencies. Dessai and Hulme (2007) state that these methods aid in making informed judgments regarding flood risk reduction by providing vital information on the reliability of flood frequency estimates. Analyzing the statistical likelihood of devastating floods is the standard procedure in flood frequency investigations. conventional approaches, such maximum likelihood estimations or moments, use parametric probability distributions when applied to past flood data. But when the terrain is complex or the water flow patterns are changing, these methods may fail to capture crucial features. Because distributions are not normally distributed, most methods for fitting them factor on uncertainty. Among these techniques are Bayesian inference, quantile regression, and L-moments. With these methods, flood risk management choices may be more consistently made. Assessing flood risk, susceptibility, and hazard necessitates an understanding of the factors contributing to a possible catastrophe. Ken Granger (2002) and UNCHS-HABITAT (1981) have delineated the concepts of hazard, vulnerability, element at risk, and risk as follows:

Hazard (H) refers to the likelihood of a potentially catastrophic natural catastrophe occurring inside a certain area over a particular timeframe.

Vulnerability (V) is the extent of loss experienced by a specific at-risk element or a collective of such components due to the manifestation of a natural event of a particular magnitude.

The element at risk (E) encompasses public services, infrastructure, utilities, civil engineering projects, economic activities, individuals, buildings, and similar entities within a specified region.

Risk, denoted as R, is the anticipated extent of loss resulting from a certain natural event. Risk analysis is the systematic use of available data to assess the likelihood of particular occurrences and their potential severity of outcomes.

Flood risk arises from the interplay of hazard and susceptibility.

"Risk = Hazard x Vulnerability" (Ologunorisa, 2001).

Figure 2 illustrates that the flood risk evaluation procedure primarily comprises three stages: risk identification, risk analysis, and risk assessment. The selection of risk analysis indicators is crucial; so, using suitable evaluation indicators to establish an index system will enhance the quality of flood risk assessment (Huang et al 2020). The theory of disaster science posits that a circular feedback mechanism including three elements governs the catastrophic process: the disaster environment, the causative cause, and the affected entity. The severity of the catastrophe is contingent upon the stability of the catastrophic environment, the risks posed by disaster-inducing causes, and the susceptibility of the impacted entities.

Figure 2. Overview of flood disaster risk assessment processes

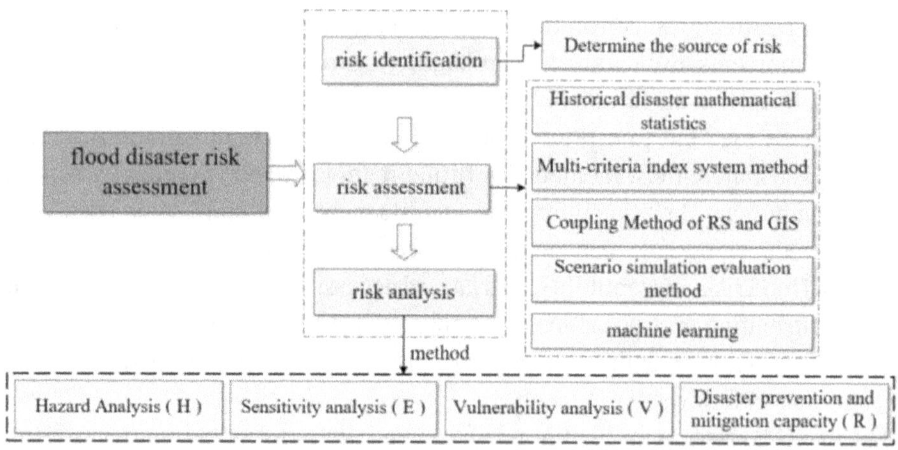

Climate change affects hydrological systems, altering rainfall, snowmelt, land use, and precipitation patterns. Hydrologists use climate models and downscaling methodologies to predict future water conditions based on greenhouse gas emissions. However, uncertainty in climate projections complicates decision-making due to model inaccuracies and concerns over economic development, societal impacts, and governmental actions. To make sound decisions, it's essential to evaluate multiple forms of uncertainty and employ effective adaptation measures across different future scenarios.

3 GEO-SPATIAL MODELING OF FLOOD EXTENT

Employing GIS and supplementary data, geo-spatial modelling of flood dissemination accurately delineates and quantifies inundated areas, encompassing terrain configuration, waterways, extremes, and historical flooding events. Water signals are identified by satellite imagery, which also offers accurate flood maps that incorporate shadows and vegetation. This information is essential for flood prediction, disaster preparation, and recovery support. It optimizes resource use, safeguards property and individuals, and identifies deficient regions.

3.1 Remote Sensing and GIS Techniques for Flood Mapping and Flood Inundation Models

GIS and remote sensing have revolutionized flood mapping by providing precise and up-to-date information on flood mitigation measures, depth, and severity. Utilizing data gathered from terrestrial sensors, drones, and satellites, these systems monitor land development, hydrological processes, and alterations in land cover. A primary method employed for flood mapping involves the analysis of optical and radar imagery transmitted from satellites (Sabir et al. 2024). Satellite imagery from sources like Landsat and Sentinel provides extensive data on surface water levels, floodplain topography, and flooding trends. Optical cameras provide the identification and quantification of submerged objects, akin to Landsat and Sentinel. Flood monitoring during nighttime or under obscured conditions is optimally performed using radar cameras, particularly those employing Synthetic Aperture Radar (SAR) (Hawkins and Sutton, 2011). Flood risk maps are created by integrating geospatial data, including satellite imagery, topographical maps, digital elevation models, land use databases, and hydrological infrastructure information.

This process enables modeling of flood propagation and intensity evaluation. GIS-based flood mapping often uses citizen science efforts, crowd-sourced data, and social media feeds for enhanced monitoring and response. Scenario simulation assessment is a method for dynamically modeling and evaluating disaster processes, based on rainfall frequency and intensity. Advancements in hydrological and hydrodynamic research have improved scenario modeling in flood risk assessment. The one-dimensional pipe network hydraulic model is the most used initial numerical model, while two-dimensional numerical models are popular for simulating surface flooding progression following precipitation events.

Table 1 illustrated that Enhanced realism in simulation outputs has garnered interest in three-dimensional numerical models such as Delft3D, FLOW-3D, water quality simulations, and flow field simulations (Renn et al. 2008; Randall et al. 2019; Djordjevic´ et al. 2005; Kadam et al. 2012). As highlighted in Table 1 Researchers employed hydrological and hydrodynamic models to simulate scenarios and assess flood risk in a designated study region. Their two-dimensional flood simulation approach integrates empirical formulas with artificial intelligence, merging a one-dimensional pipe network model with a two-dimensional surface model. The simulation technique delineates the trajectory of future risk assessment. Risk zoning, based on submerged depth and duration requirements, was analyzed using empirical calculations inside the sponge city demonstration area of Jinan City. Suarez utilized scenario simulations to investigate the impact of climate change and flooding on Boston's urban transportation infrastructure. Su developed a two-dimensional hydraulic simulation to emulate the spatial and temporal dynamics of waterlogging caused by rainstorms. Bisht determined that the two-dimensional Mike Urban hydrodynamic model had superior simulation performance compared to the one-dimensional SWMM hydrological model. Zeng effectively integrated the one-dimensional SWMM hydrological model with the two-dimensional Lis Flood-FP hydrodynamic model to accurately simulate inundation extent and depth during rainstorms and floods. Field experiments (Zheng et al. 2019; Wu et al. 2017; Zeng et al. 2020; Suarez et al. 2005; Su et al. 2015) validated the precision of the coupling model.

Table 1. Flood Assessment using Geospatial modeling

Model	Model Characteristic	Input Parameters	Development Organizations
SWMM	Offers a distributed hydrological module and a one-dimensional hydrodynamic module.	Land utilization, topographical data, precipitation, precipitation intensity, pipeline network data, etc.	EAP United States Environmental Protection Agency
HEC-RAS	One-dimensional and two-dimensional hydrodynamic modules are available.	Topographical and hydrological data	U.S. Army Corps of Engineers Hydrological Engineering Center
PCSWMM	Utilizing SWMM as the foundation, it offers pre-processing and post-processing modules, hence facilitating the simplification of two-dimensional surface calculations.	Land utilization, topographical data, precipitation, precipitation intensity, pipeline network data, etc.	Canadian Hydraulic Calculation Institute
LISFLOOD-FP	Offers a two-dimensional hydrodynamic module.	Topographical and hydrological data, pipeline network data	University of Bristol
InfoWorks ICM	Highly integrated and completely functional, it allows the coupled modeling of hydrology, hydrodynamics, and water quality, while possessing robust pre- and post-processing capabilities.	Pipeline network information, topographical data	UK HR Wallingford
MIKE	Comprises MIKE URBAN, MIKE FLOOD, MIKE21, and more modules. Each module is comparatively autonomous and entirely operational, and is extensively utilized in diverse applications.	Topographic and hydrological data, surface roughness, wave dynamics, tidal elevations, etc.	Danish Hydraulic Institute DHI
Delft3D	Appropriate for three-dimensional hydrodynamic water quality simulation, capable of modeling the hydrodynamics of estuaries and ports.	Topographical and hydrological data, grid dataset	Delft Hydraulic Research Institute, Netherlands
FLO-2D	Offers a two-dimensional hydrodynamic module and an incorporated one-dimensional calculating SWMM module.	Topographical and hydrological data	FLO-2D Software, Inc.
FLOW-3D	The CFD program has a three-dimensional hydrodynamic module, appropriate for the investigation of three-dimensional flow fields.	Terrain data, hydrological data	Flow Science, USA
Flood Simulation Model	The linkage of urban ground floods and pipelines has been achieved for the first time using an unstructured grid.	Pipe network data, terrain data	China Institute of Water Resources and Hydropower Research

3.2 Validation of Flood Extent Models

When it comes to flood risk assessments and management strategies, it is crucial to validate flood extent models. When comparing the anticipated and actual flood extents, one may assess the models' ability to predict when floods will occur. When the model's settings differ, validation methods can make the necessary adjustments. To further understand how factors like precipitation, land cover, and terrain affect the precision of flood extent projections, a sensitivity analysis may be conducted. This evaluation can help in making informed suggestions for flood hazard management. In view of the dynamic nature of land use and climate change, validation and calibration are crucial for maintaining the accuracy and use of flood extent models (Dankers and Feyen, 2008). Consistent validation motivates new ways of integrating data and building models, which in turn improves people's ability to make decisions and develop solutions to mitigate flood consequences.

3.3 Future Flood Extent Mapping under Climate Change Scenarios

Storm area maps can help municipalities and lawmakers mitigate future flood threats by implementing restrictive land use regulations and identifying flood-risk zones. These maps aid in infrastructure project planning, identifying towns most susceptible to flooding, and allocating funds for flood control measures. Early warning systems in high-risk areas enable rapid evacuation, saving lives. Restoration efforts prioritizing natural barriers, such as ponds, may be more effective. Climate models are imperfect, and climate change-induced changes in land use and precipitation patterns can significantly impact flood risks. Experts use probabilistic flood extent maps to illustrate the likelihood of flooding at various severity levels, enabling leaders and communities to prepare for future flood threats. Visualization tools help identify flood-prone areas and understand associated dangers, informing them of necessary steps to mitigate risks and recover from storms. Understanding the impact of climate change on hydrology and flood risks can enhance preparedness for future inundations, protecting property and lives.

4 IMPACT OF CLIMATE CHANGE ON FLOOD FREQUENCIES

4.1 Assessing Changes in Precipitation Patterns

Precipitation analysis helps understand the effects of climate change on infrastructure, water supply, ecosystems, and agriculture. Statistical methods, such as geostatistics and spatial interpolation algorithms, generate continuous precipitation maps using data from remote sensing devices or meteorological stations. These maps display precipitation patterns, topographical elevations, and historical fluctuations across different locations, providing a comprehensive understanding of climate change's effects. Kundzewicz., et al. (2007).

Understanding future precipitation patterns is largely dependent on climate models and downscaling methodologies. These models predict future temperature, precipitation, and air circulation, simulating physical phenomena that affect Earth's climate system. Downscaling methods, such as statistical and dynamic downscaling, improve global climate models at regional or local scales, affecting precipitation. Monitoring precipitation patterns at local, regional, and global levels is facilitated by remote sensing data and precipitation analytic tools. Satellite products like the Tropical Rainfall Measuring Mission and Global Precipitation Measurement provide data on precipitation amounts, frequency, and geographic distribution. Ground-based precipitation monitoring equipment like weather radars, rain gauges, and weather stations provide reliable data for hydrological models and climate research. Integrating ground-based and remote-sensing precipitation data can improve analytical precision and understanding of precipitation patterns across temporal and spatial dimensions as demonstrated in Table 2.

Table 2. rainfall data for different regions in Pakistan from August 1 to 26, 2022

Region	Normal Rainfall (mm)	Actual Rainfall (mm)	Deviation (%)
Pakistan	50.4	176.8	251
Azad J&K	129.3	129.2	0
Baluchistan	20.9	129.7	522
Gilgit-B	12.4	40.1	225
Khyber-PK	92.9	143	54
Punjab	82.5	133.3	62
Sindh	50	442.5	784

Table 1 shows Pakistan's rainfall data, showing 251% higher than normal, with significant differences in Baluchistan and Sindh due to high moisture levels. Azad Jammu and Kashmir experienced no change in rain, while Gilgit-Baltistan, Punjab, and Khyber Pakhtunkhwa experienced different amounts. These differences could impact farming, water resource management, and infrastructure planning, making this information crucial for these areas. The data is based on standard, actual, and percentage measurements.

4.2 Evaluating Temperature Effects on Snowmelt and Hydrological Processes

Understanding the impact of temperature on snowmelt and hydrological processes is crucial for understanding water supply fluctuations, flood risk, and ecosystem functionality in regions with consistent snowfall. Models like VIC, SWAT, and SRM help simulate snowmelt runoff processes and their effects on streamflow. Remote sensing data like MODIS and AMSR2 monitors temporal snow cover fluctuations and meteorological conditions on snowmelt. Empirical observational studies, including snow surveys and meteorological station data collection, are essential for understanding temperature effects on snowmelt and hydrological processes. Accurate terrestrial observations improve snow hydrology models' precision and provide valuable insights as shown Figure 3.

The study examines future temperature variations and their impact on snowmelt and hydrological flow using statistical and dynamical downscaling methodologies. Global climate models (GCMs) forecast future temperatures based on greenhouse gas emissions, while regional climate models (RCMs) and dynamic downscaling techniques observe alterations at local scales. Statistical downscaling strategies like empirical-statistical downscaling enhance forecasting accuracy. The integration of climate model predictions with snow hydrology models allows researchers to study the effects of future temperature variations on snowmelt timing, intensity, duration, streamflow patterns, and water supply availability. Arbjerg-Nielsen et al., 2013.

Table 3 shows severe flooding patterns in Pakistan from 2003 to 2015, affecting 20 million people annually. The exact number of deaths varies from year to year. Figure 8.2 shows potential flood areas, projected to extend to the Arabian Sea and include parts of two countries. River basins and regions at risk are the most affected. The map shows different levels of flood risk, with orange areas having high risk and clear areas having low risk. Urban areas are represented by green circles, while small to medium-sized towns and cities are represented by large circles

Figure 3. Flood risk map Source:(Rana et al.,2016)

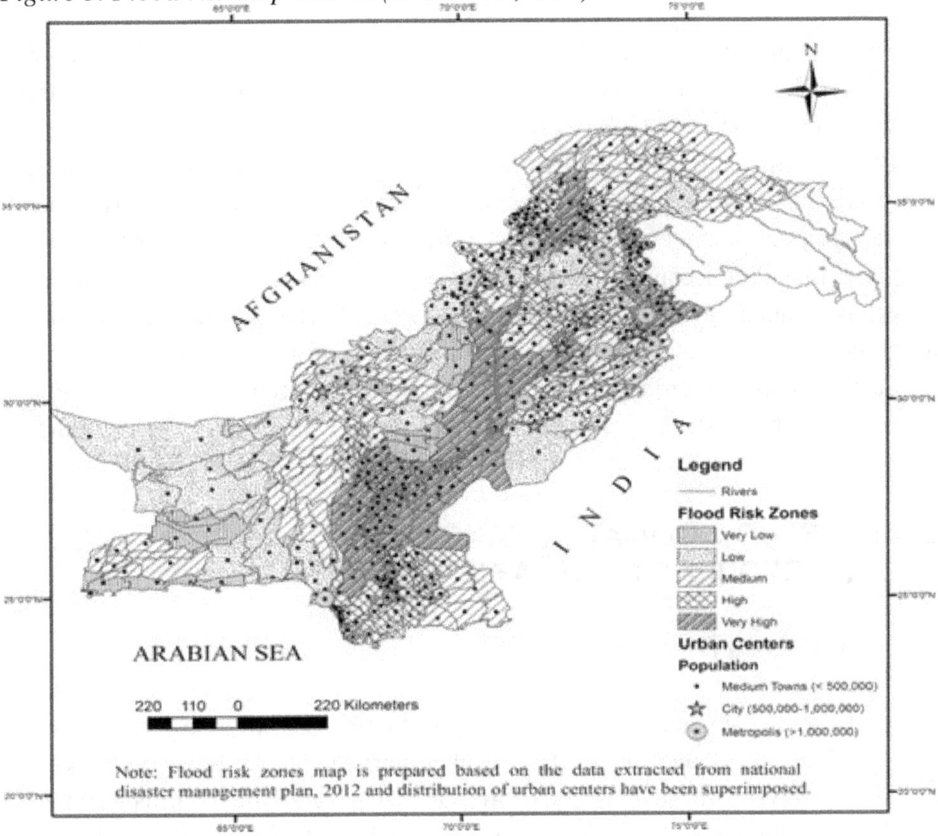

Table 3. number of people affected by floods in Pakistan from 2003 to 2015. Source: (Pakistan Weather Portal, 2023)

Flood Year	People Affected	Lives Lost	Reference/Source
2003	4376	484	Pakistan Weather Portal
2007	2 million	918	Rehman *et al.*, (2015) and Pakistan Weather Portal
2010	20 million	<1781	Pakistan Weather Portal
2011	8.9 million	434	Pakistan Weather Portal
2015	2.5 million	367	Pakistan Weather Portal

4.3 Projection of Future Flood Frequencies

Flood frequency research relies on historical data to predict future flooding occurrences. However, climate change has reduced the effectiveness of this method due to altered weather patterns. Researchers use statistical evaluations of historical storm data, climate models, and hydrological models to forecast floods. Statistical analysis explains flood patterns in specific regions, while climate models forecast future precipitation patterns. Hydrological models simulate climatic influences on flooding, including frequency, water retention, and duration. Predicting flood frequency is challenging due to limitations of climate models and potential discrepancies in real-world impacts. Changes in land use and the construction of new buildings may also increase the probability of flooding. Scientists often use multiple climate model forecasts to estimate future flood frequency (Dankers and Feyen (2008).

5 CASE STUDY: CLIMATE CHANGE AND FLOOD FREQUENCIES IN PAKISTAN

5.1 Study Area Description

Analysed is Pakistan, a vital waterway in South Asia, to assess how climate change affects flood occurrence. The basin comprising China, India, and Pakistan has varied terrain influencing its hydrological characteristics. Climate patterns including the Indian monsoon impact water supply and flow dynamics, therefore influencing the hydrology of the basin.

5.2 Historical Flood Analysis in Pakistan

Table 4 indicates that the research utilized an extensive array of data from regional water agencies and the Pakistan Meteorological Department to elucidate historical floods in Pakistan. Researchers employed GIS to analyze the geographical and temporal distribution of floods to identify trends and fluctuations in flood frequency. Their methodology encompassed an examination of snowmelt, reservoir management, and meteorological trends to ascertain the origins of these floods. Following previous floods, scholars examined the impacts on society, the economy, and the environment. Numerous disastrous floods have afflicted Pakistan since 1950, with the most lethal occurring that year and devastating the nation's infrastructure. The catastrophic flood of 2010, far more severe than the monsoon rains of 1973, affected about 20 million individuals in the Punjab and Khyber Pakhtunkhwa regions. The catastrophic floods that struck Sindh province in 2011 and 2012 obliterated entire villages and resulted in an estimated 20 million fatalities. Although 2020 was Karachi's most precipitation-laden year, the flash floods in Pakistan and Afghanistan in 2022 resulted in several fatalities and devastated agricultural production.

Table 4. Historical Flood events (Source: Government of Pakistan, Ministry of Water Resources, 2018)

Year	Event	Impact
1950	First-ever flood after Pakistan's independence	Significant damage
1973	Monsoon floods across the country	Widespread impact
2010	Massive floods affecting Khyber Pakhtunkhwa and Punjab	Over 20 million people affected
2011	Sindh province severely affected by monsoon rains	361 people killed
2012	Flooding in Khyber Pakhtunkhwa, southern Punjab, and northern Sindh	Homes destroyed; arable land affected
2020	Record-breaking rainfall in Karachi	Heaviest rain in 90 years
2022	Intense flash floods in Afghanistan–Pakistan region	Loss of life, crop damage

5.3 Geospatial Modelling and GCM Data Application

This study investigates the impact of climate change on flood patterns in a specific basin using geographic modelling tools and General Circulation Models (GCMs). GCM data, including temperature and precipitation data, predicts future meteorological conditions. Researchers developed hydrological models to assess how temperature and precipitation patterns may influence flood frequency. This comprehensive approach uses future climate models to assess potential flooding events. The study emphasizes the need for significant climate adaptation strategies and international cooperation in Pakistan to mitigate flood risks, highlighting the interrelated impacts of climate change on flooding frequency. as shown in table 5.

Table 5: Climate Change Impact on Flood Frequencies in the Pakistan

Impact Factors	Explanation
Increased Intensity of Rainfall Events	Increasing temperatures may lead to more frequent, intense, and substantial rainstorms, which can induce flash floods by generating abrupt influx of water into the river system.
Altered Timing of Precipitation	Alterations in monsoon patterns and precipitation timing might disrupt the regular flow regime, perhaps leading to unforeseen flood occurrences.
Glacier Melt and GLOFs	Glacial lakes may overflow due to accelerated glacier melting from elevated temperatures, leading to Glacial Lake Outburst Floods (GLOFs) that discharge substantial volumes of water downstream.
Sea Level Rise and Coastal Flooding	The absence of a coastline in the Indus Basin means that sea level rise can influence delta regions and exacerbate tidal surges, potentially leading to floods.
Changes in Snowmelt Dynamics	Temperature fluctuations can alter the timing and rate of snowmelt, hence affecting river flow. Accelerated snowmelt can elevate river discharge and heighten the danger of floods.
Landslides and Sediment Transport	Alterations in precipitation patterns may induce landslides, potentially obstructing rivers and leading to downstream flooding in the case of a catastrophic dam failure. Alterations in sediment movement may influence flood risk and river morphology.

6 ADAPTATION AND MITIGATION STRATEGIES

6.1 Flood Risk Management and Adaptation Measures

Flood risk management is essential in contemporary environmental planning and catastrophe preparedness, given the escalating frequency and severity attributed to climate change. Methods encompass meticulous planning, delineating floodplains, evaluating infrastructure damage, and scrutinising historical flood data. Governments can implement planning restrictions, land-use strategies, and construction standards to mitigate susceptibility to flood hazards. Implementing disaster response strategies and early warning systems is crucial to mitigate fatalities and property damage resulting from floods as shown in Figure 4. Investments in infrastructure, such as flood barriers and green infrastructure, can significantly reduce flooding risk by controlling water flow, preventing flooding from affecting populated areas, and enhancing biodiversity and ecosystem services. Effective flood risk management involves community involvement, skill development, public education, storm insurance, evacuation plans, and resilience applications. Cooperation and information exchange across borders are crucial for effective flood risk control. A comprehensive plan involving engineering solution, government actions, community involvement, and ecosystem-based methods is needed to lower flooding risks (Ghate *et al.* 2020).

Figure 4. Flood Risk Management Strategies and exemplary measures. P. Matczak et al. 2015

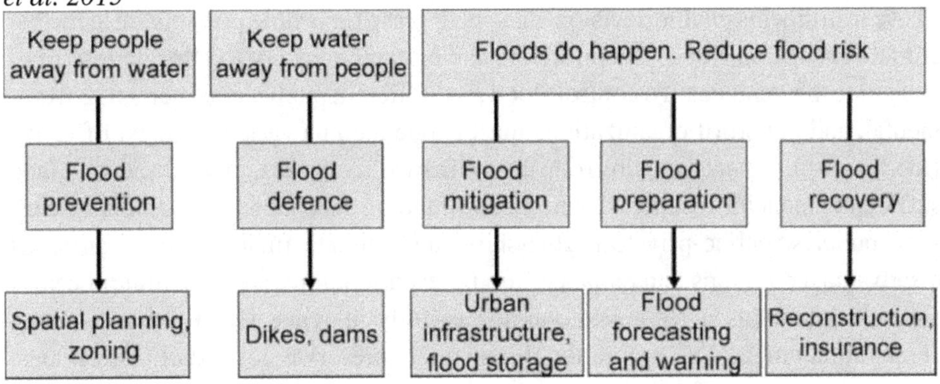

6.2 Integration of Climate Change Projections in Flood Management Plans

Extreme weather events, like floods, are becoming more often and more destructive, making climate change predictions more relevant to flood management strategies. People can make better judgments and come up with more adaptable solutions to upcoming problems when climate models are integrated. By integrating climate change forecasts into flood control strategies, authorities may increase their capacity to forecast future floods, claims AghaKouchak et al. (2014). Thus, they identified locations vulnerable to floods because to altered precipitation patterns, increased sea levels, or other climate-related variables so that governments may allocate monies for methods of flood prevention. By means of climate change forecasts, which provide information on long-term climatic patterns and uncertainty, water management strategy may be improved, and catastrophic risks can be reduced. This method increases community resilience to floods and helps to minimize negative choices. Future climate forecasts help flood forecasting models to identify at-risk locations with more accurate and timely warnings.

6.3 Policy Implications and Decision Support

Flood risk management is mostly dependent on climate change, hence flexible solutions are necessary to reduce susceptibility and raise resistance. Tools for comprehensive risk analysis and planning are very vital as decision-makers must integrate forecasts of climate change into their procedures. This will help us to identify high-risk locations, concentrate adaptation efforts, and create sensible regulations.

Cost-benefit studies, risk modelling, and scenario analysis—all of which help to manage flood risk in a changing environment—are critical decision-support tools. These instruments enable decision-makers to grasp the whole possible influence of climate change on floods, therefore enhancing their capacity to allocate resources and guide their choices. To control flood risk, different public, commercial, governmental, and nonprofit organizations must cooperate in a varied strategy. Effective flood control projects require robust governance structures, clear directions, and sufficient financial resources. Alternative funding mechanisms, including insurance policies, public-private partnerships, and climate finance tools, can assist disadvantaged regions and enhance flood prevention initiatives. Natural processes and ecosystem-based strategies can successfully alleviate flood risks, fostering cooperation among risk mitigation, biodiversity preservation, and climate change adaption, as seen in Figure 5.

Figure 5. The main concept of the proposed decision support system (DSS) for developing flood risk reduction policies, and the main technologies used. (Pantalona et al. 2021)

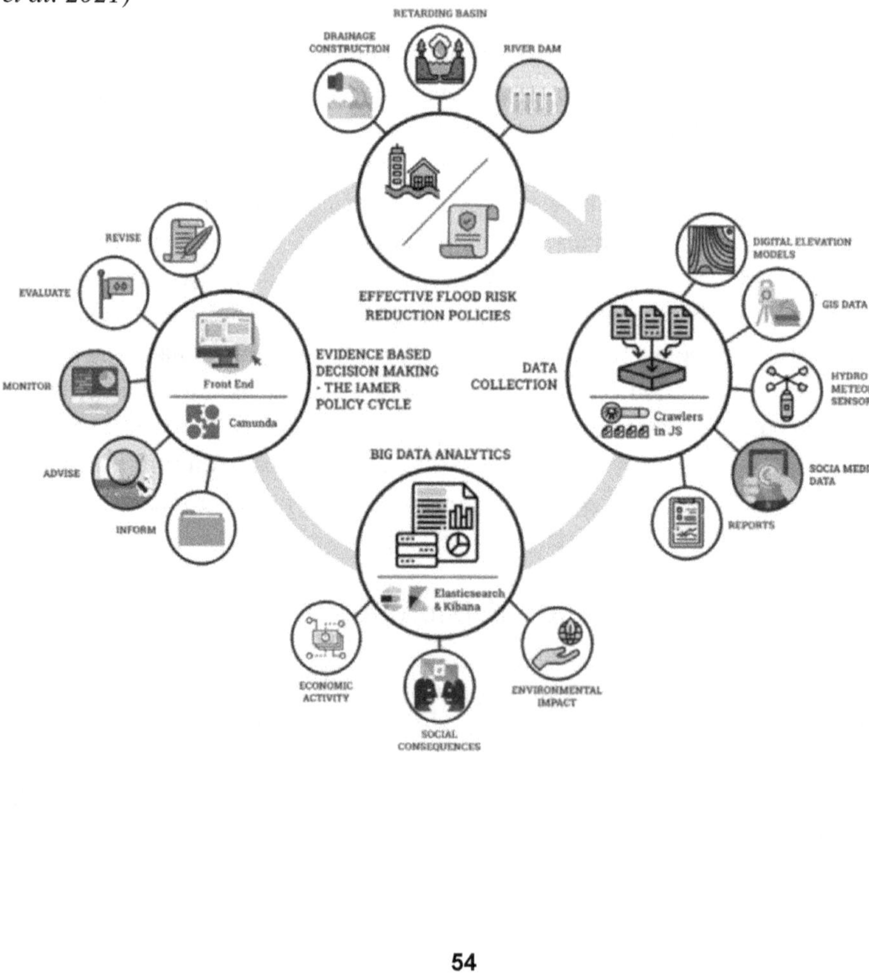

7 CONCLUSIONS AND FUTURE DIRECTIONS

7.1 Conclusions:

The frequency of floods may increase because of climate change-related changes in precipitation patterns and the melting of snow. This highlights the need for improved methods of threat assessments and modelling. Research building upon this one should concentrate on more complex climate models and how they interact with one another. Communities that are vulnerable to flooding could benefit from societal studies and climate predictions that aim to lessen the risks. Alterations to legislation, decision-support systems, and climatic predictions are also essential for a comprehensive strategy.

Estimates of climate change may be greatly improved for flood management and disaster planning by include future climatic scenarios into flood risk calculations. Finding susceptible locations, allocating funds for adaptation strategies, and drafting robust legislation are all guided by this approach. To build institutions, rein in government spending, and develop technological solutions, we need innovative approaches. Appropriate leadership, well-defined procedures, and sufficient funding are necessary for carrying out flood control measures. By thinking forward in this innovative way, lawmakers may be able to anticipate problems and find solutions.

To identify the best adaptation strategies for certain climate circumstances, decision-assistance systems reduce vulnerability and increase population resilience. To completely reduce the danger of floods, they focus on management tactics, alternate funding sources, and institutional frameworks. Figure 6 shows how this approach, which considers not only political and economic factors but also technical and financial ones, improves watershed recovery.

Figure 6. Disaster management using GIS (Source: Paul, 2018).

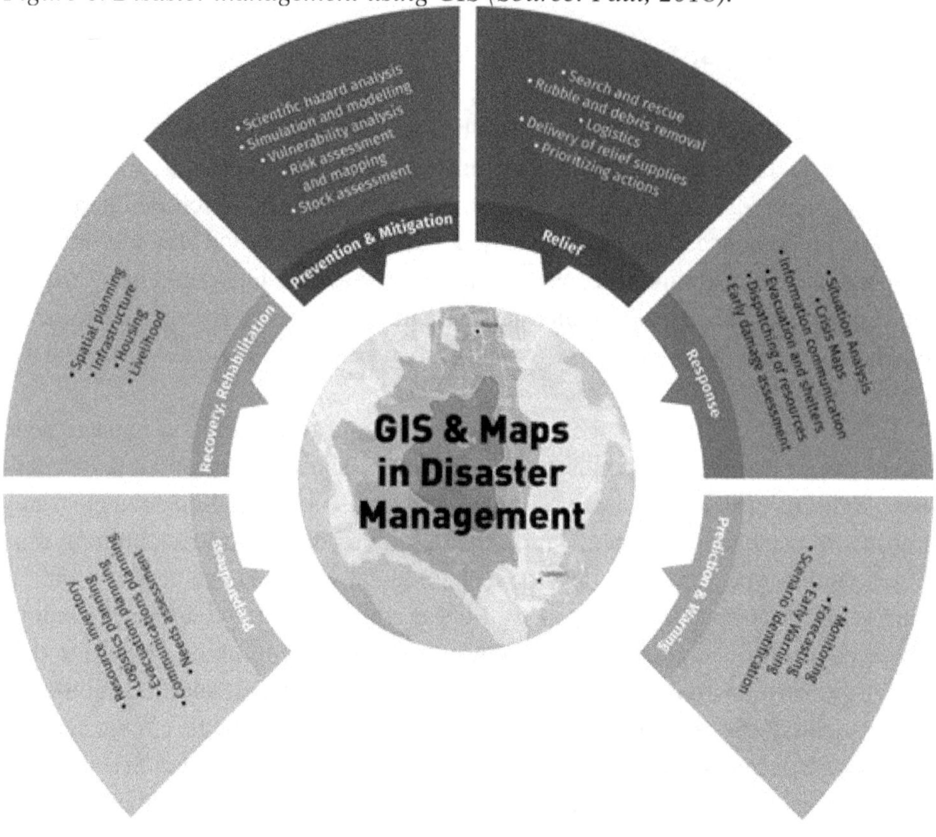

7.2 Limitations and Recommendations for Future Research

To address these challenges, studies suggest improving climate models by incorporating sophisticated methods, high-resolution data, and understanding the interaction between flood hazards, land-use patterns, and climate change. Analyzing how urbanization and deforestation affect flood dynamics can help create more effective strategies, ultimately leading to stronger flood control techniques. Floods seriously affect people's lives as well as companies; they cause relocation and disturbance of livelihoods. Knowing these consequences will let one create policies and projects to protect social and physical infrastructure. By means of evaluation of current flood risk management techniques, one may find areas for progress and thereby promote the creation of more flexible systems. Understanding the limits of present understanding emphasizes the requirement of thorough strategies to lower flood hazards in a climate under change. Future studies should give local-specific concerns a priority as well as honing climate models and looking at the social and financial effects of floods. By means of such initiatives, communities will be able to better equip themselves for increasingly frequent and severe floods, therefore guaranteeing increased resilience to future difficulties.

REFERENCES

AghaKouchak, A., Huning, L. S., Chiang, F., Mazdiyasni, O., & Vahedifard, F. (2014). Global warming and changes in risk of concurrent climate extremes: Insights from the 2014 California drought. *Geophysical Research Letters*, *41*(24), 8847–8852. DOI: 10.1002/2014GL062308

Arnbjerg-Nielsen, K., Willems, P., Olsson, J., Beecham, S., Pathirana, A., Bülow Gregersen, I., Madsen, H., & Nguyen, V.-T.-V. (2013). Impacts of Climate Change on Rainfall Extremes and Urban Drainage Systems: A Review. *Water Science and Technology*, *68*(1), 16–28. DOI: 10.2166/wst.2013.251 PMID: 23823535

Bates, B. C.(2008). *Climate Change and Water.* Intergovernmental Panel on Climate Change.

Chang, H. (2016). Exploring the Relationship between Land Use Change and Flood Damage through an Agent-Based Model. *The Science of the Total Environment*, *550*, 711–721.

Dankers, R., & Feyen, L. (2008). Climate Change Impact on Floods in the United Kingdom Due to Changes in Extremes. *Climate Research*, *32*(1), 163–172.

Dessai, S., & Hulme, M. (2007). Assessing the robustness of adaptation decisions to climate change uncertainties: A case study on water resources management in the East of England. *Global Environmental Change*, *17*(1), 59–72. DOI: 10.1016/j.gloenvcha.2006.11.005

Djordjevic´, S., Prodanovic´, D., & Maksimovic´, C. (2005). ˇ.; Maksimovic´, Cˇ.; Ivetic´, M.; Savic´D. SIPSON—Simulation of Interaction between Pipe flow and Surface Over-land flow in Networks. *Water Science and Technology*, *52*, 275–283. DOI: 10.2166/wst.2005.0143 PMID: 16248205

Dutta, D. (2013). A 2D Flood Inundation Model for Brisbane River: Model Development and Sensitivity Analysis. *Environmental Modelling & Software*, *43*, 55–66.

Fowler, H. J., Blenkinsop, S., & Tebaldi, C. (2007). Linking Climate Change Modelling to Impacts Studies: Recent Advances in Downscaling Techniques for Hydrological Modelling. *International Journal of Climatology*, *27*(12), 1547–1578. DOI: 10.1002/joc.1556

Ghate, S., Kolhe, M., & Narkhede, V. (2020). Climate change induced flood risk assessment and adaptation in urban areas. *The Science of the Total Environment*, *703*, 135558.

Government of Pakistan, Ministry of Water Resources. (2018). Annual Flood Report 2018. Office of the Chief Engineering Advisor & Chairman Federal Flood Commission, Islamabad. https://ffc.gov.pk/wp-content/uploads/2020/06/Annual-Flood-Report2018.pdf

Hawkins, E., & Sutton, R. (2011). The Potential to Narrow Uncertainty in Regional Climate Predictions. *Bulletin of the American Meteorological Society*, *90*(8), 1095–1107. DOI: 10.1175/2009BAMS2607.1

Hirabayashi, Y., Mahendran, R., Koirala, S., Konoshima, L., Yamazaki, D., Watanabe, S., Kim, H., & Kanae, S. (2013). Global Flood Risk under Climate Change. *Nature Climate Change*, *3*(9), 816–821. DOI: 10.1038/nclimate1911

Huang, G., Luo, H., Lu, X., Yang, C., Wang, Z., Huang, T., & Ma, J. (2020). Urban flood risk analysis and zoning methods. *Water Resources. Prot.*, *36*, 1–17.

IPCC. (2014). Synthesis Report. Contribution of Working Groups I, II and III to the Fifth Assessment Report of the Intergovernmental Panel on Climate Change. *Climatic Change*, 2014.

Kadam, P., & Sen, D. (2012). Flood inundation simulation in Ajoy River using MIKE-FLOOD. *ISH Journal of Hydraulic Engineering*, *18*(2), 129–141. DOI: 10.1080/09715010.2012.695449

Kundzewicz, Z. W. (2007). Precipitation as an Indicator of Climate Change. *Climate Research*, *34*(3), 199–207.

Matczak, P., Lewandowski, J., Choryński, A., Szwed, M., & Kundzewicz, Z. W. Flood risk governance arrangements in Europe, *Proc. IAHS*, 369, 195–199, , 2015. DOI: 10.5194/piahs-369-195-2015

Mustafa, S., Sabir, R. M., Sarwar, A., Safdar, M., Al Ansari, M. S., & Hussain, S. (2024). Precision Agriculture and Unmanned Aerial Vehicles (UAVs). In Khang, A. (Ed.), *Agriculture and Aquaculture Applications of Biosensors and Bioelectronics* (pp. 83–108). IGI Global., DOI: 10.4018/979-8-3693-2069-3.ch005

Muzammil, H., Zaman, M., Shahid, M. A., Safdar, M., Majeed, M. D., & Sabir, R. M. (2023). The Impacts of Climate Change on Monsoon Flood Situations in Pakistan. *Environmental Sciences Proceedings.*, *25*(1), 63. DOI: 10.3390/ECWS-7-14255

Nofal, O. M., & van de Lindt, J. W. (2020). Understanding flood risk in the context of community resilience modeling for the built environment: Research needs and trends. *Sustainable and Resilient Infrastructure*. Advance online publication. DOI: 10.1080/23789689.2020.1722546

Ologunorisa, E. T. (2001) An Assessment of Flood Risk in the Niger Delta, Nigeria. Unpublished Ph. D. Thesis, University of Port-Harcourt, Nigeria 303pp.

Pakistan Meteorological Department. (2020). Climate Data Processing System.

Pakistan Weather Portal. (2023, Apr 30). Exploring the complexity of El Niño and its effects on rainfall in Pakistan. Retrieved from https://pakistanweatherportal.com/2023/04/30/exploring-the-complexity-of-el-nino-and-its-effects-on-rainfall-in-pakistan/

Pantalona, G., Tsalakanidou, F., Nikolopoulos, S., Kompatsiaris, I., Lombardo, F., Norbiato, D., Ferri, M., Kovats, L., & Haberstock, H. (2021). Decision support system for flood risk reduction policies: The case of a flood protection measure in the area of Vicenza. *Data & Policy*, *3*, e26. DOI: 10.1017/dap.2021.23

Paul, S. (2018, Jul 4). Managing the environment using GIS. Geospatial World. Retrieved from https://www.geospatialworld.net/article/managing-the-environment-using-gis/

Prudhomme, C. (2002). Hydrological Modelling of the Upper Thames Basin Using the Environment Agency's Precipitation-Runoff Model. *Hydrology and Earth System Sciences*, *6*(2), 193–212.

Rana, I. A., & Routray, J. K. (2016, Jul). Risk assessment of flood-prone urban areas of Pakistan. Paper presented at the International Conference on Disaster Management: From Polar Region to the Local Communities, National Institute of Development Administration (NIDA), Bangkok, Thailand.

Randall, M., Sun, F., Zhang, Y., & Jensen, M. B. (2019). Evaluating Sponge City volume capture ratio at the catchment scale using SWMM. *Journal of Environmental Management*, *246*, 745–757. DOI: 10.1016/j.jenvman.2019.05.134 PMID: 31226528

Raza, A. (2024). Climate Change Impacts on Crop Productivity and Food Security: An Overview. In Kanga, S., Singh, S. K., Shevkani, K., Pathak, V., & Sajan, B. (Eds.), *Transforming Agricultural Management for a Sustainable Future. World Sustainability Series*. Springer., DOI: 10.1007/978-3-031-63430-7_8

Renn, O. (2008) White Paper on Risk Governance: Toward an Integrative Framework//Global Risk Governance; Springer: Dordrecht, The Netherlands, pp. 3–73.

Sabir, R. M. (2024). Managing Water Resources for Sustainable Agricultural Production. In Kanga, S., Singh, S. K., Shevkani, K., Pathak, V., & Sajan, B. (Eds.), *Transforming Agricultural Management for a Sustainable Future. World Sustainability Series*. Springer., DOI: 10.1007/978-3-031-63430-7_3

Sabir, R. M. (2024). Remote Sensing and Precision Agriculture: A Sustainable Future. In Kanga, S., Singh, S. K., Shevkani, K., Pathak, V., & Sajan, B. (Eds.), *Transforming Agricultural Management for a Sustainable Future. World Sustainability Series*. Springer., DOI: 10.1007/978-3-031-63430-7_4

Safdar, M., Shahid, M. A., Zaman, M., Rasul, F., Muzammal, H., Raza, A., Sabir, R. M., & Zafar, U. (2023). Drought Monitoring with Multiple Indices and Management through Various Techniques: A Review. *Engineering Proceedings*, *56*(1), 307. DOI: 10.3390/ASEC2023-16602

Srinivas, H., & Nakagawa, K. (2015). Urbanization and Informalization of Waste Management in Asian Developing Countries: Challenges and Opportunities in the Twenty-First Century. *Habitat International*, *47*, 22–31.

Su, B., Huang, H., & Zhang, N. (2015). Dynamic risk assessment method for urban waterlogging based on scenario simulation. *J. Tsinghua Univ. Sci. Technol.*, *55*, 684–690.

Suarez, P., Anderson, W., Mahal, V., & Lakshmanan, T. R. (2005). Impacts of flooding and climate change on urban transportation: A systemwide per-formance assessment of the Boston Metro Area. *Transportation Research Part D, Transport and Environment*, *10*(3), 231–244. DOI: 10.1016/j.trd.2005.04.007

Ullah, K., & Zhang, J. (2020). GIS-based flood hazard mapping using relative frequency ratio method: A case study of Panjkora River Basin, eastern Hindu Kush, Pakistan. *PLoS One*, *15*(3), e0229153. DOI: 10.1371/journal.pone.0229153 PMID: 32210424

UNCHS. (1981) Settlement Planning for Disasters, Nairobi, World Meteorological Organization (W.M.O) (1975) Drought and Agriculture.

Wu, X., Wang, Z., Guo, S., Liao, W., Zeng, Z., & Chen, X. (2017). Scenario-based projections of future urban inundation within a coupled hydrodynamic model framework: A case study in Dongguan City, China. *Journal of Hydrology (Amsterdam)*, *547*, 428–442. DOI: 10.1016/j.jhydrol.2017.02.020

Zeng, Z., Lai, C., Wang, Z., Wu, X., Huang, G., & Hu, Q. (2020). Rapid simulation of urban rainstorm and flood based on WCA2D and SWMM models. *Adv. Water Sci.*, *31*, 29–38.

Zheng, J., Zhang, C., Li, Y., You, Q., Liu, J., Wang, L., & Wu, Y. (2019). Research on prevention and control of waterlogging in subway stations based on InfoWorks ICM model. *J. Beijing Norm. Univ. Nat. Sci.*, *55*, 648–655.

Chapter 4
Environmental Control: Types, Forms, and Methods of Enforcement

Luminita Diaconu
https://orcid.org/0000-0002-6736-6344
Academy of Economic Studies of Moldova, Moldova

Yassine Mouniane
https://orcid.org/0000-0002-6037-5301
Faculty of Sciences, Ibn Tofail University, Morocco

ABSTRACT

This chapter focuses on ecological control methods as critical instruments to enforce legal compliance with environmental standards, and for sustainable management of our natural resources. Primary mechanisms of compliance such as regulatory sanctions, market instruments, technological monitoring and voluntary compliance. Various methodological approaches are also discussed and these include scientific and analytical methods. Six types of control are addressed - inspection, investigation, administrative supervision, monitoring, environmental assessment and audit, the last of which is presented as the most comprehensive method. Similarly there is attention in the chapter to the preparatory phase of control which includes information collection and analysis, comparison of data to find anomalies which directs the planning of verification actions. General controls aim to guarantee the legality, effectiveness, efficiency and transparency of environmental action, while detecting any malfunctions or abuses that may ultimately be necessary to protect the environment.

DOI: 10.4018/979-8-3693-9132-7.ch004

1. INTRODUCTION

The rapid decline in ecosystems, the worrisome loss of biodiversity, and the growing evidence of climate change have all led the world to the pressing need to protect natural resources and achieve sustainability (Wang et al., 2024; Nimma et al., 2025). The combined effects of a range of human activities, including industrial pollution, deforestation, overconsumption, and unplanned urbanization have caused significant stress to the earth's ecological balance in recent decades (Tazerji et al., 2022; Nair et al., 2023). These realities have revealed the inefficiencies of traditional approaches to environmental management, leading to the growth of more integrated and scientifically based approaches to address complex and moving environment issues (Chopra et al., 2024; Pasupuleti et al., 2024). Under this umbrella, a significant strategy that has emerged within the last decade that addresses the disparity between environmental laws and their implementation has been ecological control (Deguine et al., 2021). Ecological control goes beyond enforcement or monitoring, and is multi-dimensional, and integrated in its consideration of how humans interact with the environment (Qu et al., 2024). Ecological control serves as an important tool within the wider framework of environmental governance that focuses on transparency, accountability, and always changing plans for improvement across sectors and levels of intervention (Muccitelli et al., 2023). Essentially, ecological control seeks to achieve that human actions/conducts comply with legal regulations and environmental standards while advancing practices that foster long-term sustainability of ecosystems (Khanam et al., 2023). It is not static nor punitive, but rather a dynamic, adaptive, participatory process that draws on stakeholder feedback and evidence-based decision-making. This contemporary view of ecological control is consistent with how we think about sustainable development, where a comprehensive monitoring and regulatory framework brings together the social, economic, and ecological aspects of sustainability (Ruggerio, 2021; Zhang et al., 2022).

This chapter will provide a deeper examination of the concept, spatial and temporal scope, and method of application of ecological control. The main goal of the chapter is to articulate the meaning of ecological control, and articulate the roles is plays, the value of ecological control today, and its various applications in environmental contexts. Through this lens, the chapter is intended to build on the conceptual and operational treatment of ecological control to offer environmental professionals, policy makers, and researchers a comprehensive, but practical, conceptual underpinnings of ecological control that will guide their work and their efficacy in accomplishing positive environmental outcomes.

This chapter is organized around a few specific goals. It first defines ecological control in its broadest sense, distinguishing it from other forms of environmental oversight, including administrative inspection or judicial enforcement. The chapter also classifies the generally recognized types of ecological control, including legal-regulatory, technical-scientific, economic-incentive, and community-based approaches. In examining these mechanisms, the reader will recognize that ecological control needs to be contextualized and often involve a mix of instruments to achieve the desired results. In addition, the chapter addresses the methodological tools and processes that accompany ecological control in socio-ecological systems. EIA, ecological audits, systems for monitoring data, performance indicators, and GIS are all important aspects of ecological control, as they can help to identify environmental risks, assess compliance, and recommend remedial or preventive action. More specifically, we focus on the preparatory phase of ecological control, which includes collecting relevant information, defining control strategies, and choosing and measuring performance indicators.

Moreover, this chapter emphasizes the critical need for ecological audits, which are defined here as diagnostic tools developed to systematically evaluate the environmental performance of activities, programs, or institutions. Audits enable the identification of non-compliance, gaps in implementation, and potential opportunities to improve performance. Audits are regulatory instruments as well as educational instruments that promote environmental learning within an institution and a community. In conclusion, this chapter argues for a comprehensive, transdisciplinary, and forward-looking form of ecological control that is based on scientific rigor and ethical responsibility. As humanity faces unprecedented environmental challenges, ecological control becomes a compelling and necessary component in the defense against climate change and shaping a sustainable future. Ecological control is not only a tool for compliance and enforcement but also for transformation that helps societies move beyond compliance and contribute to ecological responsibility.

2. LITERATURE REVIEW

The literature review will begin by examining the development of environmental law and the increasing importance of effective enforcement mechanisms (Pan et al., 2024). It will examine different theoretical approaches to compliance behavior, including deterrence theory, social norms theory and behavioral economics, to understand the motivations behind non-compliance and the factors that promote compliance with environmental regulations. The review will analyze different international and national environmental enforcement frameworks, including those used by the Environmental Protection Agency (EPA) in the United States and the European Union (EPA, 2021). It will also assess the effectiveness of different enforcement tools, such as civil penalties, criminal sanctions, injunctions and supplementary environmental projects (SEPs), in achieving environmental objectives

(He et al., 2024). In addition, the literature review will consider the role of compliance monitoring, including self-reporting, inspections and audits, in ensuring that regulated entities meet their environmental obligations (Awewomom et al., 2024). The review will also consider the challenges and limitations of current enforcement approaches, such as resource constraints, regulatory complexity and the difficulty of detecting and prosecuting environmental offences.

The literature review for the chapter synthesizes existing research on environmental control mechanisms, their effectiveness, and the role of enforcement in achieving environmental objectives. This review will focus on various aspects of environmental management systems (EMS), regulatory frameworks, and enforcement strategies. It provides an overview of existing research on environmental control and enforcement methods (Andeobu et al., 2021). Key themes explored include:

2.1. Types and Forms of Environmental Control

- Preventive vs. reactive control (OECD, 2020).
- Voluntary vs. mandatory environmental compliance (Porter & van der Linde, 1995).

2.2. Methods of Environmental Law Enforcement

- Administrative, civil, and criminal enforcement (Faure & Svatikova, 2012).
- The role of environmental agencies and regulatory bodies (Kiss & Shelton, 2021).
- Market-based enforcement tools (e.g., carbon taxes, emissions trading) (Stavins, 2003).

2.3. Comparative Analysis of Global Best Practices

- EU's environmental directives and compliance monitoring.
- The U.S. Environmental Protection Agency's (EPA) enforcement model.
- China's evolving environmental enforcement mechanisms.

Despite significant regulatory advancements, enforcement challenges remain, including lack of institutional capacity, corruption risks, and weak penalties for non-compliance (UNEP, 2022) At the same time, a significant body of literature discusses models of environmental control that link organizational performance to effective environmental management. For instance, a study proposes a conceptual model of environmental control that highlights how enhanced control mechanisms can lead to improved individual, group, and organizational performance. This model emphasizes the importance of workplace capabilities that provide control and correlation with positive behavioral outcomes and business performance (O'Neill, 2010; He et al., 2024).

We must mention that the implementation of EMS has gained traction globally, particularly within defense sectors. A comprehensive review identifies various EMS concepts, applications, and evaluation methods. It notes that while there are no significant differences in EMS adoption across countries or sectors, the evaluation methods remain inconsistent. The study underscores the need for tailored approaches to EMS evaluation that consider specific organizational contexts (Delaney et al., 2025). Additionally, ISO 14001 serves as a benchmark for organizations aiming to implement effective EMS practices. Research indicates that organizations adopting ISO 14001 often experience improvements in environmental performance, although results can vary based on how "environmental performance" is defined (Arocena et al., 2021; Mosgaard et al., 2022).

The effectiveness of different enforcement mechanisms is crucial for ensuring compliance with environmental regulations. Studies have shown that a combination of civil, criminal, and administrative enforcement strategies can enhance compliance rates among regulated entities. The literature suggests that effective monitoring and evaluation practices are essential for understanding the impact of enforcement actions on environmental outcomes (National Audit Office, 2022; Choy et al., 2024; AlMarri et al., 2025).

Despite the extensive research on environmental control and management systems, several gaps remain. For instance, there is a lack of empirical studies that directly link specific enforcement methods to measurable improvements in environmental performance. Additionally, inconsistencies in definitions and measurement scales across studies hinder comparative analyses. Future research should focus on developing standardized metrics for assessing environmental performance and exploring innovative enforcement strategies that can adapt to changing regulatory landscapes.

The literature notes that the types of environmental control are not mentioned separately in the legislation governing the conditions and manner of carrying out inspections. In fact, the legislation refers to the types, conditions and manner of carrying out control in various fields, in such a way that it also applies to environmental protection. The activities of environmental authorities, compliance monitoring, environmental use beneficiaries, environmental tax payers, economic agents, whose activity is related to environmental uses, as well as the correct choice of forms, methods and types of environmental control, we consider that it should be a separate subject of regulations in the field of state control. This issue arises because forms, methods and types of environmental control constitute the basic mechanism of realization of environmental control. Therefore, to increase the efficiency of ecological control, the regulations concerning the characteristics (elements) of this mechanism, namely organizational and technical, economic and ecological control, should be adjusted as much as possible.

The literature suggests taking into account the following aspects, which determine the classification of the types of environmental control, such as: the time of implementation, the volume of problems examined and solved (subject to verification), the nature of the control acts, the method of inspection, the location where the control is carried out, the sources of information examined and inspected, the frequency and character of the controls and others. The following paper will address each of these in turn. Depending on the moment of implementation, environmental monitoring is divided into three categories: preliminary, current (operational) and subsequent. It should be noted that in the practice of environmental authorities, preliminary control is of an auxiliary nature. The application of preliminary control is limited to the planning stage of environmental verifications and is aimed at ensuring that optimal decisions are taken. In this regard, the preliminary environmental control aims to address the issue of granting environmental benefits to the beneficiary of the use of environmental resources. This may include setting or extending the terms of use of environmental resources, identifying the size or amount of payment for the use of natural resources, and so on. To illustrate this point, it is imperative to note that a mandatory condition for the conferral of the right to undertake an economic activity that involves the utilization of natural resources is the implementation of prior control and verification of the ecological state of the environmental factors, as well as the potential beneficiary's capacity to ensure the protection and the use of environmental resources. This is to ensure that the cessation or inability to realize the use of environmental resources is not influenced by the financial or other constraints of the beneficiary.

Externally, the results of preliminary environmental monitoring can be formalized in the form of expert opinions on draft agreements on the granting of environmental use rights facilities, and so forth. Routine (operational) control is carried out on a daily basis in order to prevent violations of “discipline” in the exercise of environmental use rights. In the field of environmental law, such control is part of the regular work of environmental authorities, which is often its essence. This type of control involves assessing the accuracy of activities involving the utilization of environmental factors, including various transactions, and the reliability of data provided by beneficiaries regarding the level of utilization or pollution of environmental factors. For instance, control may encompass the volume of wood mass cut and its sales, the type and area of falling. In this regard, it would be advantageous to introduce the concept of 'environmental evasion', given that the excessive utilization of natural resources by authorized beneficiaries constitutes an act of evasion. The peculiarity of the current ecological control regime is that it is a precondition for the implementation of the concept of realization of economic transactions with environmental resources.

Therefore, the current environmental control is such an operational type of control, which is based on the study of permits and authorizations, primary documents, stocks of recovered resources (such as stocks of deposits, wood, etc.), analysis of the procedure of carrying out recoveries and other such activities. All of them together allow the supervisory authorities, as well as the entities subject to control, to respond quickly to operational changes, which are required in the activity of environmental protection, including prevention of violations of environmental legislation and thus ensure the exclusion of environmental damage. As for the subsequent control, we consider that it is limited to checking the level of realization of the right of environmental benefit only after its realization. It is the post factum verification of economic activities, such as natural resource utilizations, transactions for the past period. This is carried out based on the analysis of reports and balance sheets, as well as through verifications, including audits. These checks are carried out directly on the spot (at enterprises, institutions and organizations) or because of reports provided by beneficiaries. Subsequent control is characterized by an in-depth study of all aspects of the economic activity already carried out. Such an audit is more comprehensive in nature than a preliminary or current audit, but its beneficial effect is more limited. This is because the subsequent control no longer makes it possible to prevent the occurrence of damage resulting from the economic activity carried out.

Considering the frequency of ecological control, it must be recognized that the most complete should be recognized namely the subsequent ecological control. It is carried out at the end of a certain reporting period. This subsequent control is carried out primarily by the method of documentary verification against the de facto situation. It is characterized by an in-depth study of the economic activities of environmental resource use by the recipient of environmental resource use in comparison with those indicated or permitted in the authorization or permit documentation. This control makes it possible to reveal existing deficiencies. These are usually desk checks. In addition to verifying the legality of carrying out the activity of utilization of environmental factors, such controls serve as a basis for the correct calculation of the payment for the environmental utilization already carried out, as well as for assessing the correctness of the documentation of the process of environmental resources utilization. Thus, the main purpose of subsequent environmental control is "to assess the appropriateness and completeness of the fulfillment of obligations related to the use of environmental resources by the beneficiaries of environmental use. It should be noted that with the development of the legislative framework in the context of integration into the European Union, the need for essential changes in the field of environmental control regulations is dictated.

3. MATERIAL AND METHODS

This chapter will use a mixed-methods approach combining legal analysis, comparative case studies and a review of the empirical literature to examine the types, forms and methods of environmental enforcement. The legal analysis will involve a review of environmental laws, regulations and judicial decisions in selected jurisdictions to identify the range of enforcement tools available and the legal standards governing their application. Comparative case studies will be conducted to analyze the implementation and effectiveness of different enforcement approaches in specific environmental contexts, such as air pollution control, water quality management and hazardous waste management. These case studies will draw on publicly available data, government reports and academic research to assess the impact of enforcement on environmental outcomes. In addition, a review of the empirical literature will be undertaken to synthesize findings from quantitative and qualitative studies on the effectiveness of different enforcement methods. This will involve searching academic databases and government websites for relevant studies and critically evaluating their methodologies and findings.

4. RESULTS AND DISCUSSIONS

The results will be organized around key themes such as the effectiveness of different enforcement instruments in deterring non-compliance, the factors influencing the severity of enforcement actions, and the impact of enforcement on environmental quality. The discussion will also address the challenges and limitations of environmental enforcement, such as resource constraints, regulatory complexity and the difficulty of detecting and prosecuting environmental offences (Kiss & Shelton, 2021). In addition, the discussion will explore the role of different actors in the enforcement process, including government agencies, the private sector and civil society organizations. Indeed, the discussion section presents a comparative analysis of the different types and forms of environmental enforcement identified in the literature review and case studies. It will examine the strengths and weaknesses of civil, criminal and administrative enforcement mechanisms, as well as the role of economic instruments such as fines and penalties in promoting compliance.

4.1. Types of Environmental Control Mechanisms

At the same time, based on their enforcement approach, environmental control mechanisms can be categorized into three main types:

4.1.1. Regulatory (Command-and-Control) Mechanisms

Governments set legal limits on pollutants and impose penalties for non-compliance. Example: The EU Industrial Emissions Directive (2010/75/EU) mandates strict emission limits for industries, requiring companies to adopt Best Available Techniques (BAT). Strengths: Legally binding, ensures direct compliance. Weaknesses: Can be rigid and costly for businesses.

4.1.2. Market-Based Instruments (MBIs)

Economic tools that incentivize compliance through taxes, subsidies, and tradable permits. Example: The EU Emissions Trading System (ETS) allows companies to trade carbon credits, creating a financial incentive to reduce emissions. Strengths: Encourages innovation and cost-effective compliance. Weaknesses: Requires strong regulatory oversight to prevent market manipulation.

4.1.3. Voluntary and Self-Regulation Approaches

Companies adopt sustainability standards voluntarily (e.g., ISO 14001, environmental audits). Example: The Eco-Management and Audit Scheme (EMAS) in the EU promotes voluntary compliance. Strengths: Fosters corporate social responsibility (CSR), flexible. Weaknesses: Relies on corporate goodwill, weaker enforcement. These approaches often work together—for instance, regulatory limits are complemented by financial incentives to improve compliance.

4.2. Forms of Environmental Enforcement

Enforcement methods vary by jurisdiction and depend on institutional capacity, technological resources, and governance structures. The key enforcement mechanisms include:

4.2.1. Administrative Enforcement (Civil Sanctions & Fines)

Governments impose financial penalties for violations (Faure & Svatikova, 2012). Example: In Germany, the Federal Environment Agency (UBA) enforces heavy fines for exceeding pollution limits.

Challenge: Some companies factor fines into business costs, reducing deterrent effects.

4.2.2. Criminal Enforcement (Legal Action Against Violators)

Applied in severe cases like illegal dumping, hazardous waste disposal, or environmental disasters. Example: The U.S. Environmental Protection Agency (EPA) prosecuted Volkswagen in the Dieselgate scandal (2015), imposing $14.7 billion in penalties. Strengths: Strong deterrence, sends a clear legal message. Weaknesses: Requires lengthy litigation and proof of intent.

4.2.3. Technology-Driven Monitoring (Remote Sensing & AI-Based Enforcement)

Advances in satellite monitoring, AI, and big data have improved enforcement efficiency. Example: China uses AI-powered pollution monitoring systems to track emissions in real-time. Strengths: Reduces reliance on human inspections, improves efficiency. Weaknesses: High implementation costs, potential privacy concerns.

These methods are often used in combination for maximum effectiveness—for example, AI monitoring (technological) can trigger fines (administrative) or legal action (criminal) for persistent violations.

Effectiveness of Enforcement Methods

- **EU Model:** Strong compliance through centralized regulations and monitoring.
- **U.S. Model:** Mix of strict enforcement and market-based incentives.
- **China's Approach:** Rapid legal reforms but enforcement inconsistencies remain.
- **Developing Nations:** Struggles with weak institutions and lack of resources.

Environmental control is a multidimensional process involving regulatory, economic, and technological measures to ensure compliance with environmental laws. The effectiveness of enforcement varies across jurisdictions based on institutional capacity, legal frameworks, and socio-economic conditions. This section analyzes the different types, forms, and methods of environmental enforcement with real-world examples.

4.3. Comparative Analysis of Enforcement Models

The effectiveness of environmental enforcement depends on the institutional capacity and political will of each country. Below is a comparison of enforcement models from different regions:

Table 1. Comparative Analysis of Enforcement Models

Country/Region	Main Enforcement Strategy	Key Strengths	Challenges
European Union (EU)	Regulatory & Market-Based (ETS, strict pollution limits)	Strong legal framework, cross-border compliance	Complexity in harmonizing national laws
United States (USA)	Mixed Approach (Regulations + market incentives)	Aggressive litigation (EPA), high fines	Political shifts affect enforcement consistency
China	Technology-Driven & Centralized Enforcement	AI-powered monitoring, strict penalties for severe violations	Implementation gaps in rural areas
Developing Nations	Weak enforcement, often donor-dependent	Some progress with international aid	Corruption, lack of funding & institutional capacity

4.4. Key Findings and Results

Based on the analysis, the following results emerge:

4.4.1. Strong enforcement correlates with better environmental outcomes

Countries with strict legal frameworks and monitoring (EU, USA) have higher compliance rates than those with weak enforcement.

4.4.2. Market-based instruments encourage compliance but require oversight

ETS models in the EU and USA show that cap-and-trade systems reduce emissions, but their success depends on strong regulatory oversight.

4.4.3. Technology is transforming environmental control

AI and remote sensing (China's model) enhance real-time enforcement, reducing manual inspections.

4.4.4. Developing countries face enforcement challenges

Lack of financial resources, corruption, and institutional weaknesses hinder effective environmental control.

4.4.5. Hybrid approaches work best

The most effective models combine strict regulations, economic incentives, and technological monitoring for better compliance.

4.5. State environmental control

In specialized literature, we often find the distinction between state, departmental and public environmental control. Although, at first sight, only the controlling authority is invoked for the purposes of these types of control, we believe that it is important to bear in mind that it is not only the criterion of the authority carrying out the control that is important in this respect, but also its powers in carrying out the control, and not least the limits of such control, as well as its effects, such as the binding force of the conclusions and recommendations resulting from the control. As far as state environmental control is concerned, it is established that environmental control which is mandatory. As a rule, it is one of the basic functions of state environmental control authorities, which are bodies of the central or local executive power. Such control can, perhaps, be identified not by when it is carried out, such as preliminary, current or subsequent, but by who carries it out.

Although public environmental control is stated in the legislation, it does not "enjoy" sufficient and satisfactory legislative coverage. It must be recognized that, especially at this stage of development of our society, it is society that is complaining about most problems. This is conditioned by the fact that, in many cases, public authorities associate themselves with the users of environmental resources, either because of the major economic interest or because of the political interest, which, regrettably, should not prevail, but which is often put in a position of preference. Society, however, gives an objective assessment of the irregularities which it finds. It would therefore be necessary to regulate a separate segment relating to public environmental inspections. In this respect, we are of the opinion that the beneficiaries of environmental use should "fear" the public environmental control more than the state one.

Actually, however, the legislation does not regulate public environmental control, in particular the competences, subjects, limits of control, obligations of the beneficiaries of environmental use and other important aspects. Thus, we see that legislation does not have a uniform approach to the types of environmental control. In addition, there is no clear environmental control regime at the legislative level for the realization of public environmental control, which is an issue that needs to be addressed.

4.6. Complex and thematic environmental control

Based on the scope of the controlled problems we distinguish between complex and thematic environmental control. Complex control usually covers all aspects of environmental resource utilization activity, while thematic control covers a specific issue. Comprehensive control is an all-embracing check, covering both the issue of fair use, payment for use, and quality of environmental uses and others. They cover both basic and related environmental resource use activities. A thematic review, in turn, usually covers issues narrowly because they relate to a narrow segment of interest. For example, in legal practice thematic verification may be ordered in relation to the quality of the technological process related to the use of environmental factors.

The practice of environmental monitoring can be categorized into two distinct approaches: planned and unplanned (unannounced). Planned environmental monitoring is characterized by the advance notification to the subject of the period, nature and scope of the control. This approach enables the preparation of all necessary information and documents in advance, thereby facilitating the process for the environmental authority conducting the inspection and ensuring the efficiency of the overall outcome. In contrast, unannounced environmental inspections are typically carried out on an ad hoc basis, sporadically, and are based on a report of a specific violation. Typically, unannounced inspections are conducted on a thematic basis, though this does not preclude the possibility of complexity, contingent on the nature and content of the complaint. It is noteworthy that environmental legislation generally does not impose restrictions on the implementation of unannounced environmental inspections by environmental authorities. However, it should be recognized that the policy is that the control is predominantly planned. Based on the method of checking environmental documents the control is divided into continuous and selective. Basically, the environmental control is based on the verification of all documents and records in the (cadastral and other) registers or in the accounting records, as well as of the factual situation regarding the state of use and protection of environmental resources used by the beneficiary, for the entire period of environmental use activity. The second way involves checking the documentary segment of the period subject to control. However, if a random verification of a limited period of time establishes serious violations of environmental legislation or abuses, then environmental inspectors or other subjects with the right to control in this field of activity may indicate the need for an inspection of the entire period of activity.

In the specialized literature, environmental inspection is also differentiated according to the place where it is actually carried out. Thus, we can distinguish between office-based environmental inspections and environmental inspections carried out on the spot. The difference between these types lies in the place where the inspections are carried out. To a large extent, whether the control is carried out at the place where the use of environmental resources takes place depends on the type of control. If the control is a thematic control, aimed at verifying whether the beneficiary has the necessary permits or specific authorizations, then an on-site control is unnecessary. And conversely, if a complex control is carried out, then the need for on-the-spot checks is an immediate necessity. According to the statistics of the environmental authorities, on-site controls are more frequent, largely also due to their efficiency, because they allow the environmental authority to use various methods of control activity and consequently to obtain more information about the economic activities and the use of environmental resources.

Inspections carried out directly at enterprises and individuals naturally contribute to improving the level of organization of the process of rational use of environmental resources, ensure discipline in the execution of contracts on the part of beneficiaries of environmental use. However, such control cannot be predominant, because the relatively small staff of environmental inspectors, which cannot provide the capacity to check a significant number of beneficiaries of environmental use simultaneously, does not allow such a luxury, but also the permanence of a certain type of control can create many difficulties for the very economic activity of the beneficiaries of environmental resources.

The advantage of desk-checks is that they allow time and effort on the part of the environmental authority to check a large number of beneficiaries of environmental use on the basis of the information they have submitted, including on the basis of declarations or reports and supporting documents.

4.7. Document-based environmental control and factual environmental control

Depending on the sources of information for carrying out the monitoring, we distinguish between document-based and factual environmental monitoring.

Documentary control is based on the study of data about the checked object based on the information contained in documents. Such ecological control is based on the use of such techniques as: logical and arithmetical verification of documents; ecological control assessment of commercial transactions reflected in documents; counter checks on the volumes of resources traded or extracted or stored; assessment of the correctness of reflection of technological processes of natural resource utilization; economic analysis and others.

Factual environmental control, on the other hand, implies a de facto presence on the territory where the utilization of environmental resources takes place, with the realization of the investigation of the enterprise's activity, by applying real control techniques, such as examination, inventory, observation and others.

In practice, due to the lack of sufficient staff in the environmental control authorities, the latter generally prefer to rely on documentary control. The distinction between mandatory and optional environmental control can be made according to the degree of mandatory character. Generally, environmental control is proactive; that is to say, the authorities cannot control the beneficiary of the use of environmental resources as many times and for as long as they wish. However, in certain instances, it becomes imperative for the authorities to undertake repeated inspections or extend the duration of oversight. While environmental inspectors are obliged to ascertain whether a beneficiary of environmental resources possesses the requisite authorization, they are precluded from doing so if they have previously conducted an inspection and ascertained that the individual in question is in possession of such documentation. Conversely, the environmental inspector is obligated to ascertain whether a beneficiary of environmental use continues to commit violations identified during the inspection, particularly if the inspection authority has mandated the rectification of these deficiencies.

Depending on the frequency of environmental monitoring we distinguish between initial and repeated environmental monitoring. The initial control is the control measures carried out in relation to any given user of environmental resources for the first time. It is considered for the first time with reference to a specific object or a specific referral. Repeated environmental monitoring is the monitoring of two or more times during a given period, but which concerns the matters that were the subject of the monitoring.

4.8. Forms of ecological control

The existing literature has not undertaken a separate study of the forms and methods of environmental control. There is often a perception that forms and methods of environmental control represent a totality of techniques, which are sometimes considered as forms of control and sometimes as types of control.

As an example, there are views that traditionally there are such forms of ecological control as preliminary, current and subsequent. However, a characterization of this concept, taking into account the classification of ecological control according to the time of its realization, actually implies its types and not its forms. Discrepancies in terminology are also typical when interpreting such concepts as documentary and effective control. In this respect, it is necessary to define the form of control, including the distinction between form and type of control.

In this context, the term "form of control" is proposed to denote the manifestation or expression of specific actions undertaken by the entities exercising control. This definition is posited as the most suitable interpretation of the category of "form" within the context of philosophical inquiry.

However, it should be borne in mind that not every subject of control itself exercises direct control in one form or another. In scientific literature, the subjects of environmental control are divided into those who directly control the rational use of environmental resources - the State Environmental Inspectorate and subjects whose activity ensures environmental control. This category includes entities, in which the control function appears as one of several functions that it may have. An example would be the General Inspectorate of Emergency Situations.

However, even based on this proper general understanding of this category, we distinguish between forms of control such as quantitative, qualitative and cost, which have their own specific techniques and methods.

We should agree that when interpreting forms of environmental control, we should start with the relationship between content and form as philosophical categories, reflecting the relationship between two sides of any reality. Note that in environmental legislation there is no clear distinction between forms and methods of environmental control.

In the context of environmental governance, the legislator proposes a definition of the methods of its implementation, as evidenced by the legal wording that stipulates the conduct of environmental control by officials of environmental authorities. These authorities are entrusted with the responsibility of conducting controls and inspections, requesting explanations from beneficiaries of environmental use, verifying field data and reporting results. Additionally, they are tasked with inspecting the premises and territories utilized for the purpose of environmental resource utilization. If environmental inspections are indeed considered to fall within the purview of forms of environmental control, then it can be argued that other elements must also fall within this category. In the context of the current legislative framework, the inspection of forest areas and territories used by beneficiaries of environmental resources, with the aim of generating income from this utilization, cannot be regarded as an independent form of environmental control. Instead, it should be conducted outside the framework of environmental control.

In this case, reference is being made to the control acts in the framework of an audit carried out on the initiative of the beneficiary of the environmental audit. It is this author's opinion that this position can be applied only in relation to revenue-generating activities but cannot be excluded for other similar situations. The issue of the legislative definition of forms and methods of environmental auditing is nowadays becoming a rather relevant topic. This is due to the uncertainty and inaccuracy in the scientific definition of forms and methods of environmental control, which is confusing not only the user, but even the legislator. Although the purpose of science is to rely on research and emerging social relations to develop the directions of lawmaking, to give practical recommendations in the development and adoption of regulatory acts, as we see, science is not always able to provide simple and understandable solutions to all existing problems. We will try to offer some recommendations in this regard. In our opinion, the forms of environmental control include environmental monitoring, both state and individual and environmental operational control, including office and field and other varieties.

To ascertain the forms of environmental control, it is first necessary to discuss forms of control in general, since they in fact refer to any area of control, including environmental control. It should be noted that environmental controls inherently presuppose the application of methods of carrying them out. We could give as an example such methods as obtaining explanations from the beneficiaries of environmental use, from economic agents, whose activity is related to the use of environmental resources or their pollution, as well as requesting documents, examination of the factual situation, etc. Individual environmental control procedures may also be included, such as checking the registration of the rights to use environmental resources, checking or confirming the receipt of all permits or authorizations for use, simultaneously checking the appropriateness of environmental uses, checking the validity of the right and the correctness of the use of environmental resources with special environmental regime, checking the compliance with the procedure of natural resources use, checking the execution of the requirements of the state environmental protection agency, checking the compliance of permits.

This model of relations between environmental authorities with the beneficiaries of environmental use has been operating successfully for a long time both in our country and in foreign countries. Thus, attributing the monitoring of the activity of environmental resource use by commercial organizations and individuals to the form of ecological control in our opinion would cause quite controversial discussions. If the form is an external manifestation of the essence of a phenomenon, then the form of ecological control is an external manifestation of the essence of the concept of "ecological control". Since ecological control is an activity of environmental authorities to prevent, detect and suppress violations committed by the beneficiaries of environmental use, then monitoring commercial organizations and individuals, beneficiaries of use of environmental resources, in our opinion, is of the nature, which ensures the performance of ecological control. The monitoring of commercial organizations and individuals, beneficiaries of the use of environmental resources, is a procedure that creates special conditions for ensuring the effectiveness of environmental control by the competent authorities. In this respect, the only form of environmental control remains – is the environmental audit. It does not matter which environmental audit is in question. However, the form of environmental audit is a set of measures regulated by legal norms and which are the external expression of state management activities in the sphere of environmental protection, having as its object the organization of the procedure and methods of conducting the audit. As we said earlier, in our opinion there is only one category of forms of ecological control, which is realized in the form of office ecological audit, on-site ecological audit, environmental monitoring and environmental monitoring verification. Here's the Comparative Analysis of Environmental Enforcement Models bar chart. It visually represents the strength of regulatory enforcement, market-based mechanisms, and technology use across different regions.

Figure 1. Comparative Analysis of Environmental Enforcement Models

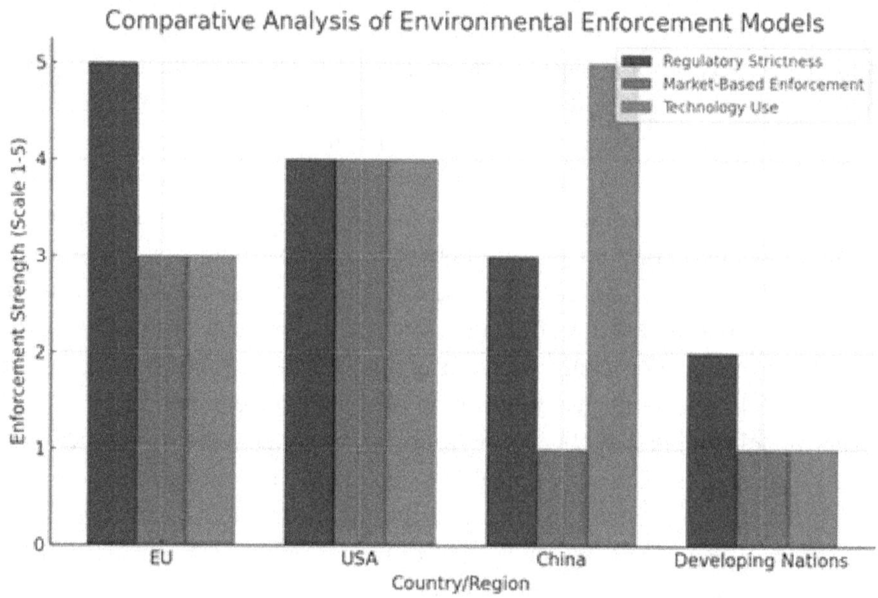

Next, I made a line chart for fines and penalties over time, followed by a pie chart for compliance enforcement methods.

Figure 2. Line chart for fines and penalties over time

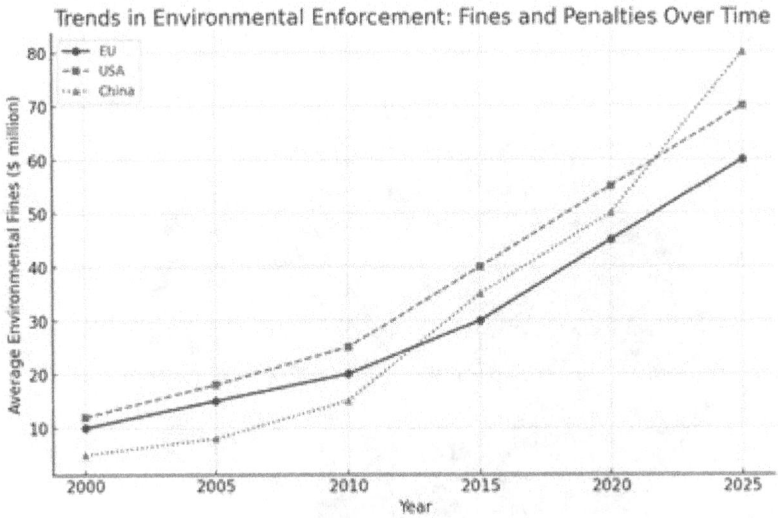

Here's the line chart showing trends in environmental fines and penalties over time. It highlights the increasing financial penalties for environmental violations, with China showing a steep rise post-2015 due to stricter laws. Next, we have a pie chart illustrating compliance enforcement methods

Figure 3. A pie chart illustrating compliance enforcement methods

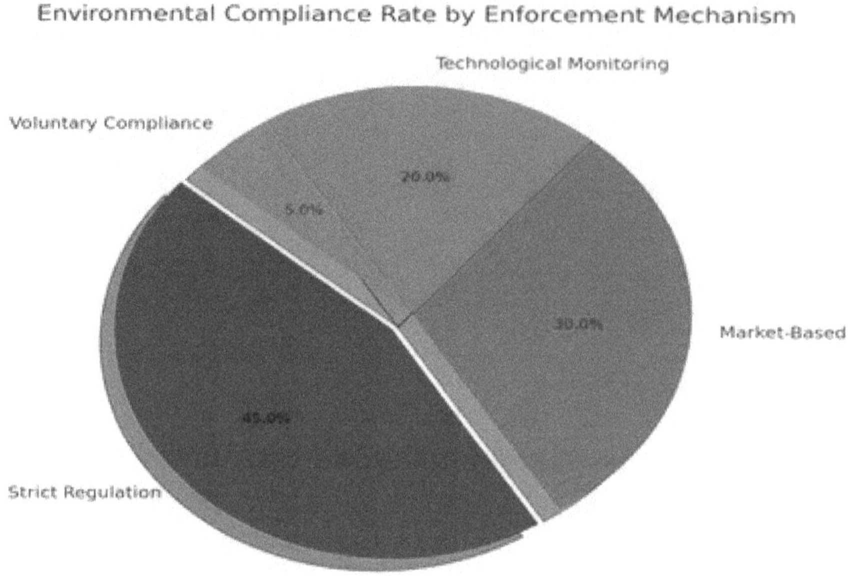

Here's the pie chart illustrating the distribution of compliance enforcement methods. It shows that strict regulatory enforcement (45%) remains the dominant mechanism, followed by market-based instruments (30%) and technological monitoring (20%), with voluntary compliance contributing the least (5%).

4.9. Ecological Control Methods

In the explanatory dictionary, "method" is defined as a set of procedures for achieving a specific goal, i.e. a way of doing something in practice. When defining the concept of environmental control methods, as with the other definitions in the environmental control literature, there is a fairly common approach to what constitutes an environmental control method. It is comprehended as a comprehensive array of techniques, procedures, and means of influence, meticulously designed to achieve a specific objective: namely, to ensure the effective utilization and conservation of the environment.

The term 'ecological control methods' refers to the comprehensive system of scientific, theoretical and practical methods of cognition, including methodological, general scientific techniques for studying the objects of control (analysis, synthesis, induction, deduction, analogy, mathematics, modelling, abstraction, statistical methods, logical method, comparative method and others); empirical methodological techniques (inventory taking, control measurements, checks and counter-checks, logic testing, scanning, written and oral enquiries, questioning of witnesses, examination of documents, examination of objects and documents, inspection of premises (territories) and objects, examination and others); techniques specific to related economic sciences (analytical techniques, theoretical methods such as application of probability (presumptions), mathematical statistics and others).

It should be noted that when environmental controls are carried out, special methods are used. Environmental legislation does not specifically refer to the concept of methods of environmental controls. Therefore, we consider it appropriate to formulate the definition of this concept. Thus, by methods of environmental control should be understood that systematized set of steps and sequential actions, aimed at the successful realization of such checks. Considering the classification of environmental control according to its types, it is necessary to proceed to the characterization of the methods of its implementation, therefore, we will specify that by methods of environmental control, we mean the techniques, methods or means of its implementation. The use of a particular method depends on a number of factors, such as the environmental control status of the control authority or the subject of the control, resulting from the characteristics of the forms of its activity. These would result from the fact, which depends on which authorities exercise control. The same may also result from the object and purpose of the control, from the grounds for the occurrence of ecological control relations of control, etc. In accordance with the methods of carrying out environmental monitoring, we can distinguish the methods of environmental monitoring into:

4.9.1. Inspections

Control acts and procedures carried out on the basis of reporting documentation. The object of verification is the documentary part of economic activity of utilization of environmental resources. Most often, they are carried out in order to eliminate violations detected from the content of the documentation submitted in the reporting process.

4.9.2. Investigations

Control acts and procedures covering a wider range of control indicators than inspections, although they are focused on a specific problem raised on an ad hoc basis;

4.9.3. Supervision

Control acts and procedures carried out by authorities entrusted with the task of monitoring compliance with environmental legislation. This refers to authorities with general powers of enforcement (police, environmental supervisory authorities, etc.), but also to authorities responsible for the management and administrative supervision of any type of environmental protection activity;

4.9.4. Analysis of environmental activities

Control acts and procedures, involving detailed periodic or annual environmental survey and reporting to determine the general assessment of the results of the state of environmental factors, as well as the activity of the use of environmental factors, and in particular, the efficiency of their use.

4.9.5. Observation (monitoring)

Control acts and procedures that involve the provision of data and information on a constant basis, such as, for example, with reference to emissions of air pollutants into the atmosphere - their ecological status, etc.

4.9.6. Audit

Control acts and procedures involving a full examination of the environmental situation in relation to the economic activity of an economic entity with a view to verifying its legality, correctness, adequacy and effectiveness. As previously mentioned, the classification of environmental control is carried out according to several criteria, including its scope.

The scope of environmental controls enables the identification of the following types of control methods formal checks, mathematical-logical checks, normative checks and direct desk checks.

A formal verification includes the verification of all forms and reporting documents established by law (e.g. documents confirming the granting of environmental benefits or confirming the export of products extracted from the natural environment, with the volumes accepted according to the device of permissive documents, etc.).

The verification process also encompasses the availability and clear fulfilment of all mandatory requirements stipulated in environmental control acts.

The mathematical-logical check makes it possible to assess the correctness of the mathematical and logical calculation of one or another indicator. For example, it compares the amount of taxable profit with the sales price and the volumes of production allowed for extraction from the natural environment. During the mathematical-logical verification, it is checked for the correctness of the calculation of the total amounts in the document, both horizontally and vertically.

Normative verification is a check of the content of a document from an ecological control point of view, i.e. in terms of compliance with the legislation in force. This method makes it possible to identify documents with illegal content, to establish facts of unjustified write-off of costs for extracted natural resources, incorrect application of quotas for the use of natural resources, etc.

Direct desk verification is a logical verification of the correspondence of digital data, according to which the beneficiary of use of environmental resources can substantiate the logic of the realized operations and according to which the amount of the volume of used environmental resources is determined.

The methods of carrying out direct desk-based environmental audits usually deal with classified information, which is the main way to establish whether a beneficiary of the use of environmental resources has underestimated the basis of control without carrying out an on-site environmental audit. In this case, the information provided by the beneficiary of the use of environmental resources (its reporting) is compared with data from controls and information from other beneficiaries of environmental use, or economic operators with which the controlled subject has business relations. Therefore, the objective of this method is the information (received) from external sources. Since the results of the analysis carried out under the chapter on the rational use of environmental resources and the environmental situation, and taking into account other available information, the final list of issues to be monitored and verified is established.

The pre-verification work is an important part of the subsequent verification process. During the pre-verification process the first impressions about the organization selected for inspection, and a strategy for a future environmental control, the question of its depth is decided. Here the judgment is made, whether it will be a comprehensive inspection, or it is enough to pay attention only to separate issues. Pre-testing, pre-verification work is aimed at familiarization with the reasons and initial data for the selection of the organization or person as the object of the inspection, verification of the necessity of the documentation to be subject to control, careful checking of errors in the reporting documentation and statements on the environmental situation, collecting additional information necessary for the organization of the work from other authorities (customs authorities, police, fire brigade and other internal affairs authorities), highlighting the main inspection issues (product release, service provision, purchase and evaluation of inventory items), strategic decision making (information storage and management, involvement of other organizations, availability of results from the first meeting, start time of the inspection, number of participants in the inspection) and others. After visiting the organization and verifying all collected data (documentation and information), it is advisable to draw up a final plan as to which issues are subject to verification.

Critical steps in the process of planning and preparing the environmental audit are:

1. Formation of an informational data table of environmental benefit recipients;
2. Desk audit of the basic situations during the control, including verification of the logical connection between individual reporting and indicators of factual use of concrete environmental resources (deposits, wood, fish, water, etc.), checking the comparability of the reporting indicators with similar indicators from the previous reporting periods, assessment of the observed situations of use and statements of use of environmental resources in terms of their conformity with the data available to the environmental control body according to the information from the economic documents from the beneficiary of use of environmental resources, as well as from other sources;
3. Data collection and identification of enterprises in similar fields of activity carried out by the beneficiary under control;
4. Analyzing the level of dynamics of the main economic indicators related to the economic activity of the enterprise, including comparative analysis with similar data of the activity of other enterprises;
5. Identification of doubtful data or inconsistencies, which may indicate possible infringements.

However, the main method of subsequent control is audit. This method of environmental control is concerned with the subject, object, purpose and basis of control. In our opinion, methods of ecological control refer to ecological control, (verifications of documentation and whether the actual state of things corresponds to what has been reported to the environmental authorities, etc.). Methods of state ecological control implementation are checks in the form of audit, examination, authorization of operations and others.

An audit means a comprehensive verification of the activities of the control unit, which is expressed in carrying out control actions on the documentary and factual study of the legality of the entire process of utilization of environmental resources and the correspondence of the final volume or turnover with the actual consumption or influence of environmental resources. No less important is the need to analyze the reliability and correctness of their reflection in the documentation of environmental uses.

The results of the inspection and audit are documented by reflecting them in a document. Examination, as a method of environmental control, means the analysis and evaluation of the condition in a certain area of activity of the object of control. The results of the examination are usually formalized in a conclusion.

Authorization of activities for the use of environmental resources means a preliminary activity of environmental use authorization. This method also implies the right to refuse to authorize the activity in question. However, the authorization to carry out activities using environmental resources is reflected in the registration of the permit in an act, but only after checking the documents submitted for the purpose of carrying out the environmental uses.

Inspections are divided into desk and field inspections, including meter checks. Desk inspections are inspections carried out on the premises of the state authority competent for environmental control.

On-site inspections are inspections carried out according to the location of the object of control, during which it is determined whether the implementation of environmental resource recovery activities complies with the provisions of the primary documents authorizing the use.

Not least important are such methods as "counter-inspections", which are those inspections that are carried out in the framework of on-site and/or office inspections, but which are aimed at establishing and/or confirming the facts related to the activities carried out by the beneficiary of the object under control.

Thus, we can conclude that the environmental control activities of the state environmental control authorities are aimed at the procedure of processing the administered information, as well as at the results of implementation of the environmental control objectives. However, it should be noted that the existing regulations indicate different levels and ways of application of the concepts of "methods" and "forms" of control, as well as their inter applicability. Thus, the legislator often assesses the quality of the methods of state environmental control, namely in terms of the forms in which it is carried out. Based on the above, noting the lack of uniformity in the theoretical understanding of the concept of methods of environmental control in environmental law, we note that this leads to a lack of consensus in the understanding of methods of environmental control. For example, environmental control methods have been understood as "specific techniques and ways of its implementation" or "the totality of techniques and methods used by the authorized authorities in environmental control activities".

Eventually, we highlight the diverse approaches to environmental control and enforcement. Our key findings include:

- Strong enforcement correlates with improved environmental outcomes, but effectiveness depends on institutional capacity.
- Market-based instruments complement traditional enforcement methods but require proper oversight.
- Technological advancements enhance monitoring but need legislative support.
- Developing nations face enforcement challenges due to weak regulatory institutions.

Thus, we recommend enhancing regulatory enforcement, increasing transparency, and integrating advanced monitoring tools to improve environmental governance worldwide.

5. CONCLUSIONS

This study highlights that environmental control mechanisms vary significantly across jurisdictions. While some countries rely on strict regulations (EU, USA), others focus on technology-driven monitoring (China) or market incentives (carbon trading). The research has provided a comprehensive analysis of environmental control mechanisms, including types, forms, and methods of enforcement across different jurisdictions. The findings reveal that effective environmental governance requires a multi-faceted approach that integrates regulatory, economic, and technological measures.

In conclusion, below are our policy recommendations:

- Enhance regulatory enforcement by increasing penalties for violations and expanding oversight agencies.
- Integrate AI-driven monitoring systems to strengthen pollution tracking and compliance verification.
- Encourage market-based solutions such as cap-and-trade programs to balance economic growth and environmental protection.
- Improve international cooperation through cross-border environmental agreements and knowledge-sharing initiatives.
- Invest in capacity-building programs in developing countries to support stronger institutional frameworks for environmental control.
- Strengthen regulatory enforcement with higher fines and stricter penalties for repeat offenders.
- Expand market-based incentives to encourage cost-effective compliance.

- Invest in AI and remote sensing technologies to improve monitoring and detection.
- Enhance international cooperation to enforce environmental regulations globally.
- Support capacity-building efforts in developing countries to improve enforcement institutions.

A hybrid enforcement model—combining legal, economic, and technological measures—is the most effective strategy for achieving long-term environmental compliance and sustainability.This study underscores the need for continuous improvements in environmental governance through legal, economic, and technological innovations. Future research should focus on evaluating long-term trends in enforcement effectiveness and the role of emerging technologies in environmentally.

REFERENCES

AlMarri, M., Al-Ali, M., Alzarooni, M., AlTeneiji, A., Al-Ali, K., & Bahroun, Z. (2025). Enterprise Resource Planning Systems for Health, Safety, and Environment Management: Analyzing Critical Success Factors. *Sustainability (Basel)*, *17*(7), 2947. DOI: 10.3390/su17072947

Andeobu, L., Wibowo, S., & Grandhi, S. (2021). An assessment of e-waste generation and environmental management of selected countries in Africa, Europe and North America: A systematic review. *The Science of the Total Environment*, *792*, 148078. DOI: 10.1016/j.scitotenv.2021.148078 PMID: 34147806

Arocena, P., Orcos, R., & Zouaghi, F. (2021). The impact of ISO 14001 on firm environmental and economic performance: The moderating role of size and environmental awareness. *Business Strategy and the Environment*, *30*(2), 955–967. DOI: 10.1002/bse.2663

Awewomom, J., Dzeble, F., Takyi, Y. D., Ashie, W. B., Ettey, E. N. Y. O., Afua, P. E., Sackey, L. N. A., Opoku, F., & Akoto, O. (2024). Addressing global environmental pollution using environmental control techniques: A focus on environmental policy and preventive environmental management. *Discover Environment*, *2*(1), 8. DOI: 10.1007/s44274-024-00033-5

Chopra, S. S., Senadheera, S. S., Dissanayake, P. D., Withana, P. A., Chib, R., Rhee, J. H., & Ok, Y. S. (2024). Navigating the Challenges of Environmental, Social, and Governance (ESG) Reporting: The Path to Broader Sustainable Development. *Sustainability (Basel)*, *2024*(16), 606. DOI: 10.3390/su16020606

Choy, S., Jiang, S., Liao, S., & Wang, E. (2024). Public environmental enforcement and private lender monitoring: Evidence from environmental covenants. *Journal of Accounting and Economics*, *77*(2-3), 101621. DOI: 10.1016/j.jacceco.2023.101621

Deguine, J. P., Aubertot, J. N., Flor, R. J., Lescourret, F., Wyckhuys, K. A. G., & Ratnadass, A. (2021). Integrated pest management: Good intentions, hard realities. A review. *Agronomy for Sustainable Development*, *41*(3), 38. DOI: 10.1007/s13593-021-00689-w

Delaney, P. G., De Vos, S., Eisner, Z. J., Friesen, J., Hingi, M., Mirza, U. J., Kharel, R., Moussally, J., Smith, N., Slingers, M., Sun, J., & Thullah, A. H. (2025). Challenges, opportunities, and priorities for tier-1 emergency medical services (EMS) development in low- and middle-income countries: A modified Delphi-based consensus study among the global prehospital consortium. *Injury*, *56*(1), 111522. DOI: 10.1016/j.injury.2024.111522 PMID: 38599953

Environmental Protection Agency (EPA). (2021). Environmental compliance and enforcement annual report. https://www.epa.gov/

Faure, M., & Svatikova, K. (2012). Environmental law enforcement: Comparative analysis. *Environmental Law Review*, *14*(1), 1–22.

He, Y., Yan, X., Fan, A., Yin, X., Meng, Q., Zhang, L., & Liu, P. (2024). Assessing the influence of actual LNG emission factors within the EU emissions trading system. *Transport Policy*, *159*, 345–358. DOI: 10.1016/j.tranpol.2024.10.030

Khanam, Z., Sultana, F. M., & Mushtaq, F. (2023). Environmental Pollution Control Measures and Strategies: An Overview of Recent Developments. In Mushtaq, F., Farooq, M., Mukherjee, A. B., & Ghosh Nee Lala, M. (Eds.), *Geospatial Analytics for Environmental Pollution Modeling*. Springer., DOI: 10.1007/978-3-031-45300-7_15

Kiss, A., & Shelton, D. (2021). *International environmental law*. Oxford University Press.

Mosgaard, M. A., Bundgaard, A. M., & Kristensen, H. S. (2022). ISO 14001 practices – A study of environmental objectives in Danish organizations. *Journal of Cleaner Production*, *331*, 129799. DOI: 10.1016/j.jclepro.2021.129799

Muccitelli, S., Pozzi, C., D'Ascanio, R., & Magaudda, S. (2023). Environmental Contract: A Collaborative Tool to Improve the Multilevel Governance of European MPAs. *Sustainability (Basel)*, *15*(10), 8174. DOI: 10.3390/su15108174

Nair, N. V., & Nayak, P. K. (2023). Exploring Water Quality as a Determinant of Small-Scale Fisheries Vulnerability. *Sustainability (Basel)*, *15*(17), 13238. DOI: 10.3390/su151713238

National Audit Office. (2022). Environmental compliance and enforcement. https://www.nao.org.uk/wp-content/uploads/2022/05/Environmental-compliance-and-enforcement.pdf

Nimma, D., Devi, O. R., Laishram, B., Ramesh, J. V. N., Boddupalli, S., Ayyasamy, R., Tirth, V., & Arabil, A. (2025). Implications of climate change on freshwater ecosystems and their biodiversity. *Desalination and Water Treatment*, *321*, 100889. DOI: 10.1016/j.dwt.2024.100889

O'Neill, M. J. (2010). A model of environmental control and effective work. *Facilities.*, *28*(3/4), 118–136. DOI: 10.1108/02632771011023104

OECD (Organisation for Economic Co-operation and Development). (2020). Environmental enforcement and compliance: Best practices. OECD Publishing. https://doi.org/DOI: 10.1787/9789264280359-en

Pan, J., Du, L., Wu, H., & Liu, X. (2024). Does environmental law enforcement supervision improve corporate carbon reduction performance? Evidence from environmental protection interview. *Energy Economics*, *132*, 107441. DOI: 10.1016/j.eneco.2024.107441

Pasupuleti, V., Thuraka, B., Kodete, C. S., & Malisetty, S. (2024). Enhancing Supply Chain Agility and Sustainability through Machine Learning: Optimization Techniques for Logistics and Inventory Management. *Logistics (Basel)*, *8*(3), 73. DOI: 10.3390/logistics8030073

Porter, M. E., & van der Linde, C. (1995). Toward a new conception of the environment-competitiveness relationship. *The Journal of Economic Perspectives*, *9*(4), 97–118. DOI: 10.1257/jep.9.4.97

Qu, Y., Wang, C., Wei, Y., Wu, L., & Zheng, N. (2024). Does Eco-Innovation of Emerging Market Firms Benefit from Knowledge Spillovers of MNC in a Multidimensional Task Environment? *MIR. Management International Review*, *64*(3), 527–565. DOI: 10.1007/s11575-024-00546-4

Ruggerio, C. A. (2021). Sustainability and sustainable development: A review of principles and definitions. *The Science of the Total Environment*, *786*, 147481. DOI: 10.1016/j.scitotenv.2021.147481 PMID: 33965820

Stavins, R. N. (2003). Market-based environmental policies: What can we learn from U.S. experience (and related research)? Resources for the Future. https://doi.org/DOI: 10.2139/ssrn.421720

Tazerji, S. S., Nardini, R., Safdar, M., Shehata, A. A., & Duarte, P. M. (2022). An Overview of Anthropogenic Actions as Drivers for Emerging and Re-Emerging Zoonotic Diseases. *Pathogens (Basel, Switzerland)*, *11*(11), 1376. DOI: 10.3390/pathogens11111376 PMID: 36422627

UNEP (United Nations Environment Programme). (2022). *Strengthening environmental law compliance and enforcement*. UNEP Publishing.

Wang, Z., Wang, T., Zhang, X., Wang, J., Yang, Y., Sun, Y., Guo, X., Wu, Q., Nepovimova, E., Watson, A. E., & Kuca, K. (2024). Biodiversity conservation in the context of climate change: Facing challenges and management strategies. *The Science of the Total Environment*, *937*, 173377. DOI: 10.1016/j.scitotenv.2024.173377 PMID: 38796025

Zhang, S., Zhang, M., Qiao, Y., Li, X., & Li, S. (2022). Does improvement of environmental information transparency boost firms' green innovation? Evidence from the air quality monitoring and disclosure program in China. *Journal of Cleaner Production, 357*, 131921. DOI: 10.1016/j.jclepro.2022.131921

Chapter 5
Harnessing AI for Climate Solutions: Tools and Techniques

Kumar Parmar
https://orcid.org/0000-0002-2502-5680
Marwadi University, India

Tejas Chandulal Chauhan
https://orcid.org/0009-0006-8718-4130
Marwadi University, India

T. Premavathi
https://orcid.org/0009-0003-0172-2021
Marwadi University, India

ABSTRACT

Subsequently, AI has been instrumental toward suppressing the climate change issue considering it sets creative techniques across the various quandaries associated with the global changes. The following areas of use have been identified: energy control, reduction of greenhouse gas emissions, climatic simulation, and prediction of storms. The strength of AI that is its suitability to work on vast, intertwined datasets can be used to explain patterns and findings which would help researchers and policymakers in eradicating climate change. Looking at the impact of climate change, AI is employed to improve energy, and decrease its usage as well as adopt to green energy. AI systems, by nature, are highly dependent on data, underline the motivating computational requirements and energy utilization, in general, and are linked with the carbon footprint. In addition, lack of data, data quality issues, and relevance and originality of professional practice could also avert the application of intelligent climate solutions based on artificial intelligence, particularly in the

DOI: 10.4018/979-8-3693-9132-7.ch005

developing world.

1. INTRODUCTION TO AI AND CLIMATE CHANGE

"Climate change is a complicated and multifarious global concern needing a thorough and integrated response involving many stakeholders including governments, businesses, and civil society. The difficulties presented by climate change call for the creation and use of sensible answers using the Fourth Industrial Revolution's technical innovations—especially artificial intelligence (AI) technology. Improving energy efficiency, increasing renewable energy systems, optimizing resource management, and enabling early warning systems for natural disasters are just a few of the several facets of climate change mitigating and adaptation strategies that AI has shown great promise in addressing. More resilient and environmentally friendly solutions can be created by using AI's powers to help worldwide adaption initiatives and lessen effects of climate change by Moreover, artificial intelligence-powered technologies are very important for tracking and evaluating climate data, forecasting future climatic patterns, and guiding policy actions meant to more successfully address climate change(Chen et al., 2023). Still, it's important to recognize that the creation and application of artificial intelligence systems themselves have environmental effects that need careful thought and resolution. Particularly deep neural networks, recent research have underlined the considerable energy consumption and carbon emissions connected with running some artificial intelligence models. This begs questions about the possible environmental effects of artificial intelligence and underlines the need of giving sustainable and energy-efficient AI methods top priority, sometimes known as "Green AI." Developing and implementing AI systems meant with environmental issues in mind would help to guarantee that their contributions to climate change mitigating and adaptation are really significant and consistent with sustainability aims. This strategy calls for giving energy efficiency top priority, using renewable energy sources, and applying sensible data management techniques front stage. By tackling these issues and using artificial intelligence's potential responsibly, its power can be used to properly fight climate change while lowering its environmental impact (Olawade et al., 2024; Taddeo et al., 2021).

Figure 1. AI-Driven Climate Action Cycle

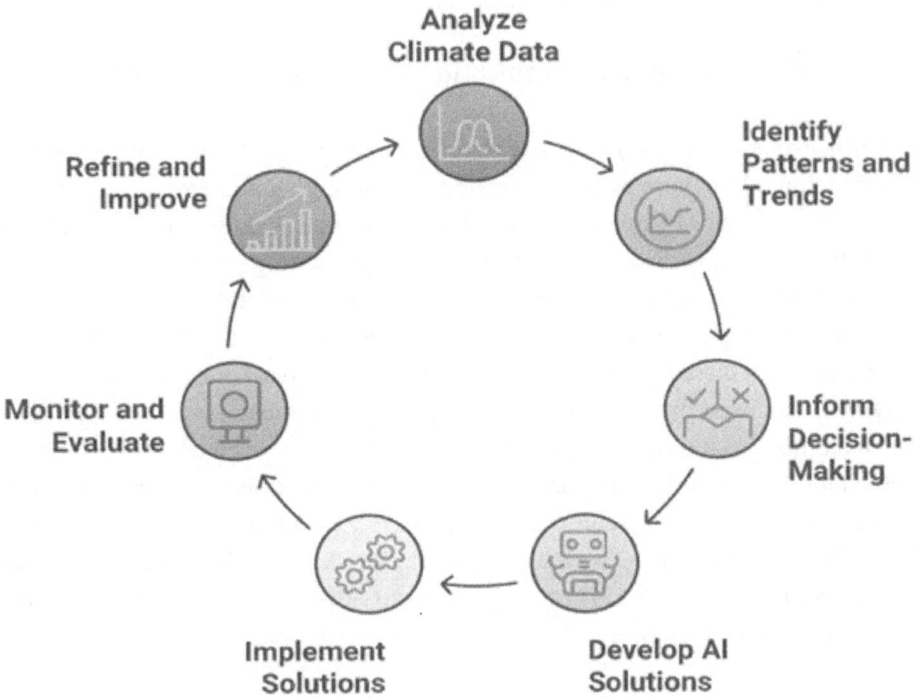

1.1 Overview of AI Technologies

Artificial intelligence refers to a combination of strong computing methods used by machines to carry out tasks that ordinarily call for intelligence. Of those, machine learning can be regarded as one of primary importance. When not being taught in each way by programmers, it uses certain computations to analyze data to make forecasts on the results they get. Cognitive learning is a subset of machine learning, typically using an artificial neural network with multiple layers to analogously process information on the model of the human brain. Through this kind of a combined structure, deep learning models can decipher hierarchical patterns and patterns from large datasets. These technologies are already in the process of revolutionizing an array of disciplines. In the field of medicine machine learning algorithms help detect diseases based on scans with high precision. There are multiple usages of machine learning where it is applied in financial institutions for fraud detection: Be it credit card payments or other, is identified instantly. These examples provide an indication of the information processing and decision-making capacities of AI which opens the door to addressing one of the greatest challenges facing the world today: climate change; (Premavathi, 2023).

1.2 The Climate Change Challenge: Causes and Impacts

Climate change refers to changes in the atmosphere's effective temperature attributable mainly to human action, particularly through the release of greenhouse gases. The emissions by burning of fossil resources like coal used in power stations and petrol or other derivatives used in vehicles generating greenhouse gases like carbon dioxide. Agriculture expansion and logging accelerate deforestation that works in the opposite direction by eliminating trees which act as carbon sinks. The repercussions of these actions are extended and progressively emerging in the world market. It has been brought out that average global temperatures are increasing, which has been resulting in frequent instances of heat waves, changes in patterns of precipitation and increased proportion of violent climatic events including hurricanes and drought. Global temperatures are increasing and leading to increased sea levels since glaciers and ice sheets are melting putting at risk of flooding coastal cities and other terrestrial ecosystems. These impacts justify the necessity for proper climate intervention (Kumar et al., 2024).

1.3 The Potential of AI to Support Global Climate Action

AI presents a set of important tools in the fight against global warming, as the technology allows to work with big data, revealing trends for both the reduction of the negative impact and the search for ways to adapt to the consequences. From the mitigation point of view, AI can help in reducing the energy consumption in the buildings, designing better transportation systems and of course, fast-tracking the use of the green technologies and products including the renewable energy sources. For instance, the adoption of AI can increase the effectiveness of the incorporation of renewable power sources such as solar and wind energy into current electric networks, as well making the current systems less dependent on fossil fuel. AI also has a great impact and can help societies adapt because they must change because of global climate change. The current advance in Artificial Intelligent can be used to come up with early warning to extreme weathers for instance floods and droughts, thus the impact can be reduced. Moreover, it can help in preparing climate smart structures for global warming for instance construction of sea walls that protect cities from floods as well as heat resistant crops that feed the people even in the event of dry spells.

1.4 Data and Climate: Why Data is Relevant to Decision-Makers

Climate action as a process requires decision-making, and decision making requires knowledge, specifically climate knowledge of the vast universe that is climatic data. This data ranges from atmospheric temperature measurements and satellite-based estimations of ice caps to detailed climate forecasts of climate change scenarios. AI performs exceptionally well in handling these large data sets to make sense of them by providing the much-needed pattern or trend analysis. AI assist Cowls et al., (2021) indicating that by turning big data into useful information, AI can support researcher, policymakers and industries in finding and implementing proper climate solutions.

1.5 Interdisciplinary Collaboration

As the climate change advocation is indeed a complex problem that cannot be solved through the lens of a single discipline, RD should embrace interdisciplinary point of view and engage experts from an array of field domains. Combining the newest AI technology with the climate science expertise of climate scientists, policy making and decision-making authority of policymakers and government officials, on-the-field experience of industries and representatives of communities affected by climatic change is vital. This integration of multiple approaches led by different faculties of studies such as computation, climate, policy, and local and indigenous knowledge is crucial for proposing diverse and interrelated and transformative solutions that can adequately respond to the complexities of global climate change. If properly utilised, the strengths and value that all these actors bring to the table, the approaches could offer fresh solutions to approaching what is arguably the world's greatest challenge – climate change. This integrated approach helps to maintain academic and practical relevance of the solutions to be developed and to consider the possibility of significant positive change.

2. AI IN CLIMATE DATA ANALYSIS AND PATTERN RECOGNITION

This includes things like the power of AI to analyze large, high-dimensional datasets, to separate signal from noise, to draw statistically valid conclusions based on observation, or to make accurate predictions in a complex environment climate science remains just as data-centric and challenging as it was before – analysts are faced with huge volumes of data from satellites, sensors, and models, for example. Because of the large data sizes that AI can cater the analysis of, it is able to be a critical aid in the deciphering of patterns, trends, and relations that will help us better understand climate change and aid in formulating valuable mitigation and adaptation strategies.

2.1 Implications of Climate Data Systems Complexity

Weather information itself is multifaceted, comprising many distinct features and indicators monitored over various spatial and temporal units. Just think about how they attempt to solve a global jigsaw with parts as temperature measurements, cloud pictures by satellites, salinity by ocean buoys, and ice cores that give one the overview of past climates (Lundstad et al., 2023). In this case, traditional approaches for data analysis do not work well with such multidimensional data. This is where AI comes into the play to provide a decisive and efficient instrumental means to deal with such a complexity. Normal human way of analyzing data cannot be able to handle data from various sources in coming up with correlations and patterns just as an AI algorithm can. Those unapparent connections uncovered by AI give a broader and complex view on Earth's climate system.

2.2 AI Techniques for Multidimensional Climate Data Processing

Two fundamental approaches in machine learning prove particularly valuable for climate data analysis: An example of a figure from a health economics assignment is shown below Figure. 2 Machine learning approach to use for climate data analysis. Supervised Learning: In this approach, algorithms are trained from labelled databases and the kind of output required is already provided. For example, the system may be taught about historical information regarding hurricanes, such as their tracks and the degrees of their intensities, and given these guides, it could forecast hurricane tracks and intensities. Unsupervised Learning: Here, algorithms search structures on virtually unknown data sets with no prior pattern recognition concept. This is valuable in finding density in climate informatics where one might not expect, for instance, connections between ocean currents and precipitation.

Figure 2. Machine Learning Approaches

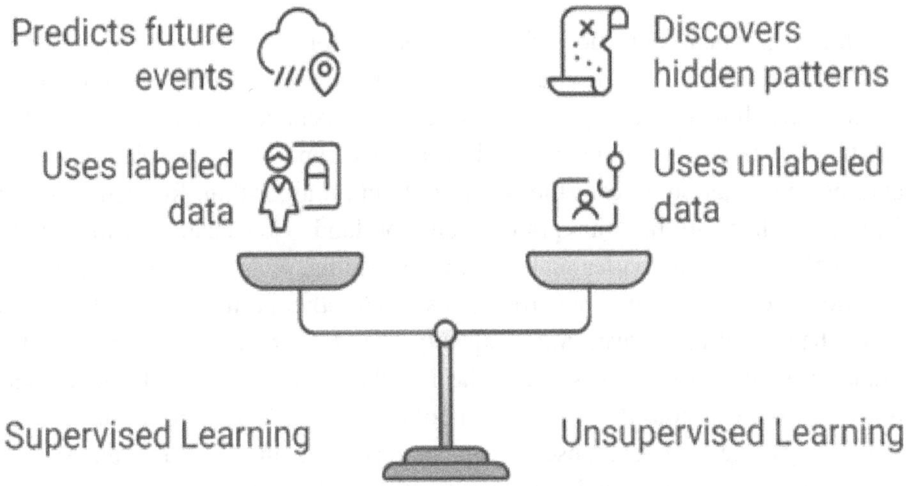

Subcategories within these broad categories include decision trees, SVMs and neural networks and each algorithm may prove advantageous within climate data analysis, depending upon the data and questions being posed.

2.3 Identifying Patterns in Climate Change Through Machine Learning

Owing to the capability of identifying complex patterns, AI has made substantial contribution in climate science. For instance, there are correlation patterns between deforestation rates in each area and alterations to temperature and precipitation in that area that have not been apparent through traditional analytical techniques but which have now been identified through the application of machine learning. In this way, AI is valuable for locating such correlations, which makes it possible to improve, for instance, the tactical arising of decisions on reforestation, according to the areas that would supposedly have the greatest influence on climatic control. Additionally, AI algorithms are employed to enhance the precision of the climate models used in analysis of climate prospects. These models are usually deficient in simulating some of the intricate relationships existent in climate system such as that between clouds, ocean and the land. The use of these methodologies can be supported by artificial intelligence methods to improve the representation of these processes and, therefore, the forecast of future climatic conditions.

2.4 The Role of AI in Revealing Root Causes of Climate Change

While the general detection of patterns is possible, AI has the significant potential of identifying specific causes of climate change by firmly relating climate information to human actions and their effects. For instance, it would otherwise take ages to analyze large amounts of satellite imagery and other geospatial data that captures deforestation all over the world with an AI algorithm. Ref. Chakraborty, 2024 It can therefore link this granular data on land cover changes with detailed records of industrial activities such as logging, mining, or agricultural expansion to measure directly the amount of carbon emission or other pollution impacts because of the changes in land cover. Such capacity to link climate changes with certain human activities on the local as well as the global level is instrumental in identifying policy-specific solutions and enforcement of the guilty. Finally, the ability of AI in establishing these basic relations between human actions and climatic changes offers helpful information that may improve mitigation and adaptation actions.

2.5 Case Studies: AI Discoveries in Climate Science

For centuries, the climate has been one of the most challenging areas for theorisation, but researchers have used the AI Godsend in this aspect. Tracking Deforestation: Moreover, with the help of satellite imagery and AI technologies, Cowls et al. 2021 have geolocated the main process of deforestation and carbon emissions. Analyzing Historical Temperature Data: Analyses done using the machine learning methods have revealed other unexplored relations between climatic temperatures and other factors that were important for understanding the long-term variations in temperature (Niyogi & Srinivasan, 2023).

3. AI FOR CLIMATE MODELLING AND PREDICTION

In addition to advancing climate science, AI is being instrumental in better climate projections and prediction systems (Gesami & Nunoo, 2024). AI can handle large volumes of data as well as find more intricate patterns that will enhance the modeling of critical climate phenomena like cloud formation, the interaction between the ocean and the atmosphere, and climatic catastrophes.

3.1 The Importance of Climate Models in Understanding Future Trends

Climate models represent complex general systems that involve climate system through atmosphere, ocean, ice and land. Climate models apply the scientific laws of physics used together with the interactions of the components in order identify how climate may change with the effects of various factors including the concentration of greenhouse gases (Nguyen et al., 2023). Contemporary climate modelling plays a central role to give accurate predictions to policymaking. They help in owning up on specific hazards like floods due to rises in sea level and other catastrophes that make governments and affected persons come up with measures for coping or minimizing such impacts. Accurate climate models and projections serve, therefore, a critical purpose in strategic decision processes related to infrastructure investments, hazard mitigation and response, and strategic provisions for the management of climate change risk. The use of AI approaches has been realized as a potential solution to optimizing climate model's precision and credibility.

3.2 AI-Driven Climate Models: Enhancing Precision and Accuracy

Using the classical models of climate one can get a qualitative and yet powerful picture of the climate of the Earth; however, this presentation can reflect only several aspects of climate determination. What makes AI useful is the possibility to refine these models and update them with better data and more precise algorithm. Machine learning algorithms can take large sets of climate data and observations as well as the simulations conducted by models and make reflections that may be difficult for other models. For instance, AI driven models have produced better estimates of the increases in global temperature, the reduction in sea ice, and the number as well as severity of storms. When integrated into climate studies AI can help scientists build models that have higher precision in dealing with future climatic change trends (Gesami and Nunoo 2024). Figure. 3 describes information about how artificial intelligence improves climatic models for enhanced forecasts.

Figure 3. AI Enhances Climate Models for Better Predictions

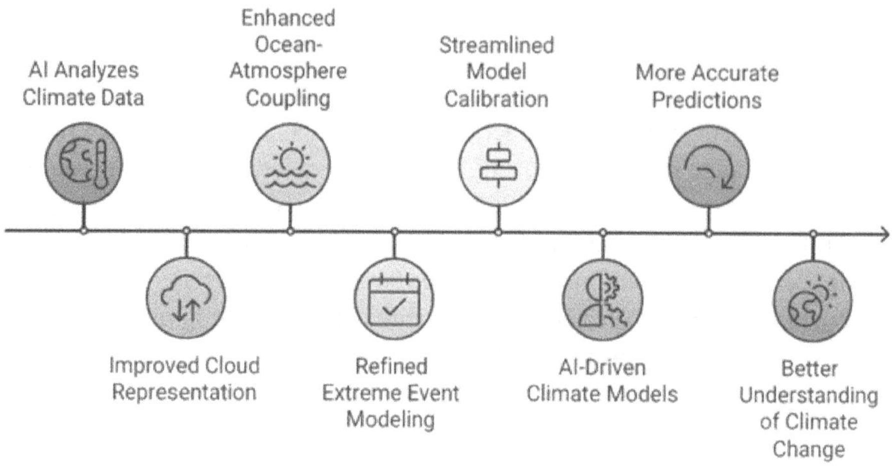

Some key ways in which AI is transforming climate modelling include: Improving the representation of clouds and other sub-grid scale processes: AI can assist the climate models to represent better the smaller-scale features, such as cloud formation or precipitation, which is difficult to achieve using large resolution models. Enhancing ocean-atmosphere coupling: Programmed models can find latent correlations between ocean and atmospheric phenomena and thus improve existing climate models. Refining extreme event modelling: We can use AI to deepen the understanding of frequency, intensity, and distribution of such extreme events by analyzing their historical characteristics, for example, hurricanes or severe heat waves. Streamlining model calibration and optimization: This can be an inconvenient task when done manually to tune and fine-tune the model to better match reality and AI can reduce the time it takes in developing models. Although climate models guided by artificial intelligence have time and again demonstrated high degrees of accuracy, a combination of the traditional physical models with AI models will provide far more profound and inclusive tools for climate analysis.

3.3 Predicting Extreme Weather Events with AI

This paper argues that knowledge in the occurrence of unfavourable weather conditions in an area is central in preventing its effects on lives and property. Machine learning is providing advanced methods for analysing these large repositories of climatic information like satellite images, weather station records or climate models concerning these climatic events such as hurricanes, floods, droughts, heat waves etc and the possibility of their occurrence is now being predicted with the help of AI. Refer Figure. 4 for AI in Extreme Weather Prediction.

Figure 4. AI in Extreme Weather Prediction

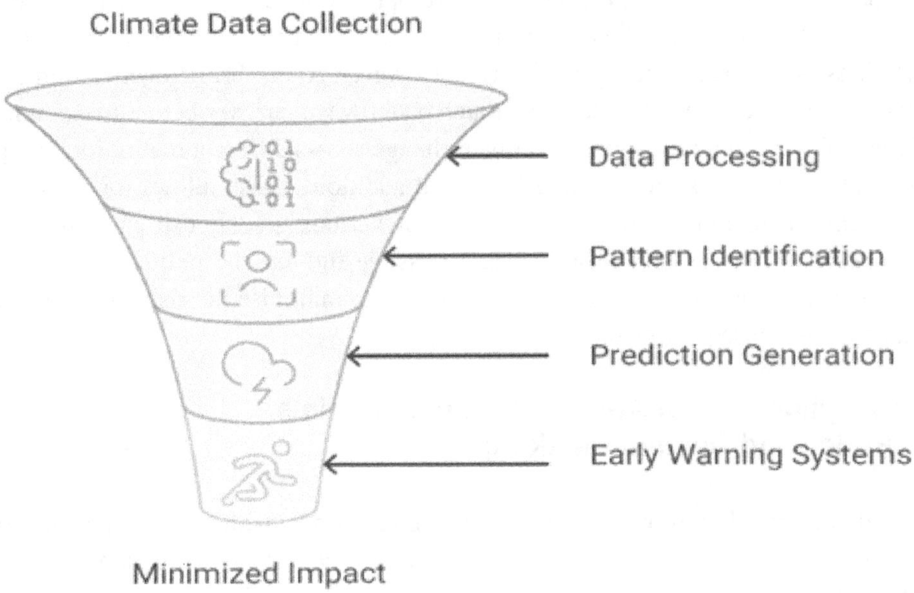

AI algorithms perform well in the analysis of real-time data inputs as weather satellites, radar and sensory data collected from the ground. This enable better and earlier prediction on the path, course and the likely outcomes to be witnessed with the extreme weather. With such sophisticated alerts, AI-based forecasting systems help governments and communities set up the early warning system, evacuate the potentially endangered, and distribute resources properly (Gesami & Nunoo, 2024). AI has also been helpful in sorting out signal from the noise in the climate data, and finding fresh patterns and links that may help to explain the causes and timing of extreme events.

3.4 Long-Term Climate Forecasting and Scenario Planning Using AI

The concerns are the anthropogenic impacts of the current climate change that EIAs require for people to understand, and the potential climate change effects that societies need to address for the foreseeable future. With the help of algorithms, AI is used to study long-term climatic indicators, discuss trends and climate forecasts within several decades or even centuries with respect to increase in temperature, changes in precipitation and sea level (Gesami & Nunoo, 2024). Digital climate tools address their uncertainty, enabling policymakers and stakeholders to model different climate scenarios, including those involving discontinuities. These tools can help decision making and guide the development of detail climate action plans that consider the effect of various emission reduction measures and adaptation activities as well as technological changes. However, AI can bring together climate models and socio-economic data in a much better way than ever before to examine the actual societal implications of climate change across different realms including agriculture, energy, structures and health. In addition, it also allows for scenario modelling in relation to the consequences of various greenhouse gas emission trajectories. Thus, by identifying several possible futures, the political decision-makers and stakeholders can analyse how they can minimise the risks of the most catastrophic consequences of climate change.

3.5 Applications of AI-Based Climate Models in Policy and Decision-Making

Climate models powered through AI, therefore are invaluable assets, especially to policy makers in the development of an efficient climate change mitigation strategy. Refer Figure. 5 Applications of AI-Based Climate Models. They can process large amount of data, draw patterns and make predictions that can guide formation of policies. For example, AI can be used to: Optimize Renewable Energy Integration: AI models can assess weather conditions, energy use and load conditions on the grid progressively for the implementation of sustainable solutions such as solar and wind energy systems (Cowls et al., 2021). Develop Carbon Pricing Strategies: AI can aid in the evaluation of the most suitable carbon pricing policies and vice versa their economic and ecological effects. Inform Climate Adaptation Planning: Using agent-based models supported by AI technologies, it is possible to estimate the isolated impacts of climate change, including rising sea levels or storm occurrence, and the subsequent design of appropriate contingency measures at the local level (Gesami & Nunoo, 2024).

Figure 5. Applications of AI-Based Climate Models

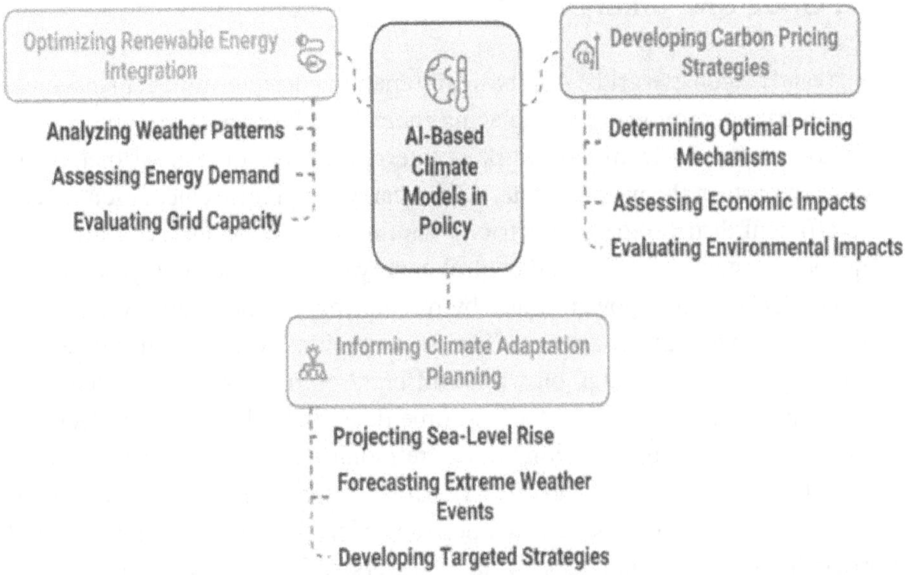

AI climate models are not yet making concrete international agreements but are ubiquitous in informing what the consequential policy choices might be. There are blogs such as Climate Change AI – Interactive Summaries where users can find examples of how AI is used in climate policymaking. However, it is crucial to mention that a policy is not an output of an AI-based model and involves many other parameters quite often hidden behind the numerous factors influencing policy-making.

4. AI-ENHANCED ENERGY SYSTEMS AND INFRASTRUCTURE

Renewable energy management and integration is another application area where the use of AI is promising, and existing energy systems can also be enhanced using AI technology. Examining weather patterns, energy load demand and usage across many sectors, AI can be used to begin to examine how the random fluctuations of wind and solar energy can best be incorporated into the 'grid' (Rashid et al., 2024). This chapter gives an insight on how AI is disrupting the energy systems in the world and a signal for a new future of energy systems.

4.1 Transforming Energy Grids with AI: Smart Grids and Smart Cities

Smart and responsive grids: AI is basically enabling the intermittent management of preindustrial energy grids by replacing them with intelligent systems. Storing huge amounts of data from a network of interrelated sensors, smart meters, and efficient forecasts of the weather state, AI can carry out precise calculations needed for the efficient distribution of electricity within the grid. AI based smart energy management systems can plan and control energy demand and supply in real time and avoid blackouts and power failure by monitoring and predicting variations in energy and green energy production. Here in smart cities AI is pivotal and versatile in regulating energy consumption within different domains such as on buildings, mobility and public utilities. As the real-time data about the usage of electricity, occupancy, and environmental conditions are monitored AI systems can discover more points to improve the efficiency and perform more advanced demand response programs mentioned above as the dynamic usage of consumption depends on supply and price. This, in turn mitigates wastage, lowers carbon footprint, and provides extremely reliable and resilient energy to communities, and part of sustainable development of cities at large.

4.2 Reducing Energy Consumption through AI-Optimized Systems

AI has the potential of using much less energy than human controlled systems in industries, residential areas and transport systems. In BYOD, intelligent systems can enhance production, automatically fine-tune equipment parameters in real-time and save energy. Smart thermostats and home appliances both detect people's habits in a house and manage to optimize energy consumption to save both, money and the environment. In transportation AI is consequently used for automating cars, smart traffic signals, and the sophisticated management of fleets. These developments can greatly reduce the usage of energy and emission outputs through decreasing the concentration of Traffic Jam, improving the route selections and encouraging the usage of electric cars. AI is being used effectively to increase the usage of renewable energy sources including wind and solar energy to address increasing needs for clean energy across the globe today (Pachot & Patissier, 2022) (Ali & Choi, 2020) (Jin et al., 2020). In the same way, it has been applied in the transportation systems with efficiency of traffic flow and congestion, and fuel consumption.

4.3 AI and Renewable Energy: Efficient Distribution and Management

This is specifically the case with renewable energies, such as solar and wind energy, where supply dispersion is problematic. These challenges are solved by AI in real-time to forecast the renewable energy generation, energy management, storage, and supply versus demand. For example, climate factors such as cloud cover and wind velocity can be forecasted to estimate changes in the efficiency of solar and wind technologies (Mohammadi & Mohammadi, 2024). Information enables grid operators to alter the energy dispatch of the grid from other sources making the electricity supply more reliable and steadier. There is also another role that AI is also playing to ensure that renewable energy is well managed and distributed. Thanks to the processing of big data of weather and electrical grids, energy demands with the help of AI systems, more effective positions for renewable sources, for example, solar panels or windmills, can be determined to create the highest output with the lowest loss. In conclusion, the employment of AI is critical for the future of the renewable energy sources and for achieving a shift to renewable energy sources (Chen et al., 2023) (Pachot, & Patissier, 2022).

4.4 AI-Driven Innovations in Energy Storage Solutions

Renewable energy is reliant on storage systems for efficient use and stability of the power grid. Due to AI, solutions for energy storage can be applied to batteries, thermal storage systems and pumped hydro power plants. As applied to energy storage, AI algorithms can forecast energy consumers' demand patterns, fine-tune charging-discharging cycles in batteries, and prolong the useful lifespan of batteries, thereby increasing the efficiency and lowering the costs of energy storage (Rashid et al., 2024). Further, AI-based EMS can further support the optimization and synchronizing of multiple energy storage applications including batteries, fly wheels, and pressurized air storage to create a more robust and adaptable grid system. Instead of being independent 'silos' they could become better aligned, more dynamic, and capable of quickly adjusting to the variability of renewable energy generation.

4.5 Real-World Applications: AI in Sustainable Energy Management

The integration of AI in energy systems and infrastructure is already yielding tangible results in various real-world applications Google's DeepMind: Google utilized its DeepMind AI to cut energy use at the firm's data centres by 40% by improving the cooling infrastructure (Adelakun, 2024). Siemens Gamesa Renewable

Energy: This company applies artificial intelligence in improving operations and productivity of wind turbines, power generation, and minimizing time duration off. Schneider Electric: A global specialist in energy management has integrated intelligent building management systems that make use of Big Data and AI to monitor energy usage, people's presence, and surrounding climate (Xin et al., 2022). This makes it possible to regulate over heating, cooling or lighting with an efficiency of reducing wastage of energy.

5. CARBON CAPTURE AND SEQUESTRATION USING AI

The constant pressure to slow the process of climate change has placed Carbon Capture and Storage technology in the spotlight. This chapter addresses the use of artificial intelligence in the enhancement of CCS techniques and their scaling for more extensive use in the fight against climate change. AI is being used effectively to advance design and implementation of carbon capture and sequestration (Jin et al., 2020) technologies, one of the vital pillars in combating climate change impacts.

5.1 Roles of Carbon Capture and Storage Technologies Explained

Carbon capture and storage comprises a range of methods intended for the capture of carbon dioxide released by industries and electricity generation plants before it is emitted to the atmosphere. The captured CO2 is then transported and it is sequestered in safe geological storage formations to eliminate it from adding to the greenhouse effect. CCS is critical for reaching considerable emission reductions and, in industries with hard-to-decarbonise processes, such as cement and steel, CCS is crucial for deep reductions in emissions. CCS thus offers the only viable route towards the decarbonisation of these industries and the consequent reduction of CO2 emissions and the impact on climate.

Figure 6. Carbon Capture and Storage Hierarchy

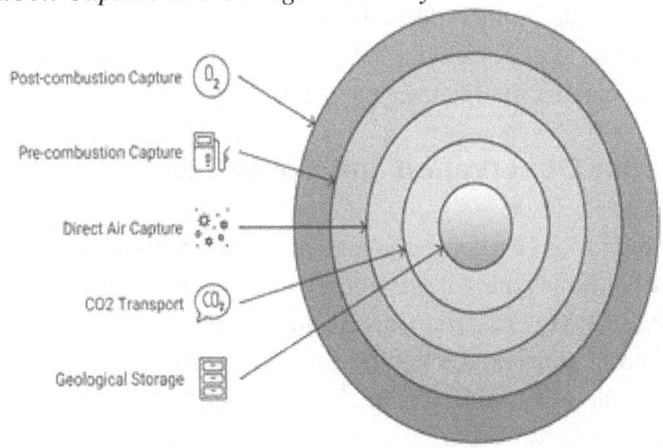

Various methods exist for capturing carbon, including - post-combustion capture: This method involves trapping of CO2 from the flue gases of power plants and industrial facilities at least after combustion processes. Pre-combustion capture: CO2 is sequestered from fuel streams before it has a chance to be burned, normally in a pre-combustion capture process whereby a fuel input is initially used to generate a H2: CO2 mixture and the released CO2 is then captured. Direct air capture: This technology captures CO2 directly from the atmosphere; it may be used to address emissions from point sources, small diffused sources or even history emissions. Bioenergy with CCS: This approach integrates the use of biomass energy production and carbon capture and storage known as CC.S. Essentially, CC.S involves trapping CO2 present during combustion or processing of biomass materials. As biomass captures CO2 while growing, BECCS can bring emissions below zero if the upstream process is taken into consideration. The captured CO2 is then transported, mostly through pipelines or by ships, to secure geological storage such as depleted oil and gas reservoirs or deep saline aquifers and it is then injected and sequestered (Jaiganesh et al., 2022).

5.2 Leveraging AI to Optimise Carbon Storage Rates

AI is revolutionising attempts at improving carbon sequestration efficiency and effectiveness. One important use is the determination of appropriate sites for CO2 sequestration. By utilizing AI algorithms, large data sets of geological information such as seismic, well log and reservoir data can be searched to pinpoint the right geological structures which can provide long term safe containment for CO2. Also, AI is applied to improving the process of carbon capture and storage. Capture plants' data on the temperatures, pressure of gases, or gas flow rates can all be captured and fed into machine learning models that can enhance the operating conditions for higher CO2 capture rates. This data-led approach could result in up to 50 percent saving in carbon capture energy usage and cost.

5.3 AI Tools for Observation and Control of Carbon Release

Implementing proper carbon management needs, examination and timely measurement of carbon emissions within multiple industries and sectors. The application of AI is helping in having real time monitoring of emissions with concentration on CO2 from different sources. Technology such as sensors and data loggers found in the industrial facilities; power plants; and transportation centres can generate huge information regarding CO2 emissions. Data collected can then be fed into an AI system to determine areas of high emissions, abnormal activities and trend analysis of emissions rates over a period. Such information is highly beneficial for companies and governments as it reveals the performance of emission reduction measures as well as guides the choice of effective strategies for carbon reduction. Due to its ability to offer an integrated analysis of carbon emissions, AI can assist stakeholders in making specific actions for changing the unfavourable trends and moving closer to a carbon-free world.

5.4 The Opportunity for AI in Declaring Industries

AI has the potential to decrease carbon emissions impact in sectors that are difficult to decarbonize, such as cement, steel and manufacturing. AI can be applied to - Optimize Industrial Processes: The sensor data can then be utilized by AI to adjust process parameters in live-time, and consequently minimize energy consumption as well as emissions. Develop Sustainable Materials: AI will enhance the search and production of substitute materials with less emission than industries such as cement and steel. Enhance Carbon Capture and Storage: AI in the context of utilization of carbon capture and storage technology consists of potential to enhance the design and operation of these technologies to increase the efficiency and reduce the cost. What is more, these industries, with the help of AI, can go a long way in the process of reduced carbon emissions and help to build a sustainable future.

6. AI-DRIVEN CLIMATE ADAPTATION AND RESILIENCE

Given the fact that the effects of climatic change are worsening, the effort to adopt suitable measures for coping with such change has never been so important. Climate adaptation and human resilience can significantly benefit by incorporating AI for predicting and mapping hazardous climates, deciding resource allocation, and designing resourceful outputs.

6.1 AI for Climate Risk Prediction and Modeling

Climate risks assessment is critical when mapping out an adaptation measure that will effectively respond to climate change. AI models can process increased information about climate conditions, the movement of oceans and other conditions to provide finer and micro-level estimations about the likelihood and impact of climate risk such as hurricanes, floods or droughts, sea level changes, and variations in precipitation (Batarseh et al., 2022). These models can call into the machine learning algorithm and learn the multiple relationships in climate data, thus assisting in improved forecasts and planning. This knowledge is essential for governments, communities and enterprises to adapt for climate change and its consequences.

6.2 Building Climate Resilience with AI: Early Warning Systems

Another area where AI is being implemented in climate adaptation is disaster risk reduction particularly early warning systems for natural disasters and other extreme climate events. AI can also learn to watch over a spectrum of different data inputs from satellite images to sensor networks to weather predictions to identify the precursors to any catastrophes (Kuglitsch et al., 2024). This information when analyzed in real-time through AI algorithms can assist communities, even provide instantaneous alarm that can help save lives and property. For instance, the use of AI in flood models help in determining the time, area and extent of floods, and hence help in planning and implementing early evacuation drills and disaster responses (Gesami & Nunoo, 2024). The types of early warning systems driven by AI include the following: The AI based early warning systems are known to drastically improve community readiness, and thus lower the effects caused by climate related disasters.

6.3 AI in Agricultural Adaptation to Changing Climate Conditions

Climate change has an impact on agricultural business by making the climate change erratic in terms of weather changes. AI brings a variety of tools which are quite effective in facilitating farmers to combat these challenges. Smart farms with the help of targeted agriculture with the help of sensors, drones, and satellite data can determine the amounts of water, fertilizer injections, and pest infestations for the highest crop yields at minimum expenses (Karunathilake et al., 2023). As well, it is helpful in identifying and creating climate-resilient crop varieties and farming practices. Artificial intelligence will be able to process big data acquired from plant genetics, climate, and agriculture yields to determine the best traits and methods that will make the transition to new types of climates. It is in this light that we believe leveraging of this technology will increase the efficacy of farmers for increasing the food systems' robustness to climate change.

6.4 Urban Planning and Infrastructure Adaptation Powered by AI

Therefore, the paper agrees with the literature as climate changes and its impacts escalate; it posits that cities and the urban context will require major alterations to improve the resilience level. As Gesami and Nunoo have indicated, AI can have a critical role here through contributing to the design of the infrastructure of the cities to enhance climate change mitigation as urban planning occurs. For instance, AI can learn about heat patterns, flood risks, or electricity usage and then decide where to install parks, white or reflective pavements and other elements to cool cities. Urban dashboards can also show conditions within the city and help city managers address new climate-related issues promptly using AI. Moreover, AI can design and build intelligent, climate-robust built environments: intelligent traffic control systems, self-driving automobiles, and microgrids of distributed energy.

6.5 Disaster Preparedness and AI enabled Crisis Response

Climate change is enhancing the frequency and intensity of disasters that are natural disasters such as hurricanes, wild fires, drought among others. Disaster is a challenging area for AI, as it enables decision makers to receive intelligence information and effective pre-programmed mechanisms that can improve the preparedness to disasters and minimize the damage, as well as save people's lives. Other applications comprise immediate weather conditions and forecasts, satellite images, and sophisticated real-time data sensors, which define possible paths and strength of forthcoming catastrophes with the help of numerous data analyzed by AI. Such information can help in allocation of emergency supplies, use of evacuation procedures, and control of disaster response procedures (Najafi et al., 2024). For instance, AI can enhance quick identification of losses and compensate for the allocation of restoration means after a disaster. Then, once a disaster has struck, post event monitoring through aerial imagery or reports from citizens, machine learning models are able to analyze the results and quickly provide maps of the damage, severity of infrastructural damage and subsequently, its repair priority. AI operated drones and robots with computer vision capability can utilize their presence in the disaster area by offering live feeds to the various agencies on the ground. It is demonstrated that through AI deployment in disaster work processes, the emergency operations performance increases at a higher rate which leads to saving more lives and achieving a faster recovery.

7. CHALLENGES OF AI IN CLIMATE ACTION

Although AI presents promising avenues for mitigating and adapting to climate change, it is essential to recognize the potential challenges and limitations inherent in its deployment (Mbuvha et al., 2024). This chapter explores the complexities of utilizing AI for climate action, addressing the environmental costs, data dependencies, and ethical considerations that warrant careful deliberation.

7.1 The Carbon Footprint of AI: Balancing Environmental Costs and Benefits

The creation, growth, and successful implementation of AI systems – especially big language models and deep learning – are very computation and energy-intensive. This energy consumption leads to utilization of energy which equals a carbon level thus causing an issue to do with the environmental friendliness or otherwise of the given AI solutions. A single large language model trained can release as much carbon dioxide as five cars in their whole life (Liu & Yin, 2024). To mitigate the environmental impact of AI, researchers and developers are exploring various strategies - Developing more energy-efficient algorithms: Improvement of techniques to minimize computational statistics and energy intake. Utilizing renewable energy sources: Using solar and wind energy as the possible source of energy for data centres and other related artificial intelligence structures. Leveraging edge computing: Moving the computations to the devices, not so much to the centralized data centres consuming a lot of energy. Managing the opportunity and risk of AI for climate change means examining the use of resources by AI systems, the energy consumption by these systems, and sustainability of such solutions.

7.2 Data Scarcity and Quality Issues in AI Climate Models

Machine learning models which are heavily harnessed by artificial intelligence require large datasets that would be used to ensure accurate predictions are made. However, depending on the area of climate change, the specificity and availability of data, as well as the quality of data, become critical barriers. Climate data, particularly historical data from some areas, sometimes can be scarce, unclear or even utterly unreliable (Gesami & Nunoo, 2024). There is a dearth of data, which can complicate the growth of both precise and statistically sound AI algorithms for forecasting climates and their consequences, as well as for determining how best to adapt to them. Deeper research shows that the quality of the data used has a strong impact on the accuracy of the AI models. Lack of comprehensive, integrated, and/or objective data can produce fallacious, consequently inefficient forecasting and decision-making. Solving data quality problems involves adequate data acquisition procedures, methods for data cleansing and early data preparation, and data verification at various stages to guarantee the effectiveness of AI-based climate solutions (Ghamisi et al., 2024).

To overcome these challenges, researchers are exploring strategies such as - Leveraging alternative data sources: Combining satellite imagery with other innovative approaches like crowdsourced data in addition to the climatic data. Developing synthetic data generation: Employing generative AI models in updating the existing datasets and synthesizing realistic data for model development. Fostering data collaboration and sharing: This means pushing for cross sector and cross international borders data sharing to establish a rich and accurate climate data pool. Since most remaining climate change AI-related difficulties are data-related, eradicating these challenges is highly significant for AI's potential to address climate change concerns.

7.3 Addressing Biases in AI Models for Climate Solutions

There is a difficulty in the fact that AI often takes input from a particular bias that it learned from the data it was trained on. In this connection, transformers can learn existing prejudices and resulting unfair climate solutions, if climate data does not contain representative selection of people and territories. For instance, if an AI model for flood prediction is trained on data of developed countries it will not be able to predict the flood risks in developing countries environments geographically and or socio economically different. Mitigating bias in AI models requires - Diverse and representative datasets: An important ingredient that proper training data must meet is valid and diverse representation of all the affected populations in all the affected regions. Bias detection and mitigation techniques: Mitigating approach to bias by addressing the issue as a part of the model building and implementation process. Transparency and accountability: Explaining how AI models work to control bias, and how they arrive at their decisions to instil confidence.

7.4 Sustainability Considerations for Energy-Intensive AI Systems

Deep learning models are characterised by their growing sophistication, which makes it more common to use hardware such as GPUs to train models and make predictions. All these GPUs are power-hungry, hence, increasing the carbon imprint of AI (Pachot & Patissier, 2022). To address this challenge, researchers are exploring - Neuromorphic computing: Building a new chip architecture targeting energy optimization which is at least one order of magnitude more efficient in terms of performance than applying GPU for neural networks computing. Quantum computing: Exploiting quantum mechanics principles to complete certain tasks which cannot be done on a conventional computer while at the same time using reduced energy. Low-power AI chips: Creating unique chips that are tailored for AI performance while having the capability to work under low power, therefore, run AI on devices with low-energy capabilities. So, putting efforts into Research and Development of more sustainable AI hardware, algorithms, etc; we can at least try to make AI a less bad instrument for climate change.

8. GOVERNANCE, ETHICS, AND POLICIES FOR AI AND CLIMATE SOLUTIONS

As climate solutions embrace the use of AI technology, the following questions of governance, ethic, and policy emerge or need to be posed. Given that AI will remain a key enabler of climate action strategies in the future, sound governance structures should be established together with consideration of the ethical concerns emanating from its use. In this chapter, the focus is on the ethical implications of AI applications for climate action, the governance of the technology in climate-related fields, and related policy prospects (Jones et al., 2023).

8.1 Ethical Considerations in AI Climate Applications

Decisions related to climate solutions based on AI elicit several ethical issues that would need redress. Such risks are: risk of bias and discrimination, privacy and data rights, AI transparency and accountability as well as benefits and costs of climate change interventions facilitated by AI technology (Coeckelbergh, 2020). The use of AI in climate change solutions raises several ethical concerns that demand careful consideration - Transparency and Explainability: Deep learning models are mainly closed to some forms of interpretation in some fields. All climate decisions made through the help of artificial intelligence should be transparent to build trust and take accountability for them. Data Privacy and Security: AI relies on data and climate solutions regularly require or imply personal data, people's groups, and ecosystems data. Data security and data anonymity or confidentiality is very important to avoid the data being misused or cause harm. Bias and Fairness: As mentioned in the previous chapter, first, the AI models can replicate and even reinforce the existing prejudices of its training information. Bias should be especially tackled in climate AI solution because such solutions impose effects on specific groups of people and communities (Corrêa et al., 2022). Responsibility and Accountability: Assigning the role of liability and obligations related to artificial intelligence in climate actions is challenging in cases where outcomes originating from it possess high degrees of social impact and the environment. Academics and policy makers must work together to solve the following ethical problems to lay down proper structures that would govern AI in climate change solutions.

8.2 AI Governance for Climate Justice and Fairness

Climate change governance using AI needs to address climate justice and ensure that all groups of people are not disadvantaged when using AI solutions. This requires - Inclusive Development and Deployment: Active involvement of various stakeholders, especially those who are more vulnerable to ill effects of climate change, in AI technologies decision-making process and implementation. Equitable Access to Benefits: Making certain that the advantages offered by use of artificial intelligence in climate change solutions ordinary pushes the social inequality or areas of living. Redressing Existing Inequalities: There are no exclusive capabilities of AI that would make existing stereotyped social and economic barriers even worse, but they should be designed to contribute to the construction of better future societies (Stahl et al., 2020).

8.3 International Policies for AI and Climate Change: Challenges and Opportunities

Climate change is a worldwide issue, and solving it cannot occur unilaterally, it also demands policy regulation which controls the application of intelligent climate solutions. Developing effective international policies for AI and climate change presents both challenges and opportunities - Challenges: As can be appreciated, there are a myriad of countries, interests, legal frameworks and ethical outlooks when it comes to AI. Opportunities: Successful cooperation between countries lets create the common approaches to AI governance, share data and experiences, and set the ethical norms for AI and climate intervention (Roberts et al., 2023). The UN and the OECD have started to develop AI guidelines and principles in relation to climate change. However, filling the gap between these principles and realistic policies that can be implemented and enforceable regulations continues to be a challenge (Nordgren, 2022). Some AI applications for Climate Change The global governance of AI and climate is needed to enable the transformation of AI to create a better tomorrow.

9. THE FUTURE OF AI IN COMBATING CLIMATE CHANGE

AI is an area of study that is expanding incredibly fast and innovation for use in climate change prevention and/or management has not yet been fully realized. This chapter again forward to the future of AI in climate which also discuss novel technologies, expected advancements, and the changing role of AI to build a climate smart environment.

9.1 Emerging AI Technologies with Potential Climate Applications

Artificial intelligence next-generation applications will play a significant role in climate change investigation and resolution. Some key areas of development include - Quantum AI: Superposition could drastically speed up AI calculations, allow insurance companies to analyze climate models quicker and come up with improved prognoses (Barnes et al., 2019). Explainable AI: XAI postulates for de-bottlenecking of AI decision making processes to gain increased trust and apply enhanced understanding of climate policies. Federated Learning: With this approach, AI models can also be trained on a decentralized set of data while preserving the privacy of the data thus promoting cross-border climate research and data analysis (Tu et al., 2022).

9.2 Future AI-Driven Innovations in Climate Research and Solutions

Stakeholders should prepare for faster growth in climate AI in the coming years as it becomes a dominant tool in the climate innovation field. Some anticipated advancements include - Hyperlocal Climate Modeling: AI can improve the climate models to give finer details of results at a local level, which will help to identify solutions to respond better to climate change. Smart Grid Optimization: AI can help to transform energy grids to allow for increased integration of renewable energy, minimize energy losses and improve the stability of the grid system. Precision Agriculture: Automating tools for agriculture have the potential to promote efficiency and productivity besides conserving resources in the future uncertain climate. AI is using the following four approaches to enhance the speed and reliability of the power grid (2023).

9.3 Preparing for the Next Decade: AI's Evolving Role in Climate Action

AI is set to become a crucial contributor to climate strategies in furthering the coming ten years. Key trends to watch for include - Increased Integration of AI into Climate Policies: AI solutions will be brought into practices in climate policies by governments and organizations in the future (Cowls et al., 2021). Growing Emphasis on Ethical and Responsible AI: With increasing integration of AI in climate solutions the emphasis will shift to ethical issues, equity, and sound AI engineering principles (Coeckelbergh, 2020). Interdisciplinary Collaboration: The climate change issues cannot be solved through blinkers of disciplinary divide. Beneficiaries: The application of AI in climate change will help scientists, policymakers, engineers, and, social scientists to come up with overall and efficient solutions. AI's role in the fight against climate change will look bright and full of opportunity in the future.

CONCLUSION

The selected key insights of this chapter outline the benefits of AI solutions for climate change mitigation but nevertheless demonstrate the necessity for proper and ethical AI development and application. Using reliable bias solutions, including achieving ensured datasets that are diverse and neutral as well as being transparent and accountable while using AI in climate-focused solutions is critical to optimally achieving the desired goals while also preserving fairness for all involved stakeholders. The increasing effects of climate change demonstrate the will be increasingly essential in addressing this problem in the future. It is possible to reduce the negative effects of AI in the environment with the help of saving new patterns of energy efficient hardware, quantum computing, low power AI chips. Future studies and innovations in this area will be crucial for AI transformation into even more significant and sustainable part of constructing a better world. The challenges to AI and climate solutions necessitate consideration of the ethical, governance, and the policy frameworks to consider, and because of this, the effort must be multi-disciplinary. If all stakeholders are to realize the potential for AI's contribution towards climate change, policymakers, researchers, and key stakeholders need to elaborate satisfactory governance structures, engage in ensuring that the benefits of AI-driven climate change are distributed equally and proportionately among countries, and work out the international policies that would meet the various national interests. By empowering this collaboration, therefore, it is possible to use AI toward accelerating climate actions, and therefore, making the world a better place with justice, resilience and sustainability.

REFERENCES

Adelakun, N. O. (2024, Jan 1). Future Trends in Artificial Intelligence for Energy Management. RELX Group (Netherlands). DOI: 10.2139/ssrn.4703931

Ali, S. S., & Choi, B. J. (2020, Jun 22). State-of-the-Art Artificial Intelligence Techniques for Distributed Smart Grids: A Review. *Electronics (Basel)*, *9*(6), 1030–1030. DOI: 10.3390/electronics9061030

Barnes, E A., Hurrell, J W., Ebert-Uphoff, I., Anderson, C W., & Anderson, D. (2019, Nov 19). Viewing Forced Climate Patterns Through an AI Lens. American Geophysical Union, 46(22), 13389-13398. DOI: 10.1029/2019GL084944

Batarseh, F. A., Donti, P. L., Drgoňa, J., Fletcher, K. M., Hanania, P., Hatton, M., Keshav, S., Knowles, B., Kotsch, R., McGinnis, S., Mitra, P., Philp, A., Spohrer, J., Stein, F., Tare, M., Volkov, S., & Wen, G. (2022, Jan 1). *Proceedings of AAAI 2022 Fall Symposium: The Role of AI in Responding to Climate Challenges*. Cornell University. DOI: 10.48550/arxiv.2212.13631

Chakraborty, S. (2024, May 22). Towards A Comprehensive Assessment of AI's Environmental Impact. Cornell University. https://doi.org//arxiv.2405.14004DOI: 10.48550

Chen, L., Chen, Z., Zhang, Y., Liu, Y., Osman, A I., Farghali, M., Hua, J., Al-Fatesh, A S., Ihara, I., Rooney, D W., & Yap, P. (2023, Jun 13). Artificial intelligence-based solutions for climate change: a review. Springer Science+Business Media, 21(5), 2525-2557. DOI: 10.1007/s10311-023-01617-y

Coeckelbergh, M. (2020, Oct 5). AI for climate: Freedom, justice, and other ethical and political challenges. *AI and Ethics*, *1*(1), 67–72. DOI: 10.1007/s43681-020-00007-2

Corrêa, N. K., Galvão, C., Santos, J. W., Pino, C. D., Pinto, E. P., Barbosa, C. K. R., Massmann, D., Mambrini, R., Galvão, L., Ugar, E. T., & Oliveira, N. D. (2022, Jan 1). Worldwide AI Ethics: a review of 200 guidelines and recommendations for AI governance. Cornell University. https://doi.org//arxiv.2206.11922DOI: 10.48550

Cowls, J., Tsamados, A., Taddeo, M., & Floridi, L. (2021, Oct 18). The AI gambit: Leveraging artificial intelligence to combat climate change—opportunities, challenges, and recommendations. *SSRN*, *38*(1), 283–307. DOI: 10.2139/ssrn.3804983 PMID: 34690449

Cowls, J., Tsamados, A., Taddeo, M., & Floridi, L. (2021, Oct 18). The AI gambit: Leveraging artificial intelligence to combat climate change—opportunities, challenges, and recommendations. *SSRN*, *38*(1), 283–307. DOI: 10.2139/ssrn.3804983 PMID: 34690449

Donnelly, B F A G. (2022, Aug 26). Here's how AI could transform the US power grid

Dresser, L. P., & Kohn, M. A. (2023, Aug 1). Artificial Intelligence and the Evaluation and Treatment of Stroke. *Delaware Journal of Public Health*, *9*(3), 82–84. DOI: 10.32481/djph.2023.08.014 PMID: 37701474

Gesami, B. K., & Nunoo, J. (2024, Jul 18). Artificial intelligence in marine ecosystem management: Addressing climate threats to Kenya's blue economy. *Frontiers in Marine Science*, *11*, 1404104. DOI: 10.3389/fmars.2024.1404104

Ghamisi, P., Yu, W., Marinoni, A., Gevaert, C., Persello, C., Selvakumaran, S., Girotto, M., Horton, B. P., Rufin, P., Hostert, P., Kumar, J., & Atkinson, P. M. (2024, May 31). Responsible AI for Earth Observation. Cornell University. https://doi.org//arxiv.2405.20868DOI: 10.48550

Heymann, F., Quest, H., Garcia, T P., Ballif, C., & Galus, M D. (2023, Nov 19). Reviewing 40 years of artificial intelligence applied to power systems – A taxonomic perspective. Elsevier BV, 15, 100322-100322. DOI: 10.1016/j.egyai.2023.100322

Jaiganesh, N., Kuo, P., Woudstra, T., Kumar, R. A., & Aravind, P. (2022, Jun 27). Negative Emission Power Plants: Thermodynamic Modeling and Evaluation of a Biomass-Based Integrated Gasification Solid Oxide Fuel Cell/Gas Turbine System for Power. *Frontiers in Energy Research*, *10*, 803756. DOI: 10.3389/fenrg.2022.803756

Jin, D., Ocone, R., Jiao, K., & Xuan, J. (2020, Apr 20). Energy and AI. Elsevier BV, 1, 100002-100002. DOI: 10.1016/j.egyai.2020.100002

Jones, A., Kuehnert, J., Fraccaro, P., Meuriot, O., Ishikawa, T., Edwards, B., Stoyanov, N., Remy, S. L., Weldemariam, K., & Assefa, S. (2023, Jun 8). AI for climate impacts: Applications in flood risk. *NPJ Climate and Atmospheric Science*, *6*(1), 63. DOI: 10.1038/s41612-023-00388-1

Kalisdha, A. (2024, Feb 1). The Impact of Artificial Intelligence and Machine Learning in Library and Information Science

Karunathilake, E. M. B. M., Le, A. T., Heo, S., Chung, Y. S., & Mansoor, S. (2023, Aug 11). The Path to Smart Farming: Innovations and Opportunities in Precision Agriculture. *Agriculture*, *13*(8), 1593–1593. DOI: 10.3390/agriculture13081593

Khandaker, T., Hossain, M. S., Dhar, P. K., Rahman, M. S., Hossain, M. A., & Ahmed, M. B. (2020, May 30). Efficacies of Carbon-Based Adsorbents for Carbon Dioxide Capture. *Processes (Basel, Switzerland)*, *8*(6), 654–654. DOI: 10.3390/pr8060654

Kim, J. (2023, Nov 22). Four ways AI is making the power grid faster and more resilient

Kuglitsch, M. M., Cox, J., Luterbacher, J., Jamoussi, B., Xoplaki, E., Thummarukudy, M., Radwan, G. S., Yasukawa, S., McClain, S. N., Albayrak, A., Oehmen, D., & Ward, T. (2024, Oct 1). AI to the rescue: How to enhance disaster early warnings with tech tools. *Nature*, *634*(8032), 27–29. DOI: 10.1038/d41586-024-03149-z PMID: 39354079

Kumar, P., Debele, S., Khalili, S., Halios, C., Sahani, J., Aghamohammadi, N., Andrade, M D F., Athanassiadou, M., Bhui, K., Calvillo, N., Cao, S., Coulon, F., Edmondson, J L., Fletcher, D., Freitas, E D D., Guo, H., Hort, M., Katti, M., Kjeldsen, T., . . . Jones, L. (2024, Mar 1). Urban heat mitigation by green and blue infrastructure: Drivers, effectiveness, and future needs. Elsevier BV, 5(2), 100588-100588. DOI: 10.1016/j.xinn.2024.100588

Lee, G. A., Shi, X., & Park, A. A. (2022, Jan 1). Technologies to Capture CO$_2$ directly from Ambient Air. Cornell University. https://doi.org//arxiv.2211.00791DOI: 10.48550

Liu, V., & Yin, Y. (2024, Apr 1). Green AI: Exploring Carbon Footprints, Mitigation Strategies, and Trade Offs in Large Language Model Training. Cornell University. https://doi.org//arxiv.2404.01157DOI: 10.48550

Lundstad, E., Brugnara, Y., Pappert, D., Kopp, J., Samakinwa, E., Hürzeler, A., Andersson, A., Chimani, B., Cornes, R., Demarée, G. R., Filipiak, J., Gates, L., Ives, G. L., Jones, J., Jourdain, S., Kiss, A., Nicholson, S. E., Przybylak, R., Jones, P., & Brönnimann, S. (2023, Jan 19). The global historical climate database HCLIM. *Scientific Data*, *10*(1), 44. Advance online publication. DOI: 10.1038/s41597-022-01919-w PMID: 36658229

Mbuvha, R., Yaakoubi, Y., Bagiliko, J., Potes, S. H., Nammouchi, A., & Amrouche, S. (2024, Jan 1). Leveraging AI for Climate Resilience in Africa: Challenges, Opportunities, and the Need for Collaboration. RELX Group (Netherlands). DOI: 10.2139/ssrn.4815919

Mohammadi, M., & Mohammadi, A. (2024, Jan 1). Empowering Distributed Solutions in Renewable Energy Systems and Grid Optimization. Springer Nature, 141-155. DOI: 10.1007/978-3-031-57567-9_7

Najafi, H., Shrestha, P. K., Rakovec, O., Apel, H., Vorogushyn, S., Kumar, R., Thober, S., Merz, B., & Samaniego, L. (2024, May 2). High-resolution impact-based early warning system for riverine flooding. *Nature Communications*, *15*(1), 3726. Advance online publication. DOI: 10.1038/s41467-024-48065-y PMID: 38698000

Nearing, G., Cohen, D., Dube, V., Gauch, M., Gilon, O., Harrigan, S., Hassidim, A., Klotz, D., Kratzert, F., Metzger, A., Nevo, S., Pappenberger, F., Prudhomme, C., Shalev, G., Shenzis, S., Tekalign, T., Weitzner, D., & Matias, Y. (2024, Mar 20). Global prediction of extreme floods in ungauged watersheds. *Nature*, *627*(8004), 559–563. DOI: 10.1038/s41586-024-07145-1 PMID: 38509278

Nguyen, T. T., Jewik, J., Bansal, H., Sharma, P., & Grover, A. (2023, Jan 1). ClimateLearn: Benchmarking Machine Learning for Weather and Climate Modeling. Cornell University. https://doi.org/DOI: 10.48550/arXiv.2307

Niyogi, D., & Srinivasan, J. (2023, Jan 1). Accurate Prediction of Global Mean Temperature through Data Transformation Techniques. Cornell University. https://doi.org//arxiv.2303.06468DOI: 10.48550

Nordgren, A. (2022, Feb 9). Artificial intelligence and climate change: ethical issues. Emerald Publishing Limited, 21(1), 1-15. DOI: 10.1108/JICES-11-2021-0106

Olawade, D. B., Wada, O. Z., David-Olawade, A. C., Fapohunda, O., Ige, A. O., & Ling, J. (2024). Artificial intelligence potential for net zero sustainability: Current evidence and prospects. In *Next Sustainability* (Vol. 4, p. 100041). Elsevier BV., DOI: 10.1016/j.nxsust.2024.100041

Pachot, A., & Patissier, C. (2022, Jan 1). Towards Sustainable Artificial Intelligence: An Overview of Environmental Protection Uses and Issues. Cornell University. https://doi.org//arxiv.2212.11738DOI: 10.48550

Pató, Z., & Mandel, T. (2022, Oct 22). Energy Efficiency First in the power sector: incentivising consumers and network companies. Springer Science+Business Media, 15(8). DOI: 10.1007/s12053-022-10062-9

Premavathi, T. (2023, Aug 21). A Comparative Analysis of ARIMA Model and LSTM Network in Predicting COVID-19 Outcomes in India. European Organization for Nuclear Research. DOI: 10.5281/zenodo.8267919

Rashid, A., Biswas, P., Biswas, A., Nasim, M A A., Gupta, K. D., & George, R. (2024, Jun 22). Present and Future of AI in Renewable Energy Domain: A Comprehensive Survey. Cornell University. https://doi.org/DOI: 10.48550/arXiv.2406

Roberts, H., Hine, E., Taddeo, M., & Floridi, L. (2023, Jan 1). Global AI governance: barriers and pathways forward. RELX Group (Netherlands). DOI: 10.2139/ssrn.4588040

Rolnick, D., Donti, P. L., Kaack, L. H., Kochanski, K., Lacoste, A., Sankaran, K., Ross, A. S., Milojevic-Dupont, N., Jaques, N., Waldman-Brown, A., Luccioni, A., Maharaj, T., Sherwin, E. D., Mukkavilli, S. K., Körding, K., Gomes, C. P., Ng, A. Y., Hassabis, D., Platt, J., & Bengio, Y. (2022, Feb 7). Tackling Climate Change with Machine Learning. *ACM Computing Surveys*, *55*(2), 1–96. DOI: 10.1145/3485128

Stahl, B C., Andreou, A G., Brey, P., Hatzakis, T., Кириченко, A M., Macnish, K., Shaelou, S L., Patel, A., Ryan, M., & Wright, D. (2020, Dec 16). Artificial intelligence for human flourishing – Beyond principles for machine learning. Elsevier BV, 124, 374-388. DOI: 10.1016/j.jbusres.2020.11.030

Taddeo, M., Tsamados, A., Cowls, J., & Floridi, L. (2021). Artificial intelligence and the climate emergency: Opportunities, challenges, and recommendations. In One Earth, *4*(6), p. 776). Elsevier BV. DOI: 10.1016/j.oneear.2021.05.018

Tok, Y C., & Chattopadhyay, S. (2023, Mar 15). Identifying threats, cybercrime and digital forensic opportunities in Smart City Infrastructure via threat modeling. Elsevier BV, 45, 301540-301540. DOI: 10.1016/j.fsidi.2023.301540

Tu, R., Roberts, N. J., Prasad, V., Nayak, S., Jain, P., Sala, F., Ramakrishnan, G., Talwalkar, A., Neiswanger, W., & White, C. (2022, Jan 1). AutoML for Climate Change: A Call to Action. Cornell University. https://doi.org//arxiv.2210.03324DOI: 10.48550

Vaughan, A., Mateo-García, G., Irakulis-Loitxate, I., Watine, M., Fernandez-Poblaciones, P., Turner, R. E., Requeima, J., Gorroño, J., Randles, C. A., Caltagirone, M., & Cifarelli, C. (2024, Aug 8). AI for operational methane emitter monitoring from space. Cornell University. https://doi.org//arxiv.2408.04745DOI: 10.48550

Xin, Z., Zhang, H., & Shi, J. (2022, Sep 1). Application Based on Artificial Intelligence in Substation Operation and Maintenance Management. *Computational Intelligence and Neuroscience*, *2022*, 1–12. DOI: 10.1155/2022/7509532

Zamani, H., Díaz, F., Dehghani, M., Metzler, D., & Bendersky, M. (2022, Jul 6). Retrieval-Enhanced Machine Learning. https://doi.org/10.114

Chapter 6
Impact of Heavy Metal Stress (Ni and Co) on Maize (Zea mays): Exploring Biochemical and Morphological Responses for Climate Resilience and Sustainability

Mahmoud Oudghiri
Faculty of Sciences, Ibn Tofail University, Morocco

Mohamed El Bakkali
Faculty of Sciences, Ibn Tofail University, Morocco

Yassine Mouniane
https://orcid.org/0000-0002-6037-5301
Faculty of Sciences, Ibn Tofail University, Morocco

Nagla Abid
Faculty of Sciences, Ibn Tofail University, Morocco

Samah Bouhassoun
https://orcid.org/0009-0003-5169-7282
Faculty of Sciences, Ibn Tofail University, Morocco

Fatima-ezzahra Jaayefar
https://orcid.org/0009-0000-0727-9356
Faculty of Sciences, Ibn Tofail University, Morocco

Fath Alah Elwahab
Faculty of Sciences, Ibn Tofail University, Morocco

Issam El-Khadir
https://orcid.org/0000-0002-3970-2865
Faculty of Sciences, Ibn Tofail University, Morocco

Ahmed Chriqui
Faculty of Sciences, Ibn Tofail University, Morocco

Mohammed Ibriz
Faculty of Sciences, Ibn Tofail University, Morocco

DOI: 10.4018/979-8-3693-9132-7.ch006

ABSTRACT

Heavy metal contamination, particularly from nickel (Ni) and cobalt (Co), is a significant issue in agricultural regions, especially those influenced by industrial activities. This study investigates the impact of Ni and Co stress on maize (Zea mays) by evaluating the plant's morphological, physiological, and biochemical responses to varying metal concentrations. We measured leaf number, leaf length, root length, fresh and dry weights of aerial and root parts, as well as chlorophyll content, proline, total sugars, polyphenols, and flavonoids. The results reveal that higher concentrations of Ni and Co caused notable reductions in growth and alterations in metabolic processes, with increased levels of stress-related markers. This research emphasizes the detrimental effects of heavy metals on maize, highlighting the need for sustainable agricultural practices to mitigate pollution. By understanding the impact of heavy metals on crop health, this study contributes to the broader goal of enhancing food security and improving resilience in regions facing soil contamination challenges.

1. INTRODUCTION

Contamination of soils by heavy metals is an increasingly serious environmental issue around the world, especially in developing countries where industrialization and urbanization often occur faster than environmental regulations can adapt (Silva et al., 2021; Adnan et al., 2022; Oudghiri et al., 2025). Trace elements such as nickel (Ni) and cobalt (Co) are naturally present in the environment but they are now being introduced into our ecosystems at an increasingly significant rate due to anthropogenic activities; including mining, industrial effluents, use of untreated wastewater in irrigation, and excessive application of phosphorous fertilizers that often contain rich trace metals (Boros-Lajszner et al., 2021; Ali et al., 2022; Khan et al., 2023). After entering the environment, these metals can persist in soils for an indefinite time without bioremediating, accumulate in plant tissues, and can finally bioaccumulate in the food chains of both humans and animals (Rajput et al., 2024), creating health and safety concerns in the long term (Kumar et al., 2019; Uddin et al., 2021; Angon et al., 2024; El-Khadir et al., 2024; Mouniane et al., 2025).

When plants are subjected to high concentrations of Ni and Co, they can initiate a vast array of physiological and biochemical perturbances (Emamverdian et al., 2015; Shakespear et al., 2024). These metals affect vital metabolic processes, such as nutrient uptake, photosynthesis, water balance, and catalytic activities. They also induce excessive production of reactive oxygen species (ROS), which induces oxidative stress, membrane damage, and ultimately inhibition of growth or death of plant tissues (Kumar et al., 2022; Liu et al., 2023; Zhao et al., 2023; Wang et al., 2024). Although low concentrations of these metals could participate in some enzymatic functions, accumulation over critical levels, would result in disruption of cellular homeostasis and reduced plant productivity (Kosakivska et al., 2021; Noor et al., 2022).

Morocco is not immune to this mounting issue. Several industrial and mining zones, including Jorf Lasfar, Guemassa, Kenitra, and Mohammedia, have already been determined to be heavy metal hotspots. In these zones, Ni, Co, Cd, and other metals are observed with elevated concentrations in agricultural soils and ecosystems in the vicinity, affecting soil fertility, microbial biodiversity, and crop yield (Hakkou et al., 2020, Zaim et al., 2021). The severity is acute for peri-urban and rural communities that rely on agriculture for economic sustenance, signifying an urgent need for research on the impact of heavy metals on relevant crops (Aslanidis et al., 2022).

In Moroccan agriculture, maize (*Zea mays* L.) is the most important of all these crops. It serves as both a dietary item and feed for livestock and is also significant in providing socio-economic stability for rural inhabitants (Zidan et al., 2024). Indeed, maize is regarded as a bio-indicator of soil pollution because it is highly sensitive to abiotic stresses. Consequently, the toxicological effects of heavy metals on maize include reduced biomass accumulation, chlorosis, stunted growth, and metabolic disruption, further affecting its potential yield (Ali et al., 2022; Khan et al., 2023).

Even though this is an important subject, there are no controlled experimental studies looking into specific effects of nickel and cobalt on maize, especially in Morocco. To fill this research gap, this particular work includes an *in vitro* experimental setup, which isolates the effect of Ni and Co without any outside variability. This *in vitro* study permits an accurate measurement of morphological traits (leaf and root growth, fresh and dry biomass), and important physiological and biochemical indicators (chlorophyll content, proline level, total sugars, polyphenols, and flavonoids).

What makes this study original is its holistic evaluation of the morphological and biochemical responses of maize to increasing concentrations of Ni and Co under controlled laboratory conditions. This dual attention gives us information on symptoms that are observable above- and below- ground associated with metal stress, as well as how maize alters its metabolism to tolerate or to respond to metal stress. By comparing the effects of two distinct metals at increasing concentrations, this study hopes to shed light on the differential toxicity and response of maize to two metals.

In this way, the aim of this chapter is to assess the phytotoxic effect of Ni and Co on maize via a series of morpho-physiological and biochemical indicators. Results are expected to provide better insight into plant responses to stress under heavy metal exposure and potentially support measures for dealing with contaminated soils, selecting tolerant genotypes, or developing phytoremediation strategies optimized to local agro-ecological conditions.

2. MATERIALS AND METHODS

2.1. Plant Material and Growth Conditions

The experiment used maize (*Zea mays* L., cv. Torro Plus) seeds, which were surface-sterilized with 5% sodium hypochlorite for 5 minutes and rinsed with distilled water. The seeds were next germinated in Petri dishes containing moistened filter paper, and the plates were wrapped in aluminum foil and incubated for 72 hours at 25°C in the dark. The seedlings were then transplanted to plastic pots containing sterilized sand and were watered with Hoagland's nutrient solution.

2.2. Experimental Design and Heavy Metal Treatments

The experiment consisted of seven different treatments where the effects of heavy metals on the studied plants were evaluated (Table 1). The first treatment (T1) was a control where no metals were applied. Treatments T2, T3, and T4 had Nickel (Ni) applied at the concentrations of 50, 100, and 150 $\mu mol \cdot L^{-1}$ for T4, respectively, while Cobalt (Co) was applied in treatments T5, T6, and T7 at the same concentrations of 50, 100, and 150 $\mu mol \cdot L^{-1}$ for T4, respectively. The design of the experiment would determine the response of the plants to increasing concentrations of Ni and Co. Each treatment was delivered via irrigation with modified Hoagland's solution containing the particular metal concentration as $NiSO_4 \cdot 6H_2O$ for nickel and $CoCl_2 \cdot 6H_2O$ for cobalt. Cultures were kept under controlled conditions (25 ± 2 °C, photoperiod 16 h light/8 h dark) for 21 days.

Table 1. Treatment conditions.

Treatment	Conditions
T1	Control
T2	Seed + 50 µmol Ni
T3	Seed + 100 µmol Ni
T4	Seed + 150 µmol Ni
T5	Seed + 50 µmol Co
T6	Seed + 100 µmol Co
T7	Seed + 150 µmol Co

2.3. Measured parameters

We performed morphological, physiological, and biochemical assessments to evaluate maize (*Zea mays* L.) tolerance to nickel (Ni) and cobalt (Co)-induced stress, enabling us to assess the effects of heavy metal on plant growth and metabolism.

2.2.1. Morphological parameters

Morphological parameters were analyzed after 7 days of exposure to different concentrations of Ni and Co, at the conclusion of the experiment. The number of leaves, per seedling, was counted manually to assess the impact of heavy metals on leaf growth. Leaf length was measured from the base to the tip of the longest leaf with a millimeter ruler. Similarly, root length was recorded for each seedling to measure root growth and nutrient adsorption capability while under stress. The fresh weight of the aerial part, as well as of the roots, was also taken directly after harvesting using a precise balance to quantify biomass accumulation under metal stress. After harvest, the samples were placed in a drying oven at 70 °C for 48 hours to obtain the dry weight for the aerial components and root. This measure was intended to determine water loss and dry matter, both of which are key parameters of plant health constraint and survival propensity in an environmental context.

2.2.2. Physiological and Biochemical parameters

The chlorophyll a, chlorophyll b, and total chlorophyll were evaluated by 80% acetone extraction and spectrophotometric analysis following the method of Arnon (1949). Chlorophyll content is an important indicator of a plant's photosynthetic state and is associated with energy metabolism being impaired during degradation in the presence of heavy metals. The proline content, a free amino acid that often accumulates in response to abiotic stress, was determined based on the method of Bates et al., 1973. Proline exerts a protective role strengthening cell membranes and protein structures, while playing a role in osmotic regulation. Total sugar content was determined using the phenol-sulfuric acid method (Dubois et al., 1956), providing an indication of the plant's energy status and metabolic reprogramming. Soluble sugars often accumulate a response to stress, as a measure to maintain osmotic pressure and contribute to antioxidant defense mechanisms. Also, total polyphenols, known to be involved in defense against oxidative stress, were determined using the Folin–Ciocalteu reagent. Their accumulation is generally also linked to the activation of antioxidant defense systems. Finally, flavonoid was determined using the aluminum chloride ($AlCl_3$) method, as a specific subpopulation of phenolic compounds known to act as strong antioxidant defense, providing a similar metric of adaptive responses to toxic metals.

2.4. Statistical Analysis

Descriptive statistics (mean and standard deviation) were calculated for each parameter. One-way analysis of variance (ANOVA) was used to evaluate the effects of heavy metal treatments on plant biomass. Significance was determined at $p < 0.05$. When appropriate, post-hoc comparisons were performed using Tukey's HSD test. The coefficient of determination (R^2) was also calculated to assess the strength of the relationship between treatments and the measured variables. The methodology used for this study includes a series of advanced statistical tests and artificial intelligence techniques aimed at assessing the impact of heavy metal treatments (Control, NI, CO) on various physiological, biochemical and secondary metabolite parameters of maize. Raw data were cleaned, normalized and subjected to correlation matrix-based principal component analysis (PCA) to reduce dimensionality and visualize relationships between samples. PCA was complemented by a correlation heatmap to assess interactions between measured variables. Next, an advanced hierarchical clustering heatmap was performed using an appropriate distance metric and an agglomeration method to identify similar groups among samples and parameters. In parallel, a one-way analysis of variance (ANOVA) was performed for each measured parameter to determine the existence of significant differences between treatments. Tukey's post-hoc test was applied to compare treatment means and identify homogeneous groups. Finally, an artificial intelligence model based on a Random Forest algorithm was applied to assess the relative importance of the different parameters in the classification of treatments. This model was trained and validated using a training and test dataset, and the results were expressed in terms of feature importance, enabling the most discriminating parameters to be identified. All these methods were applied consistently to provide a robust and comprehensive analysis of the experimental data.

3. RESULTS

3.1. Analysis of variance

Analysis of variance (ANOVA) was performed to assess the effect of heavy metal treatments (Control, Ni (Nickel), Co (Cobalt)) on physiological, biochemical and secondary metabolite parameters of maize. The results were divided into three distinct groups. The first group concerns physiological parameters including leaf number, leaf length, root length, above-ground fresh weight and root fresh weight. The results show that root and leaf length show a significant increase under nickel (Ni) treatment, as indicated by the letter A, compared with cobalt (Co) and control. The cobalt treatment shows a significant reduction in these parameters, particularly for fresh root weight, which is lower than in the other treatments. The second group includes chlorophyll pigment and biomass parameters: chlorophyll a, chlorophyll b, total chlorophyll, above-ground dry weight and root dry weight. Cobalt (Co) treatment induced a significant increase in total chlorophyll compared with nickel (Ni) and the control, as indicated by the letter B. However, heavy metal treatments decreased chlorophyll a and b levels, with cobalt (Co) showing the lowest values. Above-ground and root dry weights were also affected, with a significant reduction under cobalt treatment compared with the control. The third group includes biochemical parameters and secondary metabolites: proline, total sugars, polyphenols, flavonoids and heavy metals encoded. Proline shows a significant increase under cobalt treatment (B) compared to control (C), while nickel (Ni) shows an intermediate value (A). Total sugars, polyphenols and flavonoids show a significant increase under cobalt treatment (B), while the control shows the lowest values (C). The heavy metals encoded parameter shows significantly lower values for the control (C), showing a limited response to heavy metal treatments. These results suggest that heavy metal treatments differentially influence physiological parameters, chlorophyll pigments, biomass and secondary metabolites in maize. Cobalt appears to induce marked increases in secondary metabolites, while nickel induces more specific responses on physiological parameters such as leaf and root length.

Figure 1. ANOVA Results of Number of leaves, Length of leaves, Root length, Fresh weight of aerial part, Fresh weight of roots.

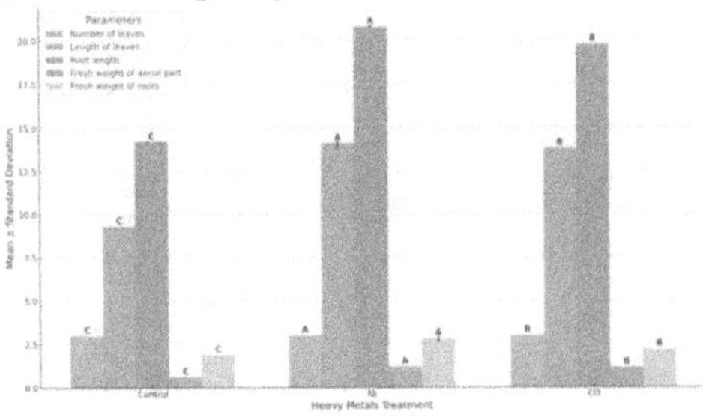

Figure 2. ANOVA Results of Dry weight of aerial part, Dry weight of roots, Chlorophyll a, Chlorophyll b, Total chlorophyll.

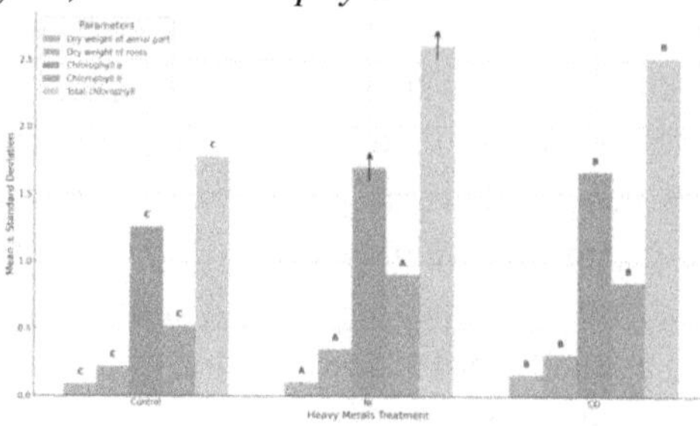

Figure 3. ANOVA Results of Biochemical Parameters (Proline, Total sugars, Polyphenols, Flavonoids) in Maize Under Heavy Metal.

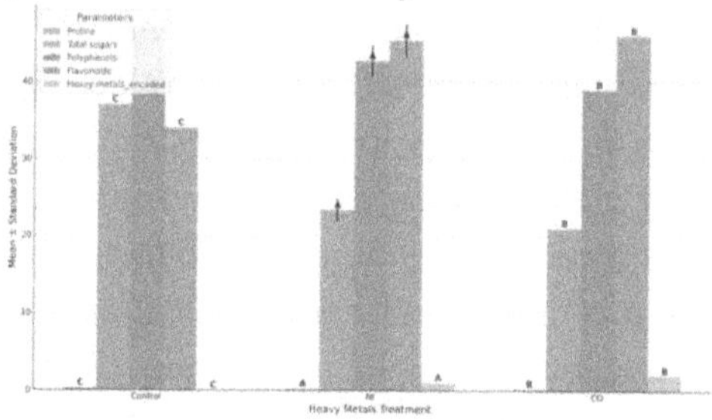

Table 2. Analysis of variance (ANOVA) of parameters evaluated under metal stress.

Source		Sum of Squares		Mean Square	F-value	Signification
Metal stress	Number of leaves	0.476	6	0.079	.833	0.564
	Length of leaves (cm)	250.665	6	41.778	366.102	0.000
	Root length (cm)	236.610	6	39.435	72.868	0.000
	Fresh weight of aerial part (g)	1.737	6	0.290	209.772	0.000
	Fresh weight of roots (g)	3.636	6	0.606	50.544	0.000
	Dry weight of aerial part (g)	0.027	6	0.005	522.532	0.000
	Dry weight of roots (g)	0.077	6	0.013	1569.615	0.000
	Chlorophyll a	1.447	6	0.241	87.300	0.000
	Chlorophyll b	0.902	6	0.150	1237.392	0.000
	Total chlorophyll	4.444	6	0.741	246.464	0.000
	Proline mg/g	0.960	6	0.160	436.225	0.000
	Total sugars mg/g	14693.333	6	2448.889	471.804	0.000
	Polyphenols mg/g	2102.286	6	350.381	40.652	0.000
	Flavonoids mg/g	19362.952	6	3227.159	125.501	0.000

3.2. Morphological parameters

All morphological parameter results are presented in table 2. The number of leaves remained more or less constant across all treatments, with an average of three leaves, even for plants treated with 50 and 100 μM of Ni or Co. There was a small reduction (2.67 leaves) with the maximum treatments (150 μM) for each metal. However, this reduction was not significantly different ($p = 0.564$), indicating that maize leaf formation is relatively insensitive to the immediate toxic effects of Ni and Co, or that the length of the experiment was insufficient for this parameter to respond. In contrast, leaf length was significantly influenced by the treatments. While the mean leaf length in control plants was 14.1 cm, it decreased to 6.3 cm (Ni) and 4.97 cm (Co) at the highest concentration of each treatment (150 μM). ANOVA test showed that there was a highly significant difference between treatments ($p < 0.001$). Cobalt was broadly toxic, compared to nickel, at lower concentrations, indicating that leaf growth was more sensitive to cobalt. This inhibition could be due to reduced photosynthesis, oxidative damage, or since the treatments added metal ions to the substrate, nutrient imbalances. Both metals also greatly decreased root growth. The mean root length of the control plants was 20.8 cm, while that in the 150 μM nickel plants was 14.9 cm, and in the cobalt plants it was 10.1 cm. Statistical analysis indicated a highly significant difference among treatments ($p < 0.001$). This indicates that root growth is sensitive to heavy metals due to their direct contact with metal ions in the substrate. The reduction in growth may be due to reduced cell division and/or elongation in the root meristems, accumulation of the toxic metals, or obstruction of essential nutrient acquisition.

An analysis of the fresh biomass of the aerial part indicates a marked difference based on the established treatments ($F = 209.772$; $p < 0.001$). For the control plants, the average fresh mass was 1.20 g. The plants displayed a slight increase with 50 μM of Ni (1.22 g), indicating a possible stimulatory effect at low dose. However, there was a marked drop with 100 μM (0.71 g) and a further decrease with 150 μM (0.54 g), indicating the toxic effects of nickel at higher doses. Cobalt treatments also caused a gradual decline in aerial biomass, which is much more prominent at 50 μM (0.65 g) through to 150 μM (0.51 g). Thus, both metals appear to have an inhibitory effect on biomass accumulation; however, cobalt appears to exhibit slightly higher toxicity especially at low concentration. For root fresh biomass, the differences in treatments are also significant ($F=50.544$; $p < 0.001$).

The fresh mass of roots decreased progressively with increasing concentrations of Ni and Co. The average mass of control plants was 2.85 g, which decreased to 1.78 g at 150 μM of Ni and to 1.42 g at 150 μM of Co. This substantial decline demonstrates an inhibition of root growth, likely due to inhibition of water and nutrient uptake processes or direct cytotoxicity of the metals at the root tissue level. Again, cobalt seems to consistently cause more deleterious effects than nickel when at the same concentrations. The analysis of dry biomass supports these trends. The dry biomass of the aerial part was significantly different among treatments ($F = 522.532$; $p < 0.001$.), where the plants exposed to 50 μM of Ni had more dry biomass than the control (0.16 g vs. 0.10 g), which may be indicative of bioaccumulation or a temporary adaptive response. At 100 and 150 μM of Ni, however, there was a major decrease in the dry weight (approximately 0.06 g). Co treatments show the same decreasing pattern (from 0.09 g to 0.05 g). The decrease in dry biomass is clear evidence for the inhibition of photosynthetic activity and restricting structural growth. Regarding the root dry biomass, the effect of treatments is extremely significant ($F = 1569.615$; $p < 0.001$). Controls show the highest value (0.35 g), followed by 50 μM of Ni (0.31 g), with a gradual reduction to 0.25 g at 150 μM. Cobalt causes a more dramatic reduction than Ni, with a reduction to 0.17 g at 150 μM. These data illustrate the sensitivity of the root system to heavy metals, which can be reliably measured with a cumulative effect of stress. The extremely high R^2 ($R^2 > 0.98$) for all parameters indicates a very strong association between the treatments and variations measured.

Table 3. Effect of heavy metals (Ni and Co) on morphological parameters.

Treatment	Number of leaves	Length of leaves (cm)	Root length (cm)	Fresh weight of aerial part (g)	Fresh weight of roots (g)	Dry weight of aerial part (g)	Dry weight of roots (g)
T1 = Temoin	3	14,10	20,80	1,20	2,84	0,10	0,34
T2 = Ni (50 µM)	3	14,16	18,85	1,22	2,20	0,16	0,31
T3 = Ni (100 µM)	3	9,55	16,80	0,71	1,90	0,05	0,28
T4 = Ni (150 µM)	2,66	6,30	14,91	0,54	1,77	0,05	0,25
T5 = Co (50 µM)	3	9,24	15,33	0,64	1,88	0,09	0,22
T6 = Co (100 µM)	3	6,28	12,53	0,55	1,77	0,05	0,19
T7 = Co (150 µM)	2,66	4,97	10,13	0,50	1,42	0,05	0,16

3.3. Physiological and biochemical parameters

All physiological and biochemical parameters results are presented in table 3. The investigation into chlorophyll a, b, and total chlorophyll concentrations show that the concentration of all three pigments decreases significantly with increasing concentrations of nickel (Ni) and cobalt (Co). In the control plants, the values were highest for chlorophyll a (1.70), chlorophyll b (0.912), and total chlorophyll (2.612). A low concentration of Ni (50 µmol) resulted in a reduced level for all three pigments with a moderate effect. The means for all three pigments were still declining from 100 µmol onwards, eventually causing significant toxicity at 150 µmol of Ni (1.1833 for chlorophyll a, 0.433 for chlorophyll b, and 1.6163 for total chlorophyll). Cobalt (Co) led to an even worse decline overall compared to Ni, particularly at 150 µmol, when the chlorophyll concentrations dropped to 1.019 for chlorophyll a, 0.290 for chlorophyll b, and 1.309 for total chlorophyll. This indicates a disturbance of the biosynthesis of photosynthetic pigments that would interfere with the overall health of the plant, possibly through the alteration of the structure of the chloroplast or the activation of oxidative stress. Comparing the two metals shows that Co is more toxic than Ni across all concentrations, especially at the higher concentrations. Most treatments exhibited low standard deviations, signifying reliable measurements;

however, the 100 µmol Co treatment had greater variability, which indicates increased sensitivity on the part of the plant at this stress level.

The breakdown of proline and total sugar demonstrated a robust increase in proline to both nickel (Ni) and cobalt (Co) levels. For the control plants, the proline content was lowest (0.170 mg/g) because there was no stress potentially induced or treated. When treated with Ni, proline content gradually increased (0.200 mg/g at the 50 µmol level, 0.460 mg/g at the 100 µmol level, and 0.790 mg/g at the 150 µmol level). Cobalt levels at 50, 100, and 150 µmol increased the proline as well, though at slightly lower amounts (0.270, 0.3667, and 0.6333 mg/g, respectively).

The results clearly show that proline is an osmoprotectant and also an indicator of stress accumulated in response to various heavy metal and/or oxidative and osmotic stresses. The trend observed in total soluble sugars mirrored that of total amino acids. Control plants had the lowest average value, 23.33 mg/g, while in response to Ni treatment, measured amounts dramatically increased, 66.67 mg/g at the 100 µmol concentration, and 92.00 mg/g at the 150 µmol concentration. Cobalt treatments, though to a lesser extent than Ni treatments, also resulted in the accumulation of sugars, with averages of 75.00 mg/g at the 100 µmol and 81.33 mg/g at the 150 µmol concentration. Overall, it is likely that sugar accumulation is an adaptive response in plants to maintain osmotic balance and protect cellular structures damaged from stress. In particular, the high levels of total soluble sugars observed for the Ni-treated plants could be reflective of the enhanced metabolism of carbohydrates or alternatively reduced utilization of sugars because of growth inhibition. Overall results suggest that both proline and soluble sugars are important physiological responses to heavy metals in plants, and that Ni has a stronger effect than Co on each factor measured. Low standard deviations for most treatments also support consistencies in values generated. Analyzing the total phenolic and flavonoid contents shows an observable increase due to nickel (Ni) and cobalt (Co) stress, indicating activation of the antioxidant defense system in response to metal stress-induced oxidative stress. The phenolic and flavonoid contents in control plants were low, 42.67 mg/g and 45.33 mg/g, respectively. The phenolic contents in plants treated with 50 µmol of Ni and Co were only slightly different from the control, indicating minimal perception of early stress. On the other hand, 100 and 150 µmol treatments of Ni resulted in phenolic levels consistently elevated to 56.33 mg/g and 72.67 mg/g, respectively. For Co treatments, the values were 58.33 mg/g at 100µmol and 60.67 mg/g at 150 µmol, with an increase likely but smaller in magnitude than Ni, which may further indicate differing capabilities of the plants to respond to different metals. Flavonoid levels varied more clearly and dynamically than the previously discussed activity, especially after Ni stress. Flavonoid levels increased to the highest levels of 69.67 mg/g and 128.00 mg/g under the two highest concentrations of Ni (100 and 150 µmol), respectively, suggesting a higher level of antioxidant response. Cobalt incubation also resulted in higher flavonoid levels (97.00 mg/g) at the highest concentration (150 µmol) than the previously discussed antioxidant activity, indicating the plant's effort to (in this case) scavenge reactive oxygen species to limit cellular damage. In conclusion, these findings indicate that phenolic compounds, particularly flavonoids, are important to the biochemical adaptation of plants under heavy metal stress and are efficient non-enzymatic antioxidants. The higher quantities under Ni treatment compared to Co infer a greater potential for inducing oxidative stress by nickel. There was some variability in standard deviation, especially in relation to flavonoid quantities, but this variability is likely a reflection of the inherent heterogeneous response of the plants under heavy metal stress when wood is available.

Table 4. Effect of heavy metals (Ni and Co) on Physiological and biochemical parameters.

Treatment	Chlorophyll a (mg/g FW)	Chlorophyll b (mg/g FW)	Total chlorophyll (mg/g FW)		Total	Polyphenols mg/g	Flavonoids mg/g
T1 = Temoin	1,70	0,91	2,61	0,17	23,33	42,66	45,33
T2 = Ni (50 µM)	1,67	0,85	2,52	0,20	23	42,33	40
T3 = Ni (100 µM)	1,58	0,59	2,17	0,46	66	56,33	69,66
T4 = Ni (150 µM)	1,18	0,43	1,61	0,79	92	72,66	128
T5 = Co (50 µM)	1,26	0,52	1,78	0,27	37,66	49,33	42
T6 = Co (100 µM)	1,10	0,50	1,60	0,36	75	58,33	79,33
T7 = Co (150 µM)	1,01	0,29	1,30	0,63	81	60,66	97

3.4. Correlation analysis

The correlation analysis indicates highly significant relationships between plant morphological growth and biochemical parameters. Leaf length is strongly and positively correlated with almost all morphological parameters (root length, fresh and dry weights of both aerial parts and rooting parts), and with chlorophyll contents, with correlation coefficients exceeding 0.9 ($p < 0.01$). A significant positive correlation is also noted between chlorophyll contents (a, b, and total) and morphological parameters illustrating that well-grown plants also have high photosynthetic activity.

On the contrary, stress-related compounds (proline, total sugars, polyphenols, flavonoids) exhibit strong and negative correlations with both morphological and chlorophyll parameters, indicating that accumulation of stress-related compounds lowers growth, likely in response to environmental stress such as salinity. Proline correlates negatively with most growth and photosynthetic parameters but exhibits strong positive correlations with total sugars, polyphenols, and flavonoids ($r>0.9$; $p<0.01$), indicating their potential synergistic role in stress tolerance.

Figure 4. Advanced Correlation Heatmap of Physiological and Biochemical Parameters in Maize Under Heavy Metal Stress with Clustered Annotations.

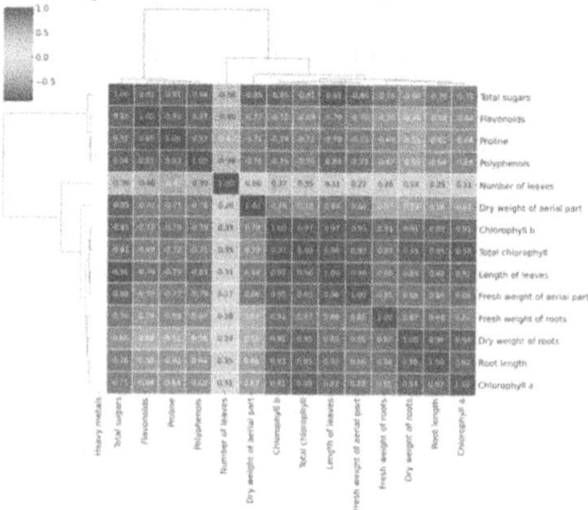

3.5. Principal component analysis

Principal Component Analysis (PCA) was carried out to assess the influence of heavy metal stress on physiological and biochemical variables in maize. The resulting biplot shows the distribution of maize samples according to the first two principal components, with vectors representing each variable measured (Biplot of Principal Component Analysis (PCA) Showing the Influence of Heavy Metal Stress on Maize Physiological and Biochemical Variables). Variables studied included leaf number, leaf length, root length, fresh and dry weights of aerial parts and roots, chlorophyll a, chlorophyll b, total chlorophyll, proline, total sugars, polyphenols and flavonoids. The proportion of variance explained by each principal component was assessed using a scree plot showing that the first principal component (PC1) explained 77.38% of the total variance, while the second principal component (PC2) explained 10.26% (Scree Plot Showing the Percentage of Explained Variance by Principal Components). Thus, most of the variability in the data is captured by PC1, while PC2 makes an additional but limited contribution. The distribution of maize samples exposed to heavy metal stress is illustrated by a PCA plot showing their distribution according to the first two principal components, allowing potential groupings to be observed (PCA Plot Showing Distribution of Maize Samples Under Heavy Metal Stress Based on Principal Components). Samples subjected to different heavy metal treatments show distinct groupings, indicating differential responses to nickel and cobalt stress. The specific contributions of each physiological and biochemical variable to the principal components (PC1 and PC2) were calculated and presented in a table showing the exact values for each of them (Contributions of Physiological and Biochemical Variables to Principal Components (PC1 and PC2)). The variables with the highest contributions to PC1 are: leaf length (1.0126), above-ground fresh weight (0.9774), root fresh weight (0.9001), above-ground dry weight (0.8438), root dry weight (0.8869), chlorophyll a (0.9397), total chlorophyll (0.9883) and chlorophyll b (1.0012). These variables are strongly correlated with PC1, indicating that they are mainly influenced by metal stress. In contrast, PC2 is mainly influenced by flavonoids (0.5096), proline (0.4727), and root dry weights (0.4955). This means that these variables make an additional contribution independent of those mainly influenced by PC1. Indeed, proline is a known stress marker, suggesting that its significant contribution to PC2 could be linked to a specific response to heavy metal stress. Biplot shows that certain variables such as flavonoids, polyphenols, and total sugars are oriented in an opposite direction to other physiological variables (e.g. leaf length, fresh and dry weight). This suggests an opposition in their variation under stress. The results also indicate that some parameters are more indicative of metal stress than others.

Figure 5. PCA Plot Showing Distribution of Maize Samples Under Heavy Metal Stress Based on Principal Components.

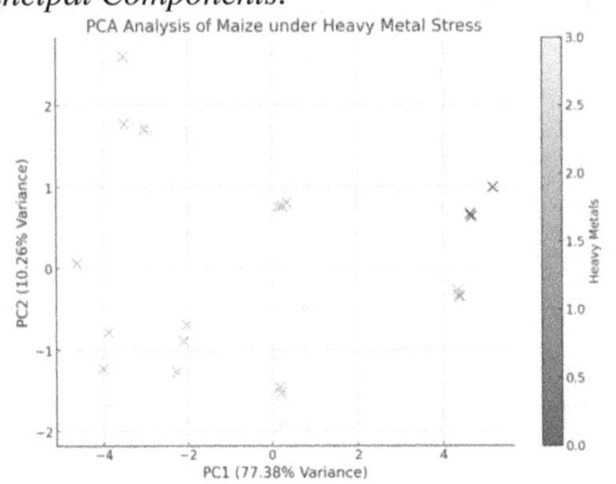

Figure 6. Biplot of Principal Component Analysis (PCA) Showing the Influence of Heavy Metal Stress on Maize Physiological and Biochemical Variables.

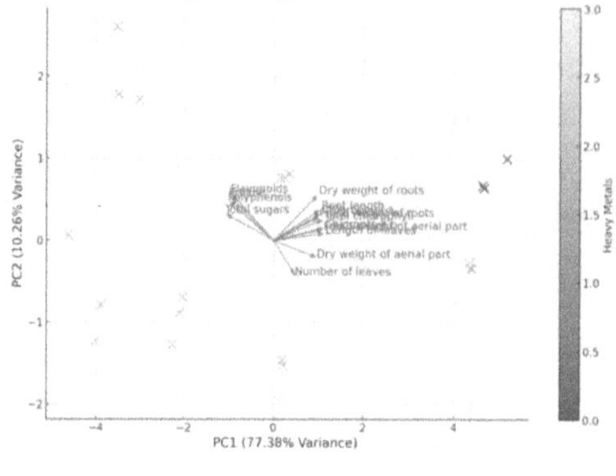

3.6. Influence of parameters

Table 4 present the relative importance of each characteristic determined by the Random Forest model applied to data from corn subjected to heavy metal treatments (Control, NI, CO). Each characteristic is ranked according to its importance, expressed as a percentage, for effectively predicting the treatments. Features with the highest importance values are considered the most influential in the classification.

Table 5. Feature Importance Analysis Using Random Forest Model.

Feature	Importance
Dry weight of roots	0.183031381
Chlorophyll a	0.111388233
Root length	0.098636221
Total chlorophyll	0.096824376
Proline	0.087072702
Chlorophyll b	0.076994443
Fresh weight of aerial part	0.076407092
Fresh weight of roots	0.062889762
Polyphenols	0.042672547
Length of leaves	0.03771868
Concentration	0.033993625
Flavonoids	0.033043832

continued on following page

Table 5. Continued

Feature	Importance
Total sugars	0.031339681
Dry weight of aerial part	0.02566015
Number of leaves	0.002327273

Comparative analysis between artificial intelligence (Random Forest) and ANOVA analysis identified the most influential parameters for discriminating heavy metal treatments in maize. The five most important characteristics identified by the AI model are: Dry weight of roots (18.30%), Chlorophyll a (11.14%), Root length (9.86%), Total chlorophyll (9.68%) and Proline (8.71%). These parameters were ranked according to their ability to correctly predict Control, NI (Nickel) and CO (Cobalt) treatments. The AI results show that Dry weight of roots is the most important characteristic, which is confirmed by the ANOVA analysis, which reveals significant differences between treatments for this parameter. In addition, the parameters Chlorophyll a, Total chlorophyll and Proline are also identified as important by AI, which is consistent with the ANOVA results showing significant differences between treatments for these parameters. Root length is identified as important by AI, which is consistent with the ANOVA results showing significant differences between treatments for this parameter. However, some discrepancies are observed between the two methods. Secondary metabolites such as polyphenols and flavonoids appear to be important in the ANOVA analysis, but were not identified among the most influential parameters by AI. This may be attributed to a complex interaction between treatments and these metabolites that is not optimally captured by the Random Forest algorithm. In summary, AI analysis using a Random Forest model identified the parameters Dry weight of roots, Chlorophyll a, Root length, Total chlorophyll and Proline as the most influential in discriminating heavy metal treatments. These results are largely consistent with those obtained by ANOVA, although some differences exist for secondary metabolites. The combined use of these two methods provides a more comprehensive and robust view of the impact of heavy metals on maize.

4. DISCUSSION

The current study clearly shows that nickel (Ni) and cobalt (Co) have negative effects on the morphological, physiological, and biochemical parameters of *Zea mays* L. plants. The results showed a negative effect of the two heavy metals in a dose-dependent manner, with cobalt at 150 μM demonstrating the highest level of toxicity for most parameters.

All morphological characteristics--including leaves' number and length, root from length, fresh and dry biomass of aerial and root parts--declined significantly as Ni and Co concentrations increased. However, while the plants applied with 50 μM of Ni or Co had a relatively similar number of leaves as the control group plants, the length of leaves revealed a marked reduction, especially at the higher concentration, 150 μM. The reduced leaf and root growth can be explained by metal-induced inhibition of cell division processes, disrupted nutrient transport and oxidative stress, all of which inhibit growth (Wang et al., 2022; Ahmed et al., 2024).

Root length is an important marker of nutrient and water uptake ability, and under Co and Ni stress, the sizes were also significantly lower, with minimal average root lengths at 150 μM Co (10.13 cm) and Ni (14.91 cm). The decreased root length could indicate damage to the root meristematic zones and root morphology changes due to excess metal. Fresh and dry weights (aerial plus root) showed a clear decline as well, with average reductions of over 50% at higher dosages. Collectively, the findings show heavy metals are directly causing poor biomass accumulation by limiting growth due to reduced photosynthate production and transport, in addition to limited water uptake (Ali et al., 2021; Pandey et al., 2022).

The content of chlorophyll (a, b and total) showed a steady reduction in response to increased concentrations of the metals under study. Compared to the control (2.61 mg/g FW, total chlorophyll), levels dropped sharply to 1.30 mg/g FW and 1.61 mg/g FW for the 150 μM Co and Ni, respectively. The decline in chlorophyll content is a well-known indicator of metal toxicity due to the inhibition of chlorophyll biosynthetic enzymes, loss of chloroplast structure, and oxidative degradation of pigments. Since chlorophyll is directly related to its efficiency in photosynthesis, its reduction mirrors the decline in biomass observed (Irfan et al., 2021; Gashi et al., 2024).

Chlorophyll levels, in contrast, did not vary significantly under stress conditions. However, proline and total soluble sugars increased dramatically, especially under the higher concentrations of metals (150 μM). Proline levels, for example, increased from 0.17 mg/g FW in the control to 0.79 mg/g FW under 150 μM Ni and 0.63 mg/g FW under 150 μM Co. This process is a known adaptive response, given that under stress conditions proline functions as an osmoprotectant, provides stabilization of proteins and membranes, and scavenges free radicals. Similarly, total sugars increased dramatically as well, rising to 92 mg/g FW under 150 μM Ni and 81 mg/g FW under 150 μM Co, relative to the control of 23.33 mg/g FW. These sugars may also act as osmolytes and as energy stores to maintain metabolism under stress conditions (Riyazuddin et al., 2021).

Exposure to metals also led to significant increases in phenolic compounds (polyphenols and flavonoids), suggesting activation of antioxidant defense systems. Flavonoids content nearly tripled compared to the control from exposure to 150 μM of Ni (128 mg/g vs 45.33 mg/g). Following the trend noted for flavonoids, cobalt treatment also led to increases in flavonoid content, with the highest level at 150 μM (97 mg/g). These secondary metabolites aid in neutralizing reactive oxygen species (ROS) that are produced during stress, thereby preventing oxidative damage to cellular structures (Hasanuzzaman et al., 2021). Proline, sugars, polyphenols and flavonoids show a strong relationship, indicating a coordinated biochemical response for increased tolerance to heavy metal induced oxidative stress (Kumar et al., 2024; Shang et al., 2024; Rao & Zheng, 2025).

Overall, these results indicate a clear physiological trade-off in maize under heavy metal stress. While plants elicit protective responses, such as accumulating osmolytes and antioxidant production, there is a cost to be paid in reduced growth and photosynthetic capacity. The stronger toxic responses noted at higher doses of cobalt than nickel suggest greater Co bioavailability or higher Co toxicity in the experimental conditions used (Rizvi et al., 2022; Gao et al., 2022).

This investigation emphasizes the sensitivity of maize to environmental contamination with heavy metals, especially in industrial areas where these contaminants can accumulate in soils (Waqas et al., 2024; Qin et al., 2024). The use of morphological, physiological, and biochemical markers offers a complete picture of the response to stress (Hafeez et al., 2023; Bai et al., 2024). This may stimulate new direction for selection or breeding or other plant improvement efforts to develop better adapted maize varieties to environments with contaminants. The study also highlights the need to survey and control heavy metal pollution in an agricultural context, as it relates to food security and human health, in Moroccan industrial areas.

5. CONCLUSION

In this study, we discuss the negative effects of heavy metals, nickel (Ni) and cobalt (Co) specifically, on the morphological, physiological, and biochemical performance of *Zea mays* L. plants. Increasing concentrations of Ni and Co lead to a significant reduction in growth parameters, chlorophyll content, and biomass production, consistent with decreased photosynthetic activity and development. The concomitant accumulation of proline, soluble sugars, polyphenols, and flavonoids showed evidence that the plants activated defense mechanisms to counter oxidative stress induced by metal exposure. The results clearly indicate that cobalt possesses a higher dose-dependent toxicity than nickel at the same concentration. This highlights the susceptibility of maize to heavy metal contamination, especially in industrial regions of Morocco where these pollutants are likely high. This study also highlights the importance of linking biochemical stress indicators with morphological markers to assess plant responses to the environment that are inclusive of all variables. In the end, this study ensures an improved understanding of plant-metal interactions while supporting the development of soil management strategies and pollution mitigation to maintain safe and productive agricultural systems in polluted environments.

REFERENCES

Adnan, M., Xiao, B., Xiao, P., Zhao, P., & Bibi, S. (2022). Heavy Metal, Waste, COVID-19, and Rapid Industrialization in This Modern Era—Fit for Sustainable Future. *Sustainability (Basel), 14*(8), 4746. DOI: 10.3390/su14084746

Ahmed, W., Sharma, G., Singh, D., & Singh, R. (2024). Ecological impact of heavy metal pollution on *Limnodrilus cervix* -abundance, bioaccumulation, and environmental dynamics in a perennial pond of Jammu, India. *Process Safety and Environmental Protection, 188*, 595–607. DOI: 10.1016/j.psep.2024.05.147

Ali, H., Khan, E., & Sajad, M. A. (2022). Phytotoxicity of heavy metals: Impacts on plants and possible remediation strategies. *Environmental Science and Pollution Research International, 29*(3), 1803–1817.

Ali, J., Ali, F., Ahmad, I., Rafique, M., Munis, M. F. H., Hassan, S. W., Sultan, T., Iftikhar, M., & Chaudhary, H. J. (2021). Mechanistic elucidation of germination potential and growth of *Sesbania sesban* seedlings with *Bacillus anthracis* PM21 under heavy metals stress: An *in vitro* study. *Ecotoxicology and Environmental Safety, 208*, 111769. DOI: 10.1016/j.ecoenv.2020.111769 PMID: 33396087

Angon, B. P., Islam, M. S., Kc, S., Das, A., Anjum, N., Poudel, A., & Suchi, S. A. (2024). Sources, effects and present perspectives of heavy metals contamination: Soil, plants and human food chain. *Heliyon, 10*(7), e28357. DOI: 10.1016/j.heliyon.2024.e28357 PMID: 38590838

Arnon, D. I. (1949). Copper Enzymes in Isolated Chloroplasts. Polyphenoloxidase in Beta Vulgaris. *Plant Physiology, 24*(1), 1–15. DOI: 10.1104/pp.24.1.1 PMID: 16654194

Aslanidis, P. S. C., & Golia, E. E. (2022). Urban Sustainability at Risk Due to Soil Pollution by Heavy Metals—Case Study: Volos, Greece. *Land (Basel), 11*(7), 1016. DOI: 10.3390/land11071016

Bai, X., Bol, R., Chen, H., Cui, Q., Qiu, T., Zhao, S., & Fang, L. (2024). A meta-analysis on crop growth and heavy metals accumulation with PGPB inoculation in contaminated soils. *Journal of Hazardous Materials, 471*, 134370. DOI: 10.1016/j.jhazmat.2024.134370 PMID: 38688214

Bates, L. S., Waldren, R. P., & Teare, I. D. (1973). Rapid determination of free proline for water studies. *Plant and Soil, 39*(1), 205–208. DOI: 10.1007/BF00018060

Boros-Lajszner, E., Wyszkowska, J., Borowik, A., & Kucharski, J. (2021). The Response of the Soil Microbiome to Contamination with Cadmium, Cobalt and Nickel in Soil Sown with Brassica napus. *Minerals (Basel), 11*(5), 498. DOI: 10.3390/min11050498

Dubois, M. K., Gils, J. K., & Hanniton, P. A. (1956). Use of phenol reagent for the determination of total sugar. *Analytical Chemistry, 28*, 350–356. DOI: 10.1021/ac60111a017

El-Khadir, I., Ktaoui, S., & Mouniane, Y. (2024). Improved Salt Stress Tolerance of *Salvia Officinalis* Grown in the Presence of a Halophytic Plant *Spergularia Maritima*: Analysis of Morpho-Physiological Parameters. In Mabrouki, J., & Azrour, M. (Eds.), *Integrated Solutions for Smart and Sustainable Environmental Conservation. Studies in Systems, Decision and Control* (Vol. 527). Springer., DOI: 10.1007/978-3-031-55787-3_2

Emamverdian, A., Ding, Y., Mokhberdoran, F., & Xie, Y. (2015). Heavy Metal Stress and Some Mechanisms of Plant Defense Response. The Scientific World Journal. *TheScientificWorldJournal, 756120*(1), 756120. Advance online publication. DOI: 10.1155/2015/756120 PMID: 25688377

Gao, C., Gao, K., Yang, H., Ju, T., Zhu, J., Tang, Z., Zhao, L., & Chen, Q. (2022). Genome-wide analysis of metallothionein gene family in maize to reveal its role in development and stress resistance to heavy metal. *Biological Research, 55*(1), 1. DOI: 10.1186/s40659-021-00368-w PMID: 35012672

Gashi, B., Buqaj, L., Vataj, R., & Tuna, M. (2024). Chlorophyll biosynthesis suppression, oxidative level and cell cycle arrest caused by Ni, Cr and Pb stress in maize exposed to treated soil from the Ferronikel smelter in Drenas, Kosovo. *Plant Stress, 11*, 100379. DOI: 10.1016/j.stress.2024.100379

Hafeez, A., Rasheed, R., Ashraf, M. A., Qureshi, F. F., Hussain, I., & Iqbal, M. (2023). Effect of heavy metals on growth, physiological and biochemical responses of plants. In *Azamal Husen. Plants and Their Interaction to Environmental Pollution*. Elsevier., DOI: 10.1016/B978-0-323-99978-6.00006-6

Hakkou, R., Benzaazoua, M., & Bussière, B. (2020). Environmental impacts of mine tailings in Morocco: A review. *Environmental Geochemistry and Health, 42*(2), 497–516. PMID: 33141377

Hasanuzzaman, M., Parvin, K., Bardhan, K., Nahar, K., Anee, T. I., Masud, A. A. C., & Fotopoulos, V. (2021). Biostimulants for the Regulation of Reactive Oxygen Species Metabolism in Plants under Abiotic Stress. *Cells, 10*(10), 2537. DOI: 10.3390/cells10102537 PMID: 34685517

Irfan, M., Mudassir, M., Khan, M. J., Dawar, K. M., Muhammad, D., Mian, I. A., Ali, W., Fahad, S., Saud, S., Hayat, Z., Nawaz, T., Khan, S. A., Alam, S., Ali, B., Banout, J., Ahmed, S., Mubeen, S., Danish, S., Datta, R., & Dewil, R. (2021). Heavy metals immobilization and improvement in maize (*Zea mays* L.) growth amended with biochar and compost. *Scientific Reports*, *11*(1), 18416. DOI: 10.1038/s41598-021-97525-8 PMID: 34531439

Khan, M. N., Ullah, R., & Abbas, T. (2023). Heavy metal stress and its impact on crop productivity: A global perspective. *Journal of Soil Science and Plant Nutrition*, *23*(1), 45–60.

Kosakivska, I. V., Babenko, L. M., Romanenko, K. O., Korotka, I. Y., & Potters, G. (2021). Molecular mechanisms of plant adaptive responses to heavy metals stress. *Cell Biology International*, *45*(2), 258–272. DOI: 10.1002/cbin.11503 PMID: 33200493

Kumar, M., Singh, S., Jain, A., Yadav, S., Dubey, A., & Trivedi, S. P. (2024). A review on heavy metal-induced toxicity in fishes: Bioaccumulation, antioxidant defense system, histopathological manifestations, and transcriptional profiling of genes. *Journal of Trace Elements in Medicine and Biology*, *83*, 127377. DOI: 10.1016/j.jtemb.2023.127377 PMID: 38183919

Kumar, S., Prasad, S., Yadav, K. K., Shrivastava, M., Gupta, N., Nagar, S., Bach, Q.-V., Kamyab, H., Khan, S. A., Yadav, S., & Malav, L. C. (2019). Hazardous heavy metals contamination of vegetables and food chain: Role of sustainable remediation approaches - A review. *Environmental Research*, *179*, 108792. DOI: 10.1016/j.envres.2019.108792 PMID: 31610391

Kumar, S., Singh, R., & Sharma, P. (2022). Nickel toxicity in plants: Mechanisms and remediation strategies. *Ecotoxicology and Environmental Safety*, *234*, 113370.

Liu, X., Zhang, Y., & Li, Q. (2023). Effects of heavy metal pollution on soil microbial communities and plant health. *Applied Soil Ecology*, *182*, 104631.

Mouniane, Y., Vibhute, A. D., Kadmi, Y. (2025). Using Phytoremediation to Treat Industrial and Pharmaceutical Wastewater: Innovations and Future Prospects. In: Biotechnology Approaches to Industrial and Pharmaceutical Wastewater Treatment. pp. 277–298. IGI Global Scientific Publishing. DOI: 10.4018/979-8-3693-8487-9.ch009

Noor, I., Sohail, H., Sun, J., Nawaz, M. A., Li, G., Hasanuzzaman, M., & Liu, J. (2022). Heavy metal and metalloid toxicity in horticultural plants: Tolerance mechanism and remediation strategies. *Chemosphere*, *303*, 135196. DOI: 10.1016/j.chemosphere.2022.135196 PMID: 35659937

Oudghiri, M., Yamani, B., Benlemlih, N., El Aammouri, S., Abid, N., Brhadda, N., Bouhassoun, S., Ziri, R., Chriqui, A., Aoujil, F. Z., El Bakkali, M., Mouniane, Y., & Ibriz, M. (2025). Effect of Heavy Metals on the Morphological and Physiological Responses of the Torro Plus Variant of *Zea mays*. *Journal of Environment and Earth Science*, *7*(4), 165–179. DOI: 10.30564/jees.v7i4.8208

Pandey, A. K., Zorić, L., Sun, T., Karanović, D., Fang, P., Borišev, M., Wu, X., Luković, J., & Xu, P. (2022). The Anatomical Basis of Heavy Metal Responses in Legumes and Their Impact on Plant–Rhizosphere Interactions. *Plants*, *11*(19), 2554. DOI: 10.3390/plants11192554 PMID: 36235420

Qin, H., Wang, Z., Sha, W., Song, S., Qin, F., & Zhang, W. (2024). Role of Plant-Growth-Promoting Rhizobacteria in Plant Machinery for Soil Heavy Metal Detoxification. *Microorganisms*, *12*(4), 700. DOI: 10.3390/microorganisms12040700 PMID: 38674644

Rajput, P., Singh, A., Agrawal, S., Ghazaryan, K., Rajput, V. D., Movsesyan, H., Mandzhieva, S., Minkina, T., & Alexiou, A. (2024). Effects of environmental metal and metalloid pollutants on plants and human health: Exploring nano-remediation approach. *Stress Biology*, *4*(1), 27. DOI: 10.1007/s44154-024-00156-y PMID: 38777953

Rao, M. J., & Zheng, B. (2025). The Role of Polyphenols in Abiotic Stress Tolerance and Their Antioxidant Properties to Scavenge Reactive Oxygen Species and Free Radicals. *Antioxidants*, *14*(1), 74. DOI: 10.3390/antiox14010074 PMID: 39857408

Riyazuddin, R., Nisha, N., Ejaz, B., Khan, M. I. R., Kumar, M., Ramteke, P. W., & Gupta, R. (2022). A Comprehensive Review on the Heavy Metal Toxicity and Sequestration in Plants. *Biomolecules*, *12*(1), 43. DOI: 10.3390/biom12010043 PMID: 35053191

Rizvi, A., Ahmed, B., Khan, M. S., Rajput, V. D., Umar, S., Minkina, T. M., & Lee, J. (2022). Maize associated bacterial microbiome linked mitigation of heavy metal stress: A multidimensional detoxification approach. *Environmental and Experimental Botany*, *200*, 104911. DOI: 10.1016/j.envexpbot.2022.104911

Shakespear, S., Sivaji, M., & Kumar, V. (2024). Navigating Through Harsh Conditions: Coordinated Networks of Plant Adaptation to Abiotic Stress. *Journal of Plant Growth Regulation*. Advance online publication. DOI: 10.1007/s00344-023-11224-4

Shang, C., Zhou, Q., Nkoh, J. N., Liu, J., Wang, J., Hu, Z., & Hussain, Q. (2024). Integrated physiological, biochemical, and transcriptomic analyses of *Bruguiera gymnorhiza* leaves under long-term copper stress: Stomatal size, wax crystals and composition. *Ecotoxicology and Environmental Safety*, *281*, 116609. DOI: 10.1016/j.ecoenv.2024.116609 PMID: 38905937

Silva, H. F., Silva, N. F., Oliveira, C. M., & Matos, M. J. (2021). Heavy Metals Contamination of Urban Soils—A Decade Study in the City of Lisbon, Portugal. *Soil Systems*, *5*, 27. DOI: 10.3390/soilsystems5020027

Uddin, M. M., Zakeel, M. C. M., Zavahir, J. S., Marikar, F. M. M. T., & Jahan, I. (2021). Heavy Metal Accumulation in Rice and Aquatic Plants Used as Human Food: A General Review. *Toxics*, *9*(12), 360. DOI: 10.3390/toxics9120360 PMID: 34941794

Wang, D., Bai, L., Huang, X., Yan, W., & Li, S. (2022). Size-dependent acute toxicity and oxidative damage caused by cobalt-based framework (ZIF-67) to *Photobacterium phosphoreum*. *The Science of the Total Environment*, *851*, 158317. DOI: 10.1016/j.scitotenv.2022.158317 PMID: 36037900

Wang, Y. L., Lee, Y. H., Chou, C. L., Chang, Y. S., Liu, W. C., & Chiu, H. W. (2024). Oxidative stress and potential effects of metal nanoparticles: A review of biocompatibility and toxicity concerns. *Environmental Pollution*, *346*, 123617. DOI: 10.1016/j.envpol.2024.123617 PMID: 38395133

Waqas, W., Yuan, Y., Ali, S., Zhang, M., Shafiq, M., Ali, W., Chen, Y., Xiang, Z., Chen, R., Ikhwanuddin, M., & Ma, H. (2024). Toxic effects of heavy metals on crustaceans and associated health risks in humans: A review. *Environmental Chemistry Letters*, *22*(3), 1391–1411. DOI: 10.1007/s10311-024-01717-3

Zaim, S., Bourrach, E., & El Alami, H. (2021). Heavy metal pollution in agricultural soils near industrial areas in Morocco. *Environmental Monitoring and Assessment*, *193*(4), 208. PMID: 33755795

Zhao, W., Chen, J., & Feng, L. (2023). Heavy metal-induced oxidative stress and its implications for sustainable crop production. *Plant Physiology and Biochemistry*, *190*, 85–97.

Zidan, K., Mandi, L., Hejjaj, A., Ouazzani, N., & Assabbane, A. (2024). Soil fertility and agro-physiological responses of maize (*Zea mays*) irrigated by treated domestic wastewater by hybrid multi-soil-layering technology. *Journal of Environmental Management*, *351*, 119802. DOI: 10.1016/j.jenvman.2023.119802 PMID: 38134502

Chapter 7
Integrating Explainable AI, Deep Learning, and Causal Inference to Unveil Salinity Stress Responses in Lavandula Dentata:
A Study Under Co-Cultivation Systems in the Context of Climate Change

Issam El-Khadir
https://orcid.org/0000-0002-3970-2865
Faculty of Sciences, Ibn Tofail University, Morocco

Yassine Mouniane
https://orcid.org/0000-0002-6037-5301
Faculty of Sciences, Ibn Tofail University, Morocco

Ahmed Chriqui
Faculty of Sciences, Ibn Tofail University, Morocco

Mohamed El Bakkali
Faculty of Sciences, Ibn Tofail University, Morocco

Driss Hmouni
https://orcid.org/0000-0001-7598-6204

DOI: 10.4018/979-8-3693-9132-7.ch007

Faculty of Sciences, Ibn Tofail University, Morocco

ABSTRACT

This chapter explores the salinity stress response of Lavandula dentata under mono-culture and co-cultivation with halophytes (Atriplex prostrata, Plantago macrorhiza) using an integrative approach combining explainable artificial intelligence, deep learning, and causal inference. Dimensionality reduction (PCA, t-SNE, UMAP) and clustering algorithms (K-Means, DBSCAN) identified stress- and culture-dependent phenotypic patterns, while supervised learning models (Random Forest, MLP) predicted cultivation conditions with high accuracy, highlighting the key role of root volume, proline, and water content. SHAP values offered model interpretability, and causal inference quantified the direct effects of co-cultivation on biomass. Results revealed that co-cultivation mitigates certain stress traits but may reduce biomass, suggesting trade-offs in physiological adaptation. This chapter demonstrates how data-driven frameworks can support agroecological strategies for salt-affected environments under climate change.

1. INTRODUCTION

The rapid growth of the global population will reach nearly 10 billion by the year 2050, which will increase more pressure on food systems globally than ever. To satiate preparing food in the 2050's global food production must increase by approximately 60-110% (Godoy et al., 2021; Noh et al., 2024). This has not been made easier because of the various environmental stressors, and abiotic stresses from climate change, are some of the most crucial. Among abiotic stresses, soil salinity is one of the most limiting factors that affect crop productivity, which impacts the yields and quality of the food produced (Ye et al., 2019; Soni et al., 2024). Approximately 20% of the world's irrigated land - an area of 62 million hectares - is affected by salinization (Etesami & Noori, 2019; Bazihizina et al., 2024). In South Asia, alone there are over 52 million hectares at risk (Mandal et al., 2018). Salt stress directly affects a plant's ability to take up water and nutrients, causes an ionic imbalance and toxicity (primarily due to Na^+ and Cl^- ions), and incurs oxidative stress, which adversely affects a plant's physiology and stunts its growth. (Singh et al., 2024).

Salinization is a global threat to food security not only through the impact on individual plants. The United Nations Environment Programme (UNEP) estimates that salinity issues affect 20% of the world's arable land and 50% of irrigated farm-land worldwide (Omuto et al., 2024). According to the World Bank, at least 24% of all irrigated land is affected by salinity and at least 50% of unproductive land in

Africa is due to salinization (Hopmans et al. 2021). The problem is expected to worsen in the coming decades, with sobering estimates suggesting that by 2050 salinity will affect as much as half of the world's arable land, resulting in an overall reduction of key crops like vegetables (Shahzad et al 2021).

Lavandula dentata L. is a perennial shrub in the family Lamiaceae. It is popular due to its essential oils, which consist of biologically active compounds like linalool, camphor, and cineole (Elbouzidi et al., 2024). These compounds provide antimicrobial, antioxidant, anti-inflammatory, and anxiolytic activities. For this reason, L. dentata is an economically important crop for the pharmaceutical, cosmetic, and perfumery industries (Habán et al., 2023). Although economically important, *Lavandula dentata* also has ecological and agronomic importance, especially in terms of its ability to adapt to Mediterranean conditions and lower water requirements than other crops (Tofah et al., 2023). However, crops like L. dentata will have limited potential in marginal and saline prone lands, especially due to their sensitivity to salt stress. Salinity has been shown to negatively affect growth, essential oil chemical composition, and biomass of other lavender species including L. dentata (Paraskevopoulou et al., 2020). Sodium chloride concentrations change the ionic balance that disrupts plant ion homeostasis, reduces photosynthetic efficiency, and decreases oil quality and yield (Akachoud et al., 2022).

Traditional solutions, for example, physical and chemical soil remedial actions (e.g. ex leaching, gypsum, lime) and transgenic solutions, have many limitations. They are typically expensive, uneconomical, and often will not work in highly saline areas (Arora et al., 2018; Morton et al., 2019). Therefore, the rapid search for alternative ecological, economic and sustainable solutions to promote soil fertility restoration and agricultural production is vitally important. In this sense, intercropping salt-sensitive species (glycophytes) with salt-tolerant plants (halophytes) represents a potential agroecological strategy. Halophytes can take up, sequester, or even release excess salts; improve rhizospheric condition; decrease toxic ions availability; and release beneficial root exudates that support growth of the neighboring plants. These beneficial mechanisms may be useful potential approaches to mitigate impacts of salinity and positively affect crop resilience (Duarte et al., 2014).

Recent sustainable agroecological strategies propose plant-plant interactions through co-cultivation with halophytes as a promising approach to alleviate salt stress effects (Atrous et al., 2025). Halophytes such as *Atriplex prostrata* and *Plantago macrorhiza* can modify rhizospheric conditions, enhance water retention, and reduce sodium uptake in companion glycophytes via nurse-plant effects, hydraulic redistribution, and rhizosphere signaling (Tarroum et al., 2021). Co-cultivation may therefore act as a bio-buffering system against salinity, but its mechanistic impact on Lavandula's physiological responses remains insufficiently characterized under field-like saline gradients.

Furthermore, climate change exacerbates the salinity problem by altering precipitation patterns, accelerating soil degradation, and expanding saline zones (Khondoker et al., 2023). Understanding the resilience mechanisms of *Lavandula dentata* within adaptive co-cultivation systems becomes urgent to guide salt-resilient agroecosystem design. However, traditional statistical tools fail to capture complex, non-linear interactions between physiological traits, cultivation context, and stress intensity (Khamidov et al., 2022).

To address this gap, artificial intelligence (AI) frameworks such as explainable machine learning, deep learning, and causal inference offer powerful tools for uncovering trait-level patterns, stress biomarkers, and cultivation effects in a data-driven and interpretable manner (Tursunalieva et al., 2024). These approaches allow for robust classification, feature attribution, and quantification of causal treatment effects, thereby reinforcing the biological interpretability and decision-making capacity for climate-resilient cultivation strategies (Zhang et al., 2024).

The objective of this study is to unravel the physiological mechanisms and causal effects underpinning salinity stress responses in *Lavandula dentata* under co-cultivation with halophytes *Atriplex prostrata* and *Plantago macrorhiza*, using a novel integrative framework that combines explainable AI, deep neural networks, and causal inference. Specifically, the study aims to (i) characterize the multivariate stress responses across culture and salinity levels, (ii) predict culture type and stress intensity based on physiological markers, (iii) identify the most influential traits using SHAP analysis, and (iv) quantify the causal impact of co-cultivation on biomass using treatment effect modeling—all within the broader context of climate change-induced soil salinization.

2. MATERIALS AND METHODS

2.1. Plant Materials

The plant species included this study consisted of *Lavandula dentata* (toothed lavender), a non-halophytic aromatic and medicinal plant, and two halophytic species: *Atriplex prostrata*, and *Plantago macrorhiza* obtained from the Oued Maleh in the Ouazzane region, known for its extreme salinity (Chriqui et al., 2024). Seedlings were grown from cuttings (for L. dentata) and seeds (for halophytes) under controlled greenhouse conditions.

2.2. Experimental Conditions

The experiment was performed in the greenhouse (experimental) at the Faculty of Science,Ibn Tofail University, Morocco. Seedlings were transplanted in pots with a homogeneous substrate of soil/sand/peat (1:1:1, v/v). Initial watering was done for 2 weeks to allow for acclimation.

2.3. NaCl Treatment (Salt Stress)

Plants were exposed to 0 mM/L (control), 34 mM/L, 86 mM/L and 172 mM/L NaCl levels of salinity. Saline solutions were delivered through irrigation every 48 h over a 30 day period to determine plant response to salt stress.

2.4. Experimental Design

The trial was carried out in a completely randomized design with each treatment replicated five times. Three growing systems were evaluated; *Lavandula dentata* in monoculture, *L. dentata* co-cultivation with *Atriplex prostrata* and *L. dentata* co-cultivation with *Plantago macrorhiza* to measure the impact of halophyte association on salt stress tolerance.

2.4. Measured Parameters

2.4.1. Morphological Parameters

Morphological parameters were evaluated to examine how salt stress and plant partners influenced *Lavandula dentata* growth. The biomass, or mass areal mass, was determined from separated parts of the aerial biomass and root biomass. In order to measure fresh biomass, all dried up stems and blades were removed from the aerial biomass, and the samples were all weighed outside of the field. Aerial biomass length was determined using a millimeter ruler, measured from collar to the tip of the main stem. Root biomass length was determined after thoroughly washed roots (to remove the soil), counting from collar to the tip of the longest root. To obtain root volume, the water displacement method was used to estimate roots volume, with the known volume of distilled water in a graduated cylinder, surrounded by roots displacement volume. The means of all these parameters provided a quantitative estimate of salt stress and interspecific interactions causing morphological development.

2.4.2. Physiological Parameters

Relative water content was calculated using fresh leaf and dry leaf weight and indicative of plants' ability to hold water under salt stress (Barrs, 1996). Chlorophyll content was measured using the extraction method with 80% acetone and subsequent spectrophotometric measurement at 645 and 663 nm using Arnon's equation to assess total chlorophyll content (Arnon, 1949).

2.4.3. Biochemical Parameters

Proline accumulation was measured by the method of Bates et al. (1973) using leaf extractions and spectrophotometric analysis to obtain proline concentration because this molecule also has osmoprotective roles in the plant defense mechanisms to salt stress. Total sugars were extracted from leaves and quantified by the method of Dubois et al. (1956) using a spectrophotometer. Because total sugars are also important osmoprotectants, the role of their investment in osmoregulatory mechanisms is equally important to the plants' adaption to stress.

2.5. Statistical Analysis

2.5.1. Advanced Artificial Intelligence Framework Combining Machine Learning, Manifold Learning, and Causal Inference for Multivariate Phenotyping in *Lavandula dentata*

The analytical workflow applied to investigate the physiological and morphological responses of *Lavandula dentata* under salinity stress and co-cultivation conditions integrated multiple steps of advanced data processing, machine learning, and manifold learning techniques. Following data collection, a rigorous preprocessing step was performed, including normalization of continuous traits, categorical encoding of cultivation treatments, and verification of data integrity. Dimensionality reduction techniques such as Principal Component Analysis (PCA) (Hotelling, 1933), Uniform Manifold Approximation and Projection (UMAP) (McInnes et al., 2020), and t-distributed Stochastic Neighbor Embedding (t-SNE) (Van der Maaten & Hinton, 2008) were applied to project high-dimensional biological data into two-dimensional space, preserving global or local structure depending on the method used (Yan & Wang, 2022). These projections enabled a comprehensive exploration of sample distribution and variance structure without prior knowledge of group labels. To explore latent groupings, unsupervised clustering was conducted using both K-Means (MacQueen, 1967) and DBSCAN (Ester et al., 1996) algorithms, where K-Means captured global centroid-based groupings while DBSCAN identified local density-based clusters and potential outliers.

Complementary to that, a correlation matrix was computed using Pearson's correlation coefficient to assess linear dependencies among physiological traits, following the formula:

$$r = \frac{\sum(x_i - \bar{x})(y_i - \bar{y})}{\sqrt{(x_i - \bar{x})^2 (y_i - \bar{y})^2}}$$

This correlation structure helped guide later interpretability steps and confirm the coherence of observed patterns. A series of intelligible visualizations were generated including enhanced boxplots stratified by culture condition for each individual trait to support feature-specific interpretation. For predictive modeling, supervised learning algorithms including Random Forest (Breiman, 2001), XGBoost (Chen & Guestrin, 2016), and Multi-Layer Perceptron (MLP) neural networks (Rumelhart et al., 1986) were trained to predict either the type of culture or salinity level based on physiological features. Cross-validation and stratified sampling ensured robustness and generalizability of classification outputs.

Post-modeling explainability was performed using SHAP (SHapley Additive exPlanations) (Lundberg & Lee, 2017) to quantify feature importance for each model. For each classifier, SHAP values were computed to estimate the marginal contribution of each feature to the prediction outcome, based on cooperative game theory principles. The contribution of a feature iii to a prediction f(x) is formally defined as:

$$\phi_i = \sum_{S \subseteq F \setminus \{i\}} \frac{|S|!(|F|-|S|-1)!}{|F|!} [f(S \cup \{i\}) - f(S)]$$

where F is the full set of features, and SSS is a subset excluding i. This approach allows for interpretable decomposition of predictions into additive feature effects.

Additionally, a causal inference framework was implemented using the DoWhy library (Sharma & Kiciman, 2020) to isolate the causal impact of cultivation type on biomass as an agronomic endpoint. Propensity score matching (Rosenbaum & Rubin, 1983) was applied to estimate the average treatment effect (ATE) between treatment groups, following the formulation:

$$ATE = E[Y1 - Y0]$$

where Y1 is the outcome under treatment, and Y0 under control. Identification of the causal effect relied on the backdoor criterion (Pearl, 2022), controlling for confounding variables through observed covariates. Causal graphs were constructed to visually represent hypothesized relationships between culture type, stress indicators, and growth parameters, consistent with the structure of graphical causal models.

This integrated methodological approach ensured that the biological insights extracted were both statistically grounded and interpretable, combining data-driven modeling, visual analytics, and theoretical frameworks of causality into a unified analytical pipeline.

3. RESULTS

The boxplot shows that *Lavandula dentata* cultivated alone exhibited the highest variability and median biomass, while co-cultivation with *Atriplex prostrata* resulted in slightly reduced median biomass but more consistent values, and association with *Plantago macrorhiza* led to the lowest biomass overall, suggesting a differential influence of companion halophytes on growth performance under the same environmental conditions (Figure 1). The chlorophyll content was lowest and most variable in monoculture, whereas both co-cultivation systems exhibited higher and more stabilized chlorophyll levels, suggesting that halophytes may help preserve photosynthetic capacity under salinity (Figure 2). Aerial part length increased significantly under co-cultivation with *Atriplex prostrata*, which showed the highest median and range values, while *P. macrorhiza* associated plants displayed moderately improved shoot growth compared to monoculture, suggesting a growth-stimulating effect from both halophytes, with *A. prostrata* exerting the strongest impact (Figure 3). Proline content was highest in monoculture, reflecting a stronger osmotic stress response, while both co-cultivation treatments—particularly with *Atriplex prostrata*—significantly reduced proline levels, indicating a potential stress-buffering effect (Figure 4). Relative water content followed the same trend, with co-cultivation enhancing hydration levels, particularly in the *Atriplex prostrata* associated treatment, highlighting improved water status likely driven by microclimatic or physiological stabilization (Figure 5). Root length showed higher variability in monoculture, while co-cultivation promoted more homogeneous root development, with *P. macrorhiza* associated samples showing marginally greater elongation, suggesting root system stabilization through halophytic interactions (Figure 6). Root volume increased significantly under both co-cultivation regimes, especially with *Atriplex prostrata*, which promoted the highest volumetric expansion, underscoring the positive role of halophytes in below-ground development and nutrient uptake (Figure 7). Total sugar content was markedly higher and more variable in monocultured *Lavandula dentata*, while both halophyte associations especially with *Atriplex prostrata* resulted in sugar suppression, indicating reduced osmotic stress and potentially altered carbon allocation under salt conditions (Figure 8). The correlation matrix revealed strong positive associations among biomass, chlorophyll, and aerial part length, while proline was strongly negatively correlated with all growth traits, particularly biomass ($r = -0.95$) and chlorophyll ($r = -0.94$), supporting its role as a key stress marker. Meanwhile, root volume and relative water content were positively associated with both biomass and photosynthetic performance, highlighting their functional contribution to salt tolerance (Figure 9). Principal Component Analysis (PCA) revealed a clear separation of cultivation types along PC1, with monocultured plants forming a distinct cluster and co-cultivated samples partially overlapping but maintaining separable multivariate profiles (Figure 10). The t-SNE projection further emphasized this separation with a nonlinear embedding, showing a distinct clustering of monoculture and moderate overlap between the halophyte-associated groups, confirming that each cultivation strategy induces a unique physiological response (Figure 11). UMAP combined with K-Means clustering revealed three well-defined groups that corresponded precisely to the three cultivation modes, reinforcing the discriminative power of unsupervised learning to capture biologically meaningful variation in response to co-cultivation (Figure 12). Applying DBSCAN clustering to the UMAP space revealed one large density-based cluster and a smaller group of edge points labeled as noise, reflecting the general coherence of physiological responses with a few outliers likely due to individual variation or micro-environmental effects (Figure 13). Finally, UMAP colored by NaCl concentration displayed a continuous physiological gradient from control (0 mM) to highly stressed plants (172 mM), indicating that salinity triggers a progressive and nonlinear reconfiguration of traits, well captured by manifold learning (Figure 14).

Figure 1. Comparative Biomass Distribution of Lavandula dentata Across Cultivation Types

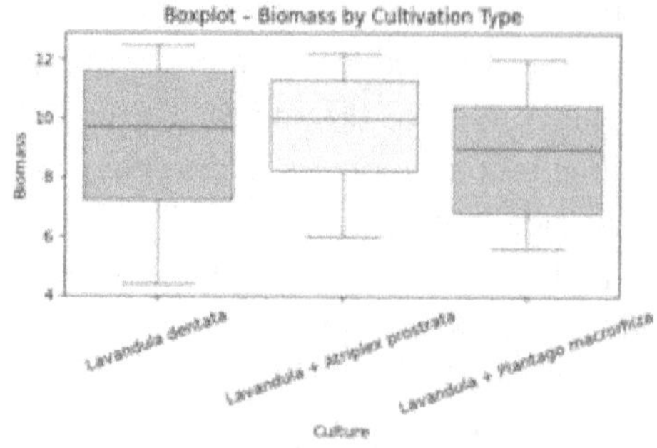

Figure 2. Variation in Chlorophyll Content of Lavandula dentata Across Cultivation Modes.

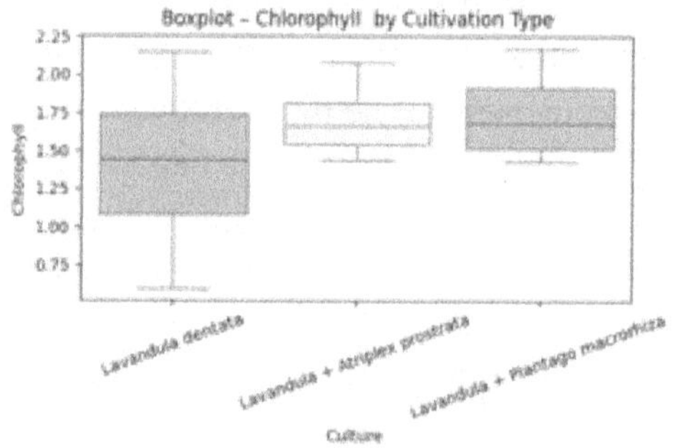

Figure 3. Effect of Cultivation Type on Aerial Part Length in Lavandula dentata.

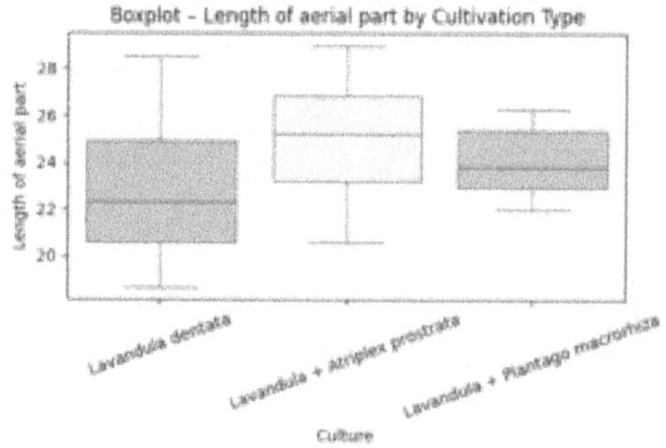

Figure 4. Proline Accumulation in Lavandula dentata Under Different Cultivation Systems.

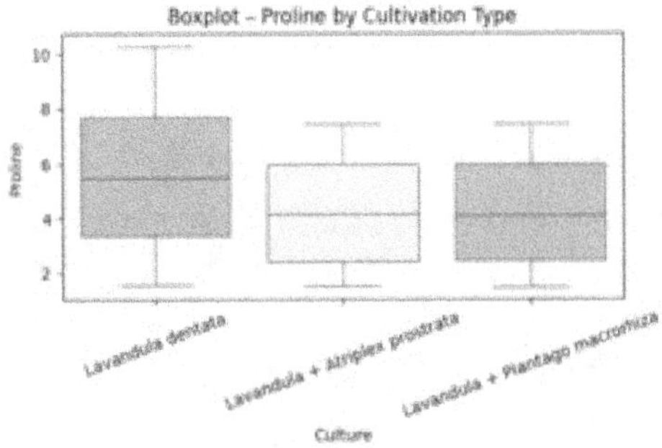

Figure 5. Relative Water Content in Lavandula dentata Across Monoculture and Halophyte Co-Cultivation.

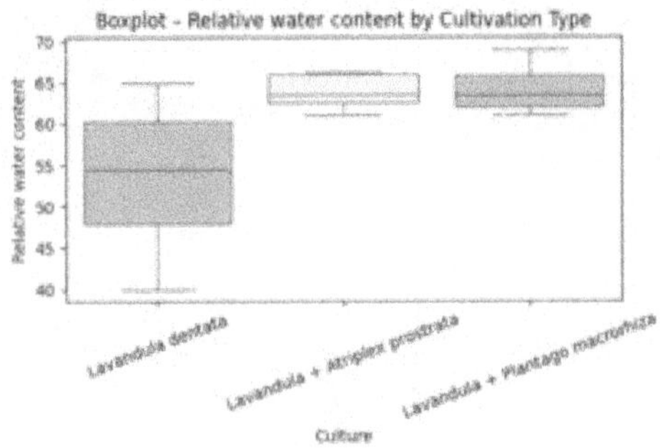

Figure 6. Root Length Response of Lavandula dentata Under Different Cultivation Regimes.

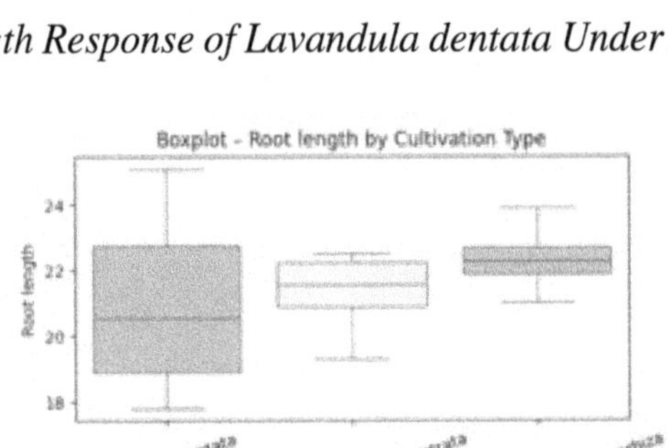

Figure 7. Root Volume Enhancement in Lavandula dentata Through Halophyte Co-Cultivation.

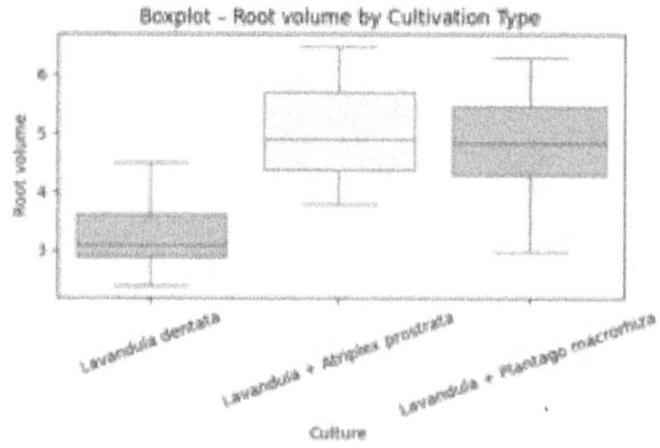

Figure 8. Total Soluble Sugar Content in Lavandula dentata Under Varying Cultivation Strategies.

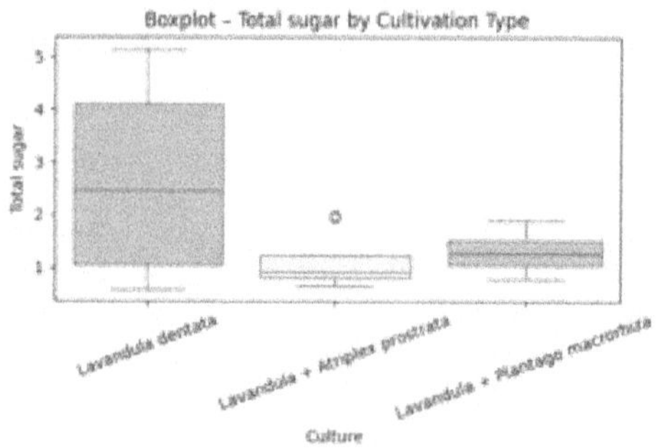

Figure 9. Correlation Matrix of Physiological Traits in Lavandula dentata Under Salinity Stress

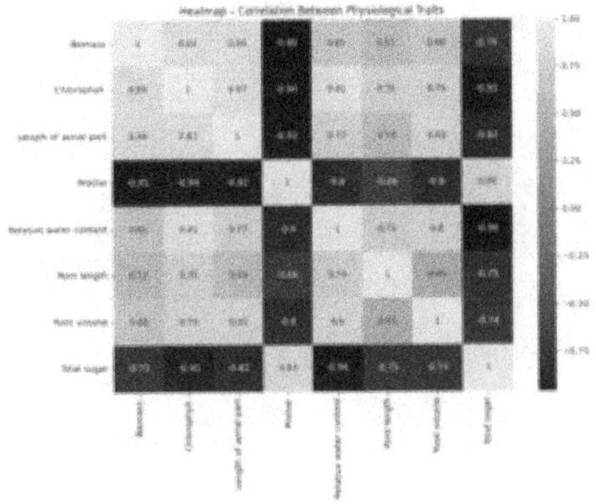

Figure 10. Principal Component Analysis (PCA) Reveals Distinct Multivariate Profiles Among Cultivation Types of Lavandula dentata

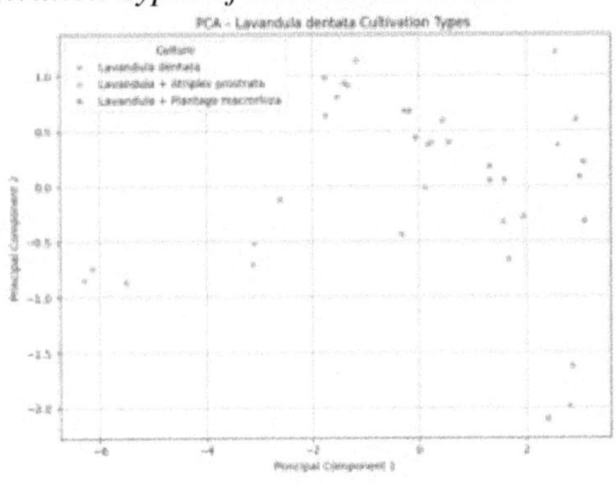

Figure 11. t-SNE Visualization of Cultivation-Dependent Physiological Patterns in Lavandula dentata.

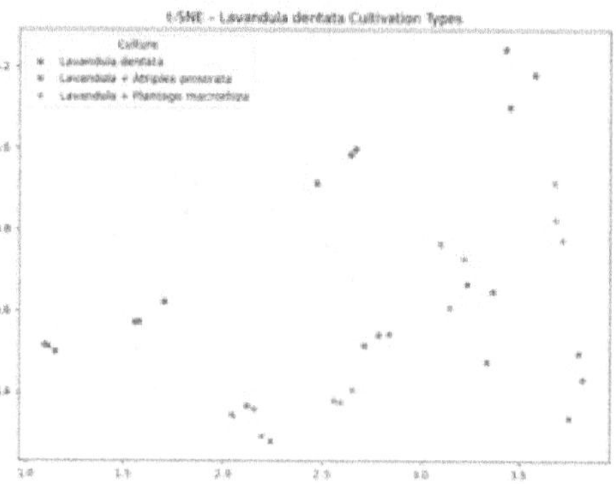

Figure 12. UMAP + K-Means Clustering Highlights Physiological Differentiation Between Lavandula dentata Cultivation Strategies.

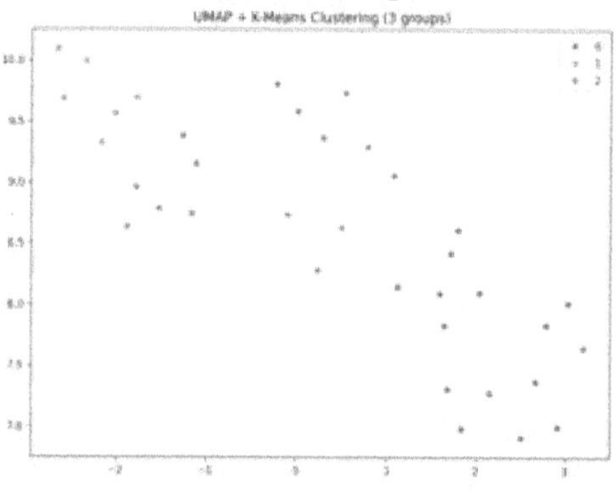

Figure 13. Density-Based Clustering (DBSCAN) Over UMAP Space Highlights Local Substructure in Lavandula dentata Responses.

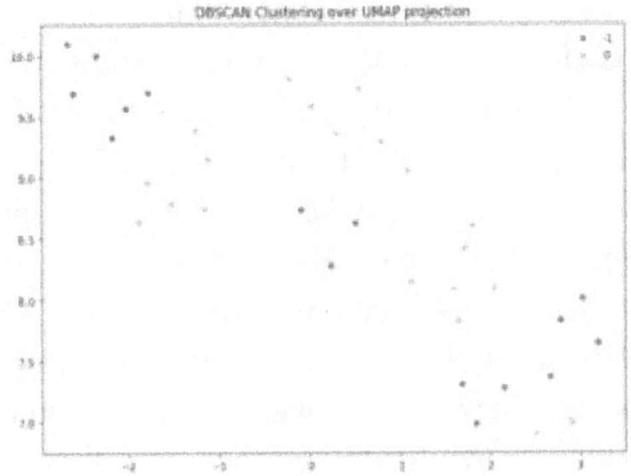

Figure 14. UMAP Projection of Lavandula dentata Colored by Salinity Stress Levels (NaCl).

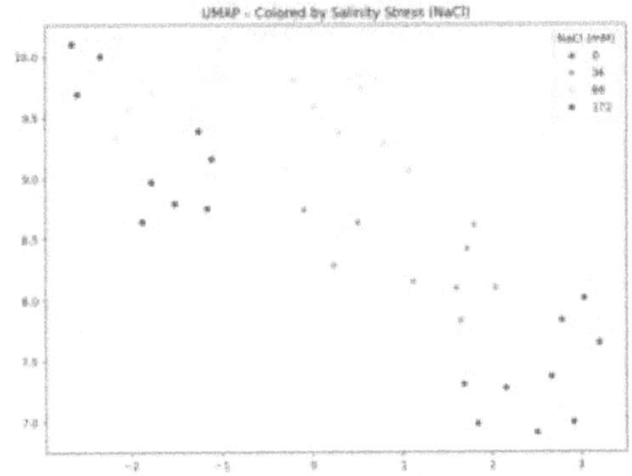

SHAP analysis provided key insights into the physiological variables driving model predictions for both culture type and salinity level classification in *Lavandula dentata*. For culture type discrimination, relative water content and root volume emerged as the most impactful features, exhibiting the highest mean SHAP values across all classes, thus enabling clear separation between monoculture and co-cultivation treatments. Total sugar and root length also contributed substantially, reflecting their sensitivity to interspecies interaction effects with companion halophytes. Intermediate contributions were observed for chlorophyll and biomass, while proline and aerial part length had the lowest discriminative power, suggesting limited influence on culture identity prediction. Class-specific SHAP distributions indicated that relative water content was particularly useful in distinguishing Lavandula grown alone, whereas root volume had more balanced contributions across co-cultivation scenarios. In contrast, the SHAP plot for salinity level prediction highlighted proline as the dominant predictive variable, confirming its role as a primary osmotic stress marker. Biomass followed closely, reflecting salt-induced growth inhibition as a robust physiological indicator. Chlorophyll degradation was also critical for discriminating stress intensity, whereas relative water content and aerial part length provided moderate predictive value. Traits such as total sugar, root length, and root volume had minimal SHAP impact in this context, suggesting a lesser role in capturing salinity-induced physiological shifts. Overall, SHAP values confirmed that different sets of physiological traits underlie culture type identification and salinity stress classification, with water retention and root traits being more relevant to cultivation identity, while osmolyte accumulation and growth depression dominated salinity response modeling (Figure 15-16).

Figure 15. SHAP-Based Feature Importance for Culture Type Classification Using XGBoost.

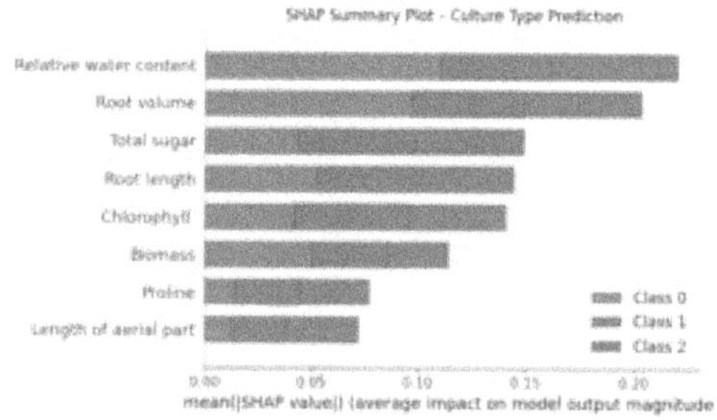

Figure 16. SHAP-Based Feature Importance for Salinity Level Classification Using XGBoost.

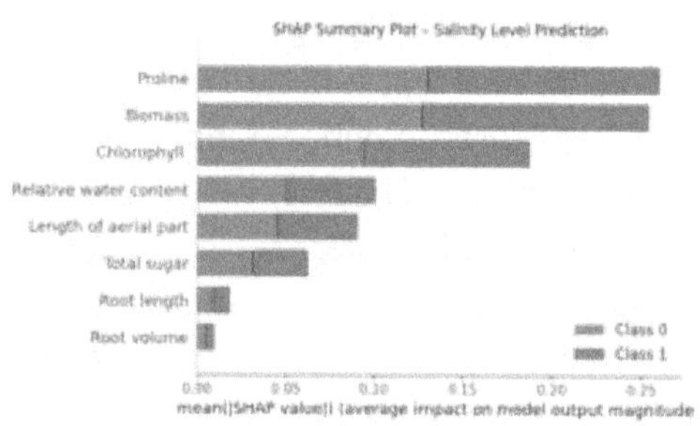

The predictive performance of supervised machine learning models was evaluated through two classification tasks: culture type prediction and salinity level classification. Model 1 aimed to distinguish between three cultivation scenarios—*Lavandula dentata* grown alone, in association with *Atriplex prostrata*, or with *Plantago macrorhiza* achieving an overall accuracy of 73% and a macro-averaged F1-score of 0.77. The highest classification performance was observed for *Lavandula dentata* grown alone, with perfect precision and recall, while lower performance was recorded for the two halophyte-associated classes due to class imbalance and moderate overlap in physiological response profiles. The corresponding confusion matrix confirmed that most misclassifications occurred between the two co-cultivation strategies. Model 2, designed to classify salinity stress levels into Low and High based on physiological parameters, achieved perfect performance (100% accuracy, precision, recall, and F1-score), demonstrating the high discriminative power of traits such as proline, biomass, and chlorophyll in salinity classification. Confusion matrices for both models illustrate the separability of classes and reinforce the robustness of the models, particularly for salinity detection. These results confirm that machine learning provides an accurate and interpretable framework for identifying physiological responses associated with cultivation practices and salt stress levels in *Lavandula dentata*.

Table 1. Classification Performance of Machine Learning Models for Predicting Culture Type and Salinity Stress in Lavandula dentata.

Model 1: Culture Type Prediction				
Class	**Precision**	**Recall**	**F1-Score**	**Support**
Lavandula dentata	1	1	1	2
Lavandula dentata + Atriplex prostrata	0,67	0,8	0,73	5
Lavandula dentata + Plantago macrorhiza	0,67	0,5	0,57	4
Accuracy	0,73	0,73	0,73	11
Macro avg	0,78	0,77	0,77	11
Weighted avg	0,73	0,73	0,72	11
Model 1: Confusion Matrix				
	Pred 0	Pred 1	Pred 2	
True 0	2	0	0	
True 1	0	4	1	
True 2	0	2	2	
Model 2: Salinity Level Prediction				
Class	Precision	Recall	F1-Score	Support

continued on following page

Table 1. Continued

Model 1: Culture Type Prediction				
Class	**Precision**	**Recall**	**F1-Score**	**Support**
High	1	1	1	6
Low	1	1	1	5
Accuracy	1	1	1	11
Macro avg	1	1	1	11
Weighted avg	1	1	1	11
Model 2: Confusion Matrix				
	Pred High	Pred Low		
True High	6	0		
True Low	0	5		

0: Lavandula dentata;1: Lavandula dentata + Atriplex prostrata ;2: Lavandula dentata + Plantago macrorhiza

A multilayer perceptron (MLP) was trained to classify culture types of *Lavandula dentata* based on a set of normalized physiological and morphological features. The model achieved a classification accuracy of 82% with a macro-averaged F1-score of 0.85, confirming the model's ability to capture relevant biological patterns. The confusion matrix showed correct classification of *Lavandula dentata* grown alone (Class 0) and good performance for both halophyte-associated cultures (Classes 1 and 2), with minor confusion between them, which reflects the shared physiological influence of the halophytes. To interpret the decision-making process of the neural network, SHAP values were computed using KernelExplainer, a model-agnostic method compatible with deep learning architectures. The SHAP summary plot (Figure 17) highlights root volume, proline concentration, and biomass as the top contributors to the model's predictions, followed by aerial part length and relative water content. These features are known to be affected by salinity and plant-plant interactions, suggesting that the MLP captured non-linear interactions between stress responses and co-culture effects. Interestingly, chlorophyll and root length, which were more prominent in linear models, appeared less influential in this deep learning setup, emphasizing the added value of complex model architectures in detecting subtle biological patterns.

Figure 17. SHAP values.

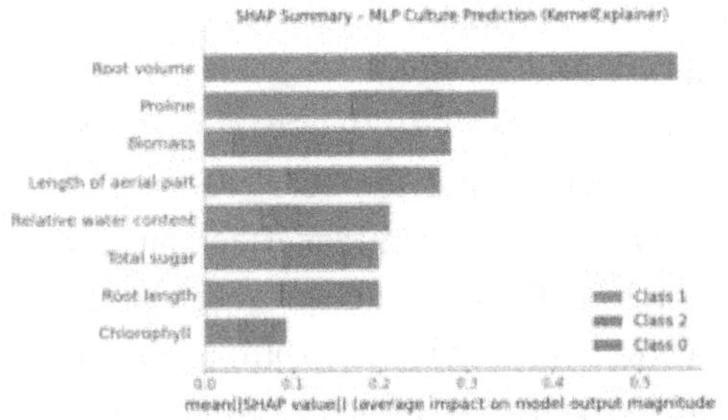

Table 2. Performance of the Multi-Layer Perceptron (MLP) Model for Culture Type Classification in Lavandula dentata Under Salinity Stress.

Model 1: Culture Type Classification				
Class	**Precision**	**Recall**	**F1-Score**	**Support**
Lavandula dentata	1	1	1	2
Lavandula dentata + Atriplex prostrata	0,8	0,8	0,8	5
Lavandula dentata + Plantago macrorhiza	0,75	0,75	0,75	4
Accuracy	0,82	0,82	0,82	11
Macro avg	0,85	0,85	0,85	11
Weighted avg	0,82	0,82	0,82	11
Model 1: Confusion Matrix				
	Pred 0	Pred 1	Pred 2	
True 0	2	0	0	
True 1	0	4	1	
True 2	0	1	3	

0: Lavandula dentata;1: Lavandula dentata + Atriplex prostrata ;2: Lavandula dentata + Plantago macrorhiza

To assess the direct causal influence of cultivation strategy on *Lavandula dentata* biomass production under salinity conditions, a formal causal inference analysis was conducted using the DoWhy framework. Three pairwise comparisons were formulated to isolate the Average Treatment Effect (ATE) of each cultivation type while adjusting for potential confounders, including morphological and physiological traits such as root length, aerial part length, relative water content (RWC), root volume, total sugar, proline content, and chlorophyll level. The first comparison quantified the causal effect of co-cultivation with *Atriplex prostrata* relative to monoculture, revealing a moderate negative effect on biomass (ATE = −0.245), suggesting that although *A. prostrata* provided some physiological buffering, it slightly reduced Lavandula growth under the tested conditions. In contrast, the second comparison showed that co-cultivation with *Plantago macrorhiza* significantly decreased biomass compared to monoculture (ATE = −1.392), indicating a stronger suppressive impact on growth. Finally, the direct comparison between the two companion halophytes confirmed that *Atriplex prostrata* had a milder effect than *Plantago macrorhiza*, with an estimated ATE of −1.101. These estimates were obtained by modeling biomass as a function of cultivation type while controlling for all relevant covariates in a backdoor-adjusted structural equation. The resulting causal graph and model-based estimation confirm that cultivation strategy exerts a quantifiable and differential causal impact on *Lavandula dentata* biomass, highlighting the superiority of *Atriplex prostrata* in mitigating stress-induced growth loss compared to *Plantago macrorhiza*. The analysis strengthens the agronomic relevance of choosing an appropriate companion species under saline environments and illustrates the value of integrating causal inference into ecological experimentation.

Table 3. Estimated Causal Effect of Co-Cultivation Strategies on Lavandula dentata Biomass Using DoWhy Framework.

Comparison	Estimand type	Model	Estimated causal effect (Δ biomass)	Interpretation
Lavandula dentata + *Atriplex prostrata* vs *Lavandula dentata* alone	Nonparametric ATE	Biomass ~ Culture type + Root length + Aerial part length + Relative Water Content + Root volume + Total sugar + Proline + Chlorophyll	-0,245	Slight drop in biomass
Lavandula dentata + *Plantago macrorhiza* vs *Lavandula dentata* alone	Nonparametric ATE	Biomass ~ Culture type + Root length + Aerial part length + Relative Water Content + Root volume + Total sugar + Proline + Chlorophyll	-1,3916	High biomass reduction
Lavandula dentata + *Plantago macrorhiza* vs *Lavandula dentata* + *Atriplex prostrata*	Nonparametric ATE	Biomass ~ Culture type + Root length + Aerial part length + Relative Water Content + Root volume + Total sugar + Proline + Chlorophyll	-1,1008	*Atriplex prostrata* > *Plantago macrorhiza*

ATE = Average Treatment Effect. The symbol "~" reads: "is modeled as a function of" or "depends on". Biomass ~ Culture_bin + Root length + ... means: Biomass is modeled as a function of crop type and physiological variables such as root length, root volume, proline, etc.

The directed acyclic graph (DAG) illustrates the assumed causal relationships between physiological traits, cultivation type, and biomass accumulation in *Lavandula dentata* under saline conditions. Cultivation type (Culture_encoded), used as a proxy for treatment assignment, is positioned as a mediator influenced by multiple upstream physiological features including root volume, relative water content, and proline concentration. These traits are considered potential confounders and are assumed to affect both the treatment allocation (i.e., cultivation with or without halophytes) and the outcome variable, biomass. Downstream, biomass is modeled as the final target node receiving directed edges from both the cultivation treatment and all physiological variables, reflecting the hypothesis that both direct and indirect pathways contribute to biomass production. The structure of the DAG is consistent with the backdoor criterion for identifying causal effects, enabling valid estimation of the Average Treatment Effect (ATE) using covariate adjustment or propensity matching. By integrating this causal model, the analytical pipeline ensures that estimated differences in biomass across cultivation conditions account for latent physiological variability, reinforcing the robustness of downstream DoWhy causal inference results.

Figure 18. Causal Graph Modeling the Direct and Indirect Influences of Cultivation Type on Biomass Production.

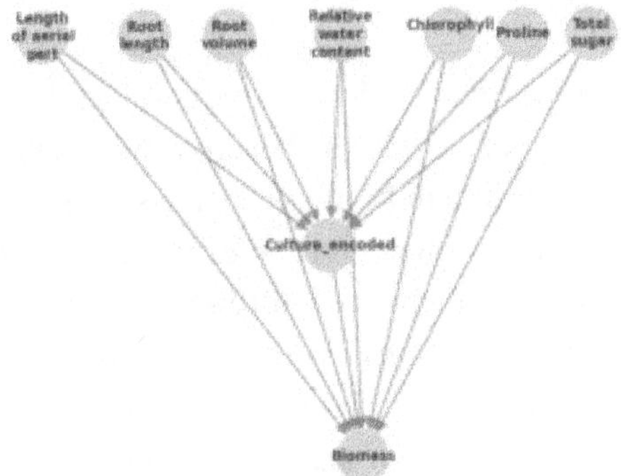

The multi-panel graph illustrates the evolution of key physiological and morphological traits in *Lavandula dentata* under increasing salinity levels (0 to 172 mM NaCl) across three cultivation systems: monoculture, co-cultivation with *Atriplex prostrata*, and co-cultivation with *Plantago macrorhiza*. Across most traits, salinity stress induced distinct physiological patterns depending on the companion species. Relative water content and root volume were better maintained under co-cultivation, especially with *A. prostrata*, indicating a buffering effect against osmotic stress. Conversely, proline—a stress biomarker—accumulated more markedly in monocultured plants, highlighting a stronger internal stress response in the absence of halophytic association. Photosynthetic pigments (chlorophyll) decreased with salinity in monocultures but remained comparatively stable in co-cultures, suggesting that halophytes may contribute to the preservation of photosynthetic machinery under saline pressure. Total sugar and aerial part length showed nonlinear responses, modulated by both salt concentration and companion identity. These patterns collectively reflect a shift in Lavandula's physiological strategy when interacting with different halophytes under saline constraints and underscore the trait-specific modulation of stress responses enabled by co-cultivation.

Figure 19. Morphological and Physiological Trait Responses of Lavandula dentata to Increasing Salinity under Three Cultivation Systems.

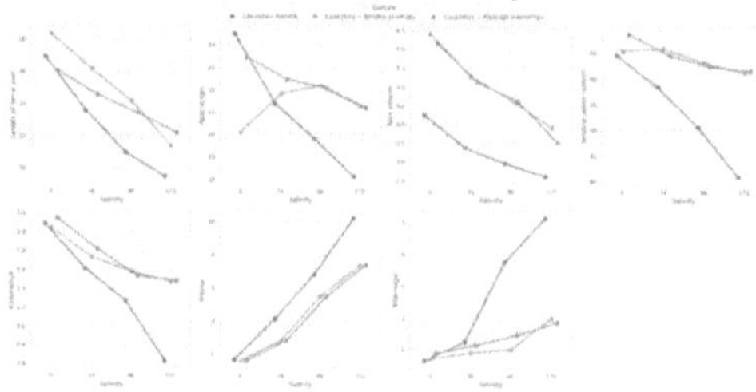

The line plot illustrates the trend of biomass reduction in *Lavandula dentata* under increasing salinity levels (0–172 mM NaCl) across three cultivation systems. The results reveal a consistent decline in biomass with rising salt concentrations for all treatments. However, co-cultivation with *Atriplex prostrata* consistently maintained higher biomass values compared to monoculture and association with *Plantago macrorhiza*, particularly at elevated salinity. This suggests that *A. prostrata* may confer a stronger buffering effect against salinity-induced growth suppression. The visual consistency of the trend confirms the impact of salt stress on biomass accumulation and supports the role of companion halophytes in modulating stress resilience.

Figure 20. Trend of Biomass under Increasing Salinity across Cultivation Systems.

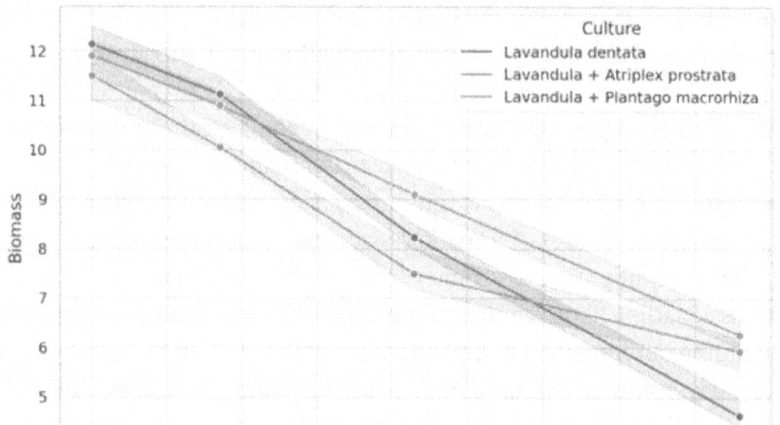

4. DISCUSSION

Climate change is one of the most serious challenges farmers/we face (Chen et al., 2024). One of the challenges of climate change is ongoing salinization of soil that is a threat to agricultural production particularly in arid and semi-arid regions like Morocco (Tang et al., 2024). High water tables, poor irrigation, and increasing evaporative loss contribute to the build-up of soluble salts in soils which can have devastating effects on the growth of non-halophytic plants (Daba & Qureshi, 2021; Tessema et al., 2023). As such, there is an immediate need to investigate new agro-ecological approaches such as crop associations with salinity tolerant species to mitigate the effects of salt stress on sensitive crops (Mishra et al., 2023).

This study purposefully investigated the effect of salt stress caused by NaCl on *Lavandula dentata,* a medicinal and aromatic plant of interest, and evaluated the ameliorating effect of two local halophytes, *Atriplex prostrata* and *Plantago macrorhiza*, grown together. The results obtained show clear trends confirming the sensitivity of *Lavandula dentata* to salt and the beneficial effect of co-culture in saline conditions (Hasegawa et al. 2014). The work presents many positive elements that justify its scientific and agronomic relevance. First, it proposes a unique methodology in the association of *Lavandula dentata* with halophyte plants as a potentially effective natural way to alleviate the effects of salt stress. This relies upon advantageous interactions between plants from an agroecological perspective, which remains one of the least studied areas of Mediterranean cropping systems (Duarte & Caçador, 2021). Then, the study examined various measurements including morphological measurement (biomass, height of the aerial part, height of the root part) as well as more biochemical or compositional responses (chlorophyll content, proline content, total sugars content), which presents a more direct measure effect of lavender's response to stress. Also, the design for the experiment was extensive in terms of many applications of increasing concentrations of NaCl and provided controlled experimental conditions with growth in a greenhouse, which contributes some reliability to or reproducibility of the initial findings presented. Lastly, these results have direct application for agricultural area dealing with increased salinity of the soil, which is a concern for greater consideration due to climate change.

The biomass of *Lavandula dentata* reduces considerably (from 12.13 g to 4.63 g) in culture alone with increasing salinity (by NaCl in the entire saline treatment). This reduction is a clear indicator of growth inhibition, as the plant would have encountered osmotic and ionic imbalance, which leads to disruption of key metabolic functions. Conversely, when *Atriplex prostrata* or *Plantago macrorhiza* are added, *Lavandula dentata* would have experienced a moderate reduction of biomass; even in high salinity, it retains a larger weight of fresh matter (6.26 g and 5.93 g, respectively.) These results demonstrate the potential protective effects of co-cultivation, presumably due to better growth conditions in the root zone.

The length of the aerial part follows the same trend as height: in monoculture, Lavandula length decreased with increasing salinity, while in co-culture, it showed a milder decrease. For instance, at 172mM NaCl, Lavandula grew on average to 21.4 cm with *Atriplex prostrata* and to 20.6 cm with *Plantago macrorhiza*, compared to 19.5 cm height when grown alone. The stability in vertical growth suggests that either root-to-root interactions and/or changes in micro-climatic conditions around the roots are influencing the plant's stress responses.

Root development is also dependent upon method of cultivation. Root length, for example, dropped to 18.13 cm at 172 mM in the case of monoculture; root length in co-culture exceeded 21 cm. Root volume exhibited the same trend, again not surprising: alone in cultivation, root volume was 2.60 ml; with *A. prostrata*, root volume was 3.90 ml; and with *Plantago macrorhiza*, root volume was 3.73 ml. By way of explanation, halophytes like *A. prostrata* and *P. macrorhiza* are oiler tolerant and have the ability to limit local salinity and soil structuring due to their influence on local ionic toxicity and stimulation of all minerals.

Relative water content, which is a reliable indicator of water status, fell sharply in Lavandula grown alone, to just 40.87% at 172 mM. With co-culture, relative leaf water content remained significantly higher at around 60%, indicating halophytes help retain hydration in plants (Flowers, 1985). This may be due to lower overall transpiration, a higher water uptake capacity, or a better soil structure that helps to retain moisture (Nikalje et al., 2018). The physiologic activity of chlorophyll levels reveal degradation of the photosynthetic system if stressed. In monoculture, chlorophyll decreases to 0.63 mg/g MF in high salinity environments. Whereas in co-culture *Atriplex prostrata* and *Plantago macrorhiza*, it was maintained at higher levels, indicative of better preservation of the photosynthetic apparatus (Farhat et al., 2021). The stability of chlorophyll is a good indicator of the plants greater resilience in the context of *Atriplex prostrata* and *Plantago macrorhiza.*

Additionally, the levels of proline, which is also an osmoprotector made by plants upon stress, greatly increases in monoculture and is found at 10.21 mg/g DM, which indicates severe stress. In co-culture, proline levels remain at a more moderate level (7.13 mg/g with *Atriplex prostrata* and 7.47 mg/g with *Plantago macrorhiza*), indicating that stress is perceived as lower due to a more favorable environment. As with proline, total sugars which are very important in osmoprotection and stabilisation of membranes, could be highly increased in *Lavandula dentata* exposed to culture only salt stress (total sugar concentrations showed 9.34 mg/g MS at 172 mM NaCl), demonstrating an osmotic stress adaptation mechanism. However, less when co-cultured with *Atriplex prostrata* (total sugars = 7.12 mg/g DM) and *Plantago macrorhiza* (7.45 mg/g DM), which again demonstrates that the plant perceives a lower level of stress in these conditions. In this case, lower sugar accumulation can be interpreted as the increased co-culture efficiency of osmotic regulation of the immediate saline environment, and thus the level of activation of the metabolic defence is not required to be as strong (Vlahos et al., 2023).

Overall, the results we obtained indicate that co-cultivation with halophytic plants is a potentially useful strategy for improving salt stress tolerance in sensitive plants. The underlying mechanisms might include halophytes sequestering Na^+ ions, halophytes either providing beneficial substances to the rhizosphere, decreasing effective salinity of the soil, or stimulating beneficial microflora (Caudal et al., 2022). Complementarity of root systems may have an important functional role in the optimization of resource uptake (Buhmann et al., 2015). Salt stress, on the other hand, is an important limiting factor for *Lavandula dentata*, but this limitation can be lessened by co-cultivation with a halophyte. There is ample possibility for these plant associations to assist in developing sustainable agricultural systems in saline areas and from a climate change perspective, the environmental stress on crops will continue to increase (Jurado et al., 2024). Such practices may help maintain the production of aromatic and medicinal plants in degraded environments.

Nevertheless, some caveats should be considered when interpreting the results. First, there is the short-term aspect of the experiment, which limits us from concluding anything about the long-term implications of co-culture, especially for the yield or quality of the essential oils. Additionally, the limited range of halophyte species limits out ability to generalize the conclusions: other species may display more beneficial characterizations. Furthermore, greenhouse conditions, while manipulated, cannot adequately reproduce the complexity of natural conditions which indicate that other features such as temperature, pathogens or interspecific competition that can have an influence on the results. Lastly, the lack of soil physico-chemical data, particularly in respect to the accumulation or mobility of toxic ions (Na^+, Cl-) do not allow us to gain a further understanding of the mechanisms implicated in the tolerance adduced.

Taking these results into consideration, there is a series of recommendations made for future experiments. The duration of the experiments should be, in time and natural conditions, to explore how stable the observed effects are and how they may permeate real conditions. Likewise, perform a more diverse number of halophyte species to try to find the best plant associations linked to the characteristics of the soil and climate conditions. In addition, conducting soil analysis before and after co-cultivation may give more visibility to the rhizospheric dynamics and possible phytoremediation effects (Mouniane et al., 2025). Finally, analyzing the effects of both methods of agriculture on *Lavandula dentata* secondary metabolites, and the oil composition of essential oils, would identify its value in terms of economic purpose. In practical terms, this research paves the way for many specific applications. It confirms the ability to have a co-cultivation with halophytes for sustainable management of saline soils potentially especially, in arid and semi-arid zones (El-Khadir et al., 2024). Thus the possibility to obtain the maximum productivity of aromatic and medicinal plants from marginal land compromised by over salinity. Since agroecological practices would make it possible such as reasonable association of crops, preservating biodiversity and improving farming systems' resilience to climate change while opening new economic opportunities to farmers (Macaluso et al., 2024).

This study presents several methodological and conceptual strengths. First, it integrates a multi-layered AI framework—including unsupervised manifold learning, supervised predictive models, explainable AI (SHAP), and causal inference—to uncover non-linear relationships between physiological traits and stress responses in *Lavandula dentata*. This comprehensive pipeline not only ensured accurate classification of salinity levels and cultivation types, but also enabled interpretable trait-level attribution, a rarely addressed gap in plant ecophysiology. Second, the inclusion of co-cultivation strategies with two contrasting halophytes (*Atriplex prostrata* and *Plantago macrorhiza*) under controlled salinity gradients offered a novel biological setting to test nurse-plant effects under real agronomic constraints. Third, the causal inference approach quantified the direct impact of cultivation practices on biomass independently of confounding traits, enabling a robust agronomic interpretation of biological interactions.

However, some limitations should be acknowledged. The relatively small sample size and the use of categorical encoding may limit the generalizability of the machine learning models in broader ecological contexts. Additionally, only biomass was used as the final causal outcome, whereas incorporating reproductive or metabolic endpoints would provide a more holistic view of plant fitness under stress. Environmental variability in natural field settings, such as soil microbiome dynamics or climatic fluctuations, were not captured in this experimental design and should be considered in future expansions.

5. CONCLUSION

This study provides a pioneering integrative framework for understanding how *Lavandula dentata* responds to salinity stress under different co-cultivation systems, within the broader context of climate change-induced soil degradation. By combining explainable machine learning, deep neural networks, and causal inference, it was possible to identify key physiological traits—such as root volume, relative water content, chlorophyll concentration, and proline accumulation—that define plant adaptation profiles. The co-cultivation with *Atriplex prostrata* emerged as a more beneficial strategy than with *Plantago macrorhiza*, exerting a milder negative impact on biomass while enhancing physiological buffering. The ability to predict stress levels with perfect accuracy and to explain these predictions using SHAP further reinforces the applicability of AI in plant science and agroecology. Most importantly, the causal quantification of treatment effects validates co-cultivation as a scalable strategy for improving stress resilience. Altogether, this work demonstrates that integrating artificial intelligence with ecological experimentation enhances our ability to interpret complex plant-environment interactions and supports the design of adaptive agricultural systems for future climatic challenges. Increasing salinization of soils due to climate change has elevated the need for sustainable solutions to maintain global agricultural output. In this regard, this paper has presented a study assessing the effect of salt stress, caused by sodium chloride, on *Lavandula dentata* as an aromatic and medicinal plant of interest, cultivated either as a monoculture or in association with the halophytes: *Atriplex prostrata* and *Plantago macrorhiza*. Overall, the results demonstrate a negative effect of salinity on plant growth and morphology, chlorophyll content, and water status in lavender plants, specifically. Nevertheless, *Atriplex prostrata* and potentially other species of halophytes reduce the negative impacts of salinity by increasing the levels of osmoprotective compounds such as proline and total soluble sugars. Overall, evidence is provided to suggest that intercropping systems hold value in contributing to salt stress tolerance in sustainable agricultural systems. It is proposed that intercropping be promoted based on the positive mechanisms of interspecific facilitation, for the rehabilitation of marginal soils, and enhance resilient agroecological practices. Further work is nevertheless warranted to establish multi environmental conditions, and in longer term under actual field conditions to validate the utility of these associations and facilitation models for practical realization.

REFERENCES

Akachoud, O., Bouamama, H., Facon, N., Laruelle, F., Zoubi, B., Benkebboura, A., Ghoulam, C., Qaddoury, A., & Lounès-Hadj Sahraoui, A. (2022). Mycorrhizal Inoculation Improves the Quality and Productivity of Essential Oil Distilled from Three Aromatic and Medicinal Plants: Thymus satureioides, Thymus pallidus, and *Lavandula dentata. Agronomy (Basel), 12*(9), 2223. DOI: 10.3390/agronomy12092223

Arnon, D. I. (1949). Copper enzymes in isolated chloroplasts. polyphenoloxidase in beta vulgaris. *Plant Physiology, 24*(1), 1–15. DOI: 10.1104/pp.24.1.1 PMID: 16654194

Arora, N. K., Fatima, T., Mishra, I., Verma, M., Mishra, J., & Mishra, V. (2018). Environmental sustainability: Challenges and viable solutions. *Environmental Sustainability, 1*(4), 309–340. DOI: 10.1007/s42398-018-00038-w

Atrous, G., Piqueras, A., Diaz-Vivancos, P., Hernández-Cánovas, A., Hamed, K. B., Hernández, J. A., & Barba-Espín, G. (2025). In Vitro-Selected Clones of the Halophyte Arthrocaulon macrostachyum Display Enhanced Salinity Stress Tolerance. *Plants, 14*(8), 1164. DOI: 10.3390/plants14081164 PMID: 40284052

Barrs, H. D. (1996). Root pressure and leaf water potential. *Science, 152*(3726), 1266–1268. DOI: 10.1126/science.152.3726.1266 PMID: 17769543

Bates, L. S., Waldren, R. P., & Teare, I. D. (1973). Rapid determination of free proline for water studies. *Plant and Soil, 39*(1), 205–208. DOI: 10.1007/BF00018060

Bazihizina, N., Papenbrock, J., Aronsson, H., Ben Hamed, K., Elmaz, Ö., Dafku, Z., Custódio, L., Rodrigues, M. J., Atzori, G., & Negacz, K. (2024). The Sustainable Use of Halophytes in Salt-Affected Land: State-of-the-Art and Next Steps in a Saltier World. *Plants, 13*(16), 2322. DOI: 10.3390/plants13162322 PMID: 39204758

Breiman, L. (2001). Random Forests. *Machine Learning, 45*(1), 5–32. DOI: 10.1023/A:1010933404324

Buhmann, A. K., Waller, U., Wecker, B., & Papenbrock, J. (2015). Optimization of culturing conditions and selection of species for the use of halophytes as biofilter for nutrient-rich saline water. *Agricultural Water Management, 149*, 102–114. DOI: 10.1016/j.agwat.2014.11.001

Caudal, F., Tapissier-Bontemps, N., & Edrada-Ebel, R. A. (2022). Impact of Co-Culture on the Metabolism of Marine Microorganisms. *Marine Drugs, 20*(2), 153. DOI: 10.3390/md20020153 PMID: 35200682

Chen, R., Su, Y., & Tran, L. (2024). Small Farmer's Perceptions of Climate Change and Adoption of Climate-Smart Practices: Evidence from Missouri, USA. *Sustainability (Basel), 16*(21), 9525. DOI: 10.3390/su16219525

Chen, T., & Guestrin, C. (2016). XGBoost : A Scalable Tree Boosting System. *Proceedings of the 22nd ACM SIGKDD International Conference on Knowledge Discovery and Data Mining*. 785-794. DOI: 10.1145/2939672.2939785

Daba, A. W., & Qureshi, A. S. (2021). Review of Soil Salinity and Sodicity Challenges to Crop Production in the Lowland Irrigated Areas of Ethiopia and Its Management Strategies. *Land (Basel), 10*(12), 1377. DOI: 10.3390/land10121377

Duarte, B., & Caçador, I. (2021). Iberian Halophytes as Agroecological Solutions for Degraded Lands and Biosaline Agriculture. *Sustainability (Basel), 13*(2), 1005. DOI: 10.3390/su13021005

Duarte, B., Sleimi, N., & Caçador, I. (2014). Biophysical and biochemical constraints imposed by salt stress: Learning from halophytes. *Frontiers in Plant Science, 5*, 746. DOI: 10.3389/fpls.2014.00746 PMID: 25566311

Dubois, M. K., Gils, J. K., Hanniton, P. A., & Smith, F. (1956). Use of phenol reagent for the determination of total sugar. *Analytical Chemistry, 28*, 350–356. DOI: 10.1021/ac60111a017

El-Khadir, I., Ktaoui, S., Mouniane, Y., Chriqui, A., Mabrouki, J., Ameziane, H., & Hmouni, D. (2024). Improved Salt Stress Tolerance of Salvia Officinalis Grown in the Presence of a Halophytic Plant Spergularia Maritima: Analysis of Morpho-Physiological Parameters. In Mabrouki, J., & Azrour, M. (Eds.), *Integrated Solutions for Smart and Sustainable Environmental Conservation. Studies in Systems, Decision and Control* (Vol. 527). Springer. DOI: 10.1007/978-3-031-55787-3_2

Elbouzidi, A., Taibi, M., El Hachlafi, N., Haddou, M., Jeddi, M., Baraich, A., Aouraghe, A., Bellaouchi, R., Mothana, R. A., Hawwal, M. F., Mesnard, F., Hano, C., Asehraou, A., Chaabane, K., El Guerrouj, B., & Addi, M. (2024). Formulation of a Three-Component Essential Oil Mixture from *Lavandula dentata, Rosmarinus officinalis*, and Myrtus communis for Improved Antioxidant Activity. *Pharmaceuticals (Basel, Switzerland), 17*(8), 1071. DOI: 10.3390/ph17081071 PMID: 39204175

Ester, M., Kriegel, H.P., Sander, J., Xu, X. (1996). A density-based algorithm for discovering clusters in large spatial databases with noise. kdd, 96(34), 226-231.

Etesami, H., & Noori, F. (2019). Soil Salinity as a Challenge for Sustainable Agriculture and Bacterial-Mediated Alleviation of Salinity Stress in Crop Plants. In Kumar, M., Etesami, H., & Kumar, V. (Eds.), *Saline Soil-based Agriculture by Halotolerant Microorganisms*. Springer. DOI: 10.1007/978-981-13-8335-9_1

Farhat, N., Kouas, W., Braun, H. P., & Debez, A. (2021). Stability of thylakoid protein complexes and preserving photosynthetic efficiency are crucial for the successful recovery of the halophyte Cakile maritima from high salinity. *Plant Physiology and Biochemistry*, *166*, 177–190. DOI: 10.1016/j.plaphy.2021.05.044 PMID: 34116337

Flowers, T. J. (1985). Physiology of halophytes. In Pasternak, D., & San Pietro, A. (Eds.), *Biosalinity in Action: Bioproduction with Saline Water. Developments in Plant and Soil Sciences* (Vol. 17). Springer., DOI: 10.1007/978-94-009-5111-2_3

Godoy, F., Olivos-Hernández, K., Stange, C., & Handford, M. (2021). Abiotic stress in crop species: Improving tolerance by applying plant metabolites. *Plants*, *10*(2), 186. DOI: 10.3390/plants10020186 PMID: 33498148

Habán, M., Korczyk-Szabó, J., Čerteková, S., & Ražná, K. (2023). Lavandula Species, Their Bioactive Phytochemicals, and Their Biosynthetic Regulation. *International Journal of Molecular Sciences*, *24*(10), 8831. DOI: 10.3390/ijms24108831 PMID: 37240177

Hasegawa, A., Oyanagi, T., Minagawa, R., Fujii, Y., & Sasamoto, H. (2014). An inverse relationship between allelopathic activity and salt tolerance in suspension cultures of three mangrove species, Sonneratia alba, S. caseolaris and S. ovata: Development of a bioassay method for allelopathy, the protoplast co-culture method. *Journal of Plant Research*, *127*(6), 755–761. DOI: 10.1007/s10265-014-0651-1 PMID: 25062702

Hopmans, J. W., Qureshi, A. S., Kisekka, I., Munns, R., Grattan, S. R., Rengasamy, P., Ben-Gal, A., Assouline, S., Javaux, M., Minhas, P. S., Raats, P. A., Skaggs, T. H., Wang, G., De Jong van Lier, Q., Jiao, H., Lavado, R. S., Lazarovitch, N., Li, B., & Taleisnik, E. (2021). Critical knowledge gaps and research priorities in global soil salinity. *Advances in Agronomy*, *169*, 1–191. DOI: 10.1016/bs.agron.2021.03.001

Hotelling, H. (1933). Analysis of a complex of statistical variables into principal components. *Journal of Educational Psychology*, *24*(6), 417–441. DOI: 10.1037/h0071325

Jurado, C., Díaz-Vivancos, P., Gregorio, B. E., Acosta-Motos, J. R., & Hernández, J. A. (2024). Effect of halophyte-based management in physiological and biochemical responses of tomato plants under moderately saline greenhouse conditions. *Plant Physiology and Biochemistry*, *206*, 108228. DOI: 10.1016/j.plaphy.2023.108228 PMID: 38043255

Khamidov, M., Ishchanov, J., Hamidov, A., Donmez, C., & Djumaboev, K. (2022). Assessment of Soil Salinity Changes under the Climate Change in the Khorezm Region, Uzbekistan. *International Journal of Environmental Research and Public Health*, *19*(14), 8794. DOI: 10.3390/ijerph19148794 PMID: 35886646

Khondoker, M., Mandal, S., Gurav, R., & Hwang, S. (2023). Freshwater Shortage, Salinity Increase, and Global Food Production: A Need for Sustainable Irrigation Water Desalination—A Scoping Review. *Earth (Basel, Switzerland)*, *4*(2), 223–240. DOI: 10.3390/earth4020012

Lundberg, S. M., & Lee, S. I. (2017). A unified approach to interpreting model predictions. *Advances in Neural Information Processing Systems*, 30.

Macaluso, D., Licciardo, F., & Carbone, K. (2024). Farming of Medicinal and Aromatic Plants in Italy: Structural Features and Economic Results. *Agriculture*, *14*(1), 151. DOI: 10.3390/agriculture14010151

MacQueen, J. (1967). Some methods for classification and analysis of multivariate observations. Proceedings of the Fifth Berkeley Symposium on Mathematical Statistics and Probability, Volume 1. *Statistics*, *5*, 281–298.

Mandal, S., Raju, R., Kumar, A., Kumar, P., & Sharma, P. C. (2018). Current status of research, technology response and policy needs of salt-affected soils in India-a review. *Journal of the Indian Society of Coastal Agricultural Research*, *36*, 40–53.

McInnes, L., Healy, J., & Melville, J. (2020). UMAP : Uniform Manifold Approximation and Projection for Dimension Reduction (No. arXiv:1802.03426). arXiv. / arXiv.1802.03426DOI: 10.48550

Mishra, A. K., Das, R., George Kerry, R., Biswal, B., Sinha, T., Sharma, S., Arora, P., & Kumar, M. (2023). Promising management strategies to improve crop sustainability and to amend soil salinity. *Frontiers in Environmental Science*, *10*, 962581. DOI: 10.3389/fenvs.2022.962581

Morton, M. J., Awlia, M., Al-Tamimi, N., Saade, S., Pailles, Y., Negrão, S., & Tester, M. (2019). Salt stress under the scalpel dissecting the genetics of salt tolerance. *The Plant Journal*, *97*(1), 148–163. DOI: 10.1111/tpj.14189 PMID: 30548719

Mouniane, Y., Vibhute, A. D., Kadmi, Y., Chriqui, A., El-Khadir, I., Soufiani, A., Doubi, M., Bautista, X. G., Cardenas, J. A. A., & Hmouni, D. (2025). *Using Phytoremediation to Treat Industrial and Pharmaceutical Wastewater: Innovations and Future Prospects.* In *Biotechnology Approaches to Industrial and Pharmaceutical Wastewater Treatment* (pp. 277–298). IGI Global. DOI: 10.4018/979-8-3693-8487-9.ch009

Nikalje, G. C., Srivastava, A. K., Pandey, G. K., & Suprasanna, P. (2018). Halophytes in biosaline agriculture: Mechanism, utilization, and value addition. *Land Degradation & Development*, *29*(4), 1081–1095. DOI: 10.1002/ldr.2819

Noh, Y., Jafarinejad, S., & Anand, P. (2024). A Review on Harnessing Renewable Energy Synergies for Achieving Urban Net-Zero Energy Buildings: Technologies, Performance Evaluation, Policies, Challenges, and Future Direction. *Sustainability (Basel)*, *16*(8), 3444. DOI: 10.3390/su16083444

Omuto, C. T., Kome, G. K., Ramakhanna, S. J., Muzira, N. M., Ruley, J. A., Jayeoba, O. J., Raharimanana, V., Owusu Ansah, A., Khamis, N. A., Mathafeng, K. K., Elmobarak, A. A., Vargas, R. R., Koetlisi, A. K., Dembele, D., Diawara, M., Mbaikoubou, M., Maria, R. M., Adam Boukary, I., Malatji, A., & Nyamai, M. (2024). Trend of soil salinization in Africa and implications for agro-chemical use in semi-arid croplands. *The Science of the Total Environment*, *951*, 175503. DOI: 10.1016/j.scitotenv.2024.175503 PMID: 39147045

Paraskevopoulou, A. T., Kontodaimon Karantzi, A., Liakopoulos, G., Londra, P. A., & Bertsouklis, K. (2020). The Effect of Salinity on the Growth of Lavender Species. *Water (Basel)*, *12*(3), 618. DOI: 10.3390/w12030618

Pearl, J. (2022). Causal Diagrams for Empirical Research (With Discussions). In H. Geffner, R. Dechter, & J. Y. Halpern (Eds.), Probabilistic and Causal Inference (p. 255-316). ACM. DOI: 10.1145/3501714.3501734

Rosenbaum, P. R., & Rubin, D. B. (1983). The central role of the propensity score in observational studies for causal effects. *Biometrika*, *70*(1), 41–55. DOI: 10.1093/biomet/70.1.41

Rumelhart, D. E., Hinton, G. E., & Williams, R. J. (1986). Learning representations by back-propagating errors. *Nature*, *323*(6088), 533–536. DOI: 10.1038/323533a0

Shahzad, A., Ullah, S., Dar, A. A., Sardar, M. F., Mehmood, T., Tufail, M. A., Shakoor, A., & Haris, M. (2021). Nexus on climate change: Agriculture and possible solution to cope future climate change stresses. *Environmental Science and Pollution Research International*, *28*(12), 14211–14232. DOI: 10.1007/s11356-021-12649-8 PMID: 33515149

Sharma, A., & Kiciman, E. (2020). DoWhy : An end-to-end library for causal inference. arXiv preprint arXiv:2011.04216. /arXiv.2011.04216DOI: 10.48550

Singh, K. M., Baksi, S., Rani, S., Jha, A. B., Dubey, R. S., & Sharma, P. (2024). NanoBoost: Maximizing crop resilience and yield via nanopriming under salt stress. *Environmental and Experimental Botany*, *226*, 105937. DOI: 10.1016/j.envexpbot.2024.105937

Soni, S., Jha, A. B., Dubey, R. S., & Sharma, P. (2024). Nanowonders in agriculture: Unveiling the potential of nanoparticles to boost crop resilience to salinity stress. *The Science of the Total Environment*, *925*, 171433. DOI: 10.1016/j.scitotenv.2024.171433 PMID: 38458469

Tang, H., Du, L., Xia, C., & Luo, J. (2024). Bridging gaps and seeding futures: A synthesis of soil salinization and the role of plant-soil interactions under climate change. *iScience*, *27*(9), 110804. DOI: 10.1016/j.isci.2024.110804 PMID: 39286506

Tarroum, M., Ben Romdhane, W., Ali, A. A. M., Al-Qurainy, F., Al-Doss, A., Fki, L., & Hassairi, A. (2021). Harnessing the Rhizosphere of the Halophyte Grass Aeluropus littoralis for Halophilic Plant-Growth-Promoting Fungi and Evaluation of Their Biostimulant Activities. *Plants*, *10*(4), 784. DOI: 10.3390/plants10040784 PMID: 33923476

Tessema, N., Yadeta, D., Kebede, A., & Ayele, G. T. (2023). Soil and Irrigation Water Salinity, and Its Consequences for Agriculture in Ethiopia: A Systematic. *Agriculture*, *13*(1), 109. DOI: 10.3390/agriculture13010109

Tofah, M. L., Mseddi, K., Al-Abbasi, O. K., Ben Yazid, A., Khechine, A., Gdoura, R., & Khannous, L. (2023). A New Lavender (*Lavandula multifida* L.) Ecotype from Arid Tunisia, with Differential Essential Oil Composition and Higher Antimicrobial Potential. *Life (Chicago, Ill.)*, *13*(1), 103. DOI: 10.3390/life13010103 PMID: 36676052

Tursunalieva, A., Alexander, D. L. J., Dunne, R., Li, J., Riera, L., & Zhao, Y. (2024). Making Sense of Machine Learning: A Review of Interpretation Techniques and Their Applications. *Applied Sciences (Basel, Switzerland)*, *14*(2), 496. DOI: 10.3390/app14020496

Van der Maaten, L., & Hinton, G. (2008). Visualizing data using t-SNE. *Journal of Machine Learning Research*, *9*, 2579–2605.

Vlahos, N., Berillis, P., Levizou, E., Patsea, E., Panteli, N., Demertzioglou, M., Morfesis, K., Voudouri, G., Krigas, N., Kormas, K., Antonopoulou, E., & Mente, E. (2023). Investigating Salinity Effects in Brackish Aquaponics Systems: Evidencing the Co-Cultivation of the Halophyte *Crithmum maritimum* with the Euryhaline *Sparus aurata. Applied Sciences (Basel, Switzerland), 13*(6), 3385. DOI: 10.3390/app13063385

Yan, J., & Wang, X. (2022). Unsupervised and semi-supervised learning : The next frontier in machine learning for plant systems biology. *The Plant Journal, 111*(6), 1527–1538. DOI: 10.1111/tpj.15905 PMID: 35821601

Ye, Y., Medina-Velo, I. A., Cota-Ruiz, K., Moreno-Olivas, F., & Gardea-Torresdey, J. L. (2019). Can abiotic stresses in plants be alleviated by manganese nanoparticles or compounds? *Ecotoxicology and Environmental Safety, 184*, 109671. DOI: 10.1016/j.ecoenv.2019.109671 PMID: 31539809

Zhang, X., Ibrahim, Z., Khaskheli, M. B., Raza, H., Zhou, F., & Shamsi, I. H. (2024). Integrative Approaches to Abiotic Stress Management in Crops: Combining Bioinformatics Educational Tools and Artificial Intelligence Applications. *Sustainability (Basel), 16*(17), 7651. DOI: 10.3390/su16177651

Chapter 8
Synergistic Effects of Halophyte Intercropping on the Salt Stress Tolerance of Rosmarinus Officinalis L:
A Sustainable Approach for Climate-Resilient Agriculture

Issam El-Khadir
https://orcid.org/0000-0002-3970-2865
Faculty of Sciences, Ibn Tofail University, Morocco

Yassine Mouniane
https://orcid.org/0000-0002-6037-5301
Faculty of Sciences, Ibn Tofail University, Morocco

Ahmed Chriqui
Faculty of Sciences, Ibn Tofail University, Morocco

Mahmoud Oudghiri
Faculty of Sciences, Ibn Tofail University, Morocco

Mohamed El Bakkali
Faculty of Sciences, Ibn Tofail University, Morocco

Driss Hmouni
https://orcid.org/0000-0001-7598-6204
Faculty of Sciences, Ibn Tofail University, Morocco

ABSTRACT

Salt stress is one of the major challenges for sustainable agriculture, particularly in areas vulnerable to the effects of climate change. This chapter explores the impact of cultivation in association with halophytes on the salt tolerance of Rosmarinus

DOI: 10.4018/979-8-3693-9132-7.ch008

officinalis, a glycophyte plant. Experience has shown that salt stress negatively affects the growth of rosemary. However, association with halophytes significantly improved morphological and physiological parameters by regulating cell hydration, increasing proline concentration, total sugars, and maintaining photosynthesis. The results suggest that this agroecological approach, by integrating salt-tolerant plants, could be a viable solution for enhancing crop resilience in the face of soil salinization and extreme climatic conditions. This study contributes to the advancement of knowledge on the integration of halophytes into agricultural practices and to the search for sustainable solutions for agriculture in the context of climate change.

1. INTRODUCTION

Salt stress is the primary abiotic factor limiting agricultural yield (Mansinhos et al., 2024), especially in arid and semi-arid areas where water is already limited. About one billion hectares (about 25% of all cultivated land) of land worldwide is currently affected by soil salinization, and salinity levels are climbing about 1% annually (Luo et al., 2017; Arora, 2019; Shokri et al., 2024). If the current trend continues, by the year 2050 more than half of all land designated for agriculture will be salt stressed (FAO, 2019; Tang et al., 2024; Rehman et al., 2025). Climate change, rising sea levels, seawater intrusion into coastal areas, and poor irrigation will exacerbate this challenge to global food security due to salt stress (El-hendawy et al., 2017; Hossain et al., 2021; Mouniane et al., 2024; Omuto et al., 2024). Soil salinity causes significant losses in crops by disrupting physiological processes in plants (Hassani et al., 2021). Salinity induces osmotic stress, ion toxicity, nutrient imbalance, and oxidative stress, decreasing the plant's capacity to germinate, grow, and produce (Kaur et al., 2024). The physiological disruption causes stunted vegetative growth, delayed photosynthesis, and a drop in both the quality and quantity of agricultural production (Ibrahimova et al., 2021; Singh et al., 2024; Oudghiri et al., 2025).

It is important that we have a clear understanding of the mechanisms involved in tolerance to salt stress in order to develop appropriate adaptation strategies (Boussora et al., 2024). Considerable strides have been made in this field, revealing mechanisms that can involve osmotic adjustment, regulation of ion homeostasis, and maintenance of photosynthetic efficiency (Pissolato et al., 2024). Plant responses to salt stress are multi-layered and operate through interconnected signalling pathways such as calcium signalling, reactive oxygen species, or phytohomones like abscisic acid at the cellular and molecular levels (Chele et al., 2021; Wang et al., 2025). Transcription factors specifically regulate the expression of stress-related genes which can improve resilience in plants (Li et al., 2024). In light of these difficulties, some complementary strategies are currently being undertaken. Significant efforts are being made in genetic improvement of salt tolerant plants using conventional breeding, modern biotechnological approaches such as genetic engineering, as well as sustainable agricultural practices (Shabbir et al., 2021; Albahri et al., 2023). Given the confluence of multiple environmental stressors, including salinity, drought, and extreme temperatures, it is clear that an integrated and interdisciplinary approach is imperative for agriculture to be resilient and sustainable in the long term (Khan et al., 2021; Ortiz & Sansinenea, 2023).

Rosmarinus officinalis L. (rosemary), a medicinal herb of the Lamiaceae family, is commonly grown and used in the Mediterranean region (Hammer & Junghanns, 2020). Its high resilience to environmental conditions, including salt stress, positions rosemary as a promising plant in the fight against desertification and rapid soil erosion (Andrade et al., 2018; Elachouri et al., 2024). Rosemary is a native herb in all coastal areas of the Mediterranean and serves a variety of purposes, such as the production of essential oil for perfumery and aromatherapy, as well as in products for pest control (González-Minero et al., 2020). Rosemary essential oil contains bioactive compounds such as 1,8-cineole, camphor, α-pinene, β-pinene, and borneol (Raffo et al., 2023). Rosemary has well-known medicinal properties in addition to its aroma, and traditional medicine has used it for antioxidant purposes. In the Mediterranean region, rosemary serves as an essential oil, but it is also a culinary herb that is appreciated for its nutritional and medicinal properties (Psarrou et al., 2020). Rosemary's volatile and non-volatile compounds have been studied for their biological activity, particularly the polyphenols in rosemary studied for antioxidant and anti-inflammatory activity (Aziz et al., 2021). Rosemary is an example of a plant that is tough against extreme environments, but it can be affected by salt stress, particularly in sites where saline irrigation and soil salinization are increasing (Christoforidi et al., 2022). While rosemary essential oil is primarily obtained through distillation, essential oil yields are low, therefore, the solid residuals created pose an environmental issue (Aulicky et al., 2025). Consequently, researchers have started working with residuals rich in antioxidant compounds (Cerdá-Gandia et al., 2024). Once the oil is removed, de-oiled rosemary leaves are valuable raw material that are odorless, thus allowing for extraction of phenolic compounds. Without the oil present, penetration of the solvent can improve which aids mass transfer using a different route of extraction (Múnera-Echeverri et al., 2024).

Intercropping with halophyte plants is an innovative and promising agroecological strategy to counteract the impacts of salinity on crops (Surówka et al., 2025). Combining salt-tolerant plants, or halophytes, with salt-sensitive crops, such as *Rosmarinus officinalis*, is designed to improve the management of saline soils and promote resilience to stress in plants. Because of the extreme levels of salinity they can tolerate, halophytes can change the chemical composition within the soil that creates a better growth environment for other plants (Abebe & Tu, 2024), either via reducing salt concentrations in the rhizosphere, adjusting water availability, or using other forms of interactions (Fouad et al., 2025). All of these benefits will reduce yield due to salinity, while also improving crop diversification and overall sustainability of agroecosystems (Zhang et al., 2024). Intercropping also maximizes the use of degraded lands that were often abandoned for use by traditional crops while promoting environmentally non-destructive agriculture (Ilakiya et al., 2024).

This research aims to assess the role of intercropping with halophytes on the salt tolerance of *Rosmarinus officinalis*, using a multifaceted methodology to assess different morphological, physiological, and biochemical parameters. From a morphological perspective, we evaluated plant growth, including height, biomass, root length, and root volume, to shed light on the effects of salt stress on development. Physiologically, we measured relative water content and chlorophyll concentration, two important physiological indicators of plant stress, to evaluate the effect of halophytes on water relations and photosynthesis. Biochemical analysis then enabled us to quantify proline, an important osmoprotectant in plants to counter salinity stress, and total sugars, which also contribute to the plant's adaptation to environmental stress. The combination of this methodology highlighted the overall and adaptive responses of *R. officinalis* under saline conditions with and without halophytes.

2. MATERIALS AND METHODS

2.1. Plant Materials

The plant materials used during the experiment were selected based upon interest in examining effects related to cultivation in association with halophytes on the salt tolerance of *Rosmarinus officinalis*. The study used *Rosmarinus officinalis* (rosemary) as a model organism to study salt stress tolerance due to its sensitivity to saline conditions. Rosemary plants were purchased from licensed nurseries and grown in pots, consisting of a substrate of maamoura soil, peat and compost for adequate drainage while supporting root growth until appropriate size was reached for the experiment. The halophyte species chosen for use in the experiment were *Spergularia salina* and *Plantago coronopus* species, collected from oued maleh located in the Ouazzane region and known for its very high salinity (Chriqui et al., 2024). *Spergularia salina* was selected because of its known ability to tolerate high salinity and improve soil condition, hence making it a great companion to rosemary plants suffering from salt stress. *Plantago coronopus* was selected for the experiment because of its ability to tolerate extreme salinity and possibly improve the growth environment of *Rosmarinus officinalis*. Both halophytes were also grown at the same time as the rosemary plants in a controlled environment greenhouse.

2.2. Cultivation Conditions and Salt Stress

The investigation was conducted in an experimental greenhouse situated at Ibn Tofail University, Faculty of Science where the environmental conditions were controlled to imitate a salt stress regime. The plants were grown in pots full of a mixture of maamoura soil, compost and peat, allowing for adequate drainage and appropriate root growth. Salt stress was induced via irrigation of plants with NaCl. Four concentrations of NaCl were used: 0 mM/L (control) and 34 mM/L, 86 mM/L, 172 Mm/L. For an experimental period of one month, plants were irrigated every other day with each of the saline solutions. With this procedure, salt tolerance of *Rosmarinus officinalis* was examined and the effects of cultivation with halophytes at various saline levels was evaluated.

2.3. Experimental Design

The experiment was conducted using a completely randomized block design (RBD) involving three main experiments, with several treatments for each experiment to evaluate the effect of intercropping with halophytes on *Rosmarinus officinalis* salt tolerance.

Rosemary alone under salt stress: *Rosmarinus officinalis* was subjected to different concentrations of NaCl with no intercropping with halophytes. The concentrations of NaCl used were: 0 mM (control), 34 mM, 86 mM, and 172 mM. This treatment allows us to view the response of the *R. officinalis* to salt stress in situ without an intercropping with halophytes.

Rosemary in association with *Spergularia salina*: For this experiment the *R. officinalis* was intercropped with *Spergularia salina*, a salt tolerant halophyte plant. We used the same concentrations of NaCl as outlined in the previous treatment to assess any positive effect of intercropping on salt stress and salt management in *R. officinalis*.

Rosemary in association with *Plantago coronopus*: This treatment of *R. officinalis* involved growing *R. officinalis* in association with *Plantago coronopus*, another halophytic or salt tolerant plant, and observational effect of this intercropping of *Plantago coronopus* on the salt tolerance of the R. officinalis. The same concentrations of NaCl was also applied 0 mM (control), 34 mM/L, 86 mM/L and 172 mM/L.

2.4. Measured Parameters

2.4.1. Morphological Parameters

Following the harvest of the crops, biomass was assessed using a precision balance. The plant height was recorded by measuring the total height of each *Rosmarinus officinalis* plant from the crown, which measures vertical growth response to salt stress and the effect of intercropping with halophytes. Root length was recorded subsequent to washing the roots to eliminate soil residue, enhancing the understanding of the impact of salt stress on the root system. The volume of root systems was recorded via water displacement in a graduated cylinder to measure the volume of water displaced by the root systems, accurately estimating the volume of the root system. Morphological parameters provide a finer assessment of plant cues in response to salt stress and the impact of associated cultivation with either of the two halophytes.

2.4.2. Physiological Parameters

Relative water content was estimated by comparing fresh leaf and dry leaf weight to determine the plants' ability to hold water under salt stress (Barrs, 1996). Chlorophyll content was measured by extracting with 80% acetone followed by spectrophotometric reading at 645 and 663 nm with Arnon's equation to measure total chlorophyll content (Arnon, 1949).

2.4.3. Biochemical Parameters

Proline accumulation was measured according to the method of Bates et al. (1973) using leaf extracts and spectrophotometric analysis to obtain proline concentration, as this molecule fulfills osmoprotective roles in plant defense mechanisms to salt stress. Total sugars were extracted from leaves and quantified by the method of Dubois et al. (1956), also using a spectrophotometer. As total sugars are also important osmoprotectants, their investment in osmoregulatory mechanisms is crucial to plant adaptation to stress.

2.4.4. Justification for the Choice of the Three Parameters Analysed

- **Biomass (g)**: Integrative performance indicator
  - Biomass is a global growth indicator integrating all the plant's physiological and biochemical responses to stress.

- Under salt stress, the reduction in biomass reflects an alteration in carbon assimilation, water balance and mineral nutrition.
- It is a widely used criterion of tolerance or sensitivity to salt stress in halophytic and glycophytic species (Munns & Tester, 2008).

- **Proline (mg/g DM)**: Osmotic stress marker
  - Proline is a classic osmoprotector: it accumulates in response to salt stress to stabilize proteins, membranes and cell osmosis.
  - Its accumulation is strongly correlated with stress levels and reflects the plant's ability to activate defense mechanisms.
  - It is often used as a biomarker of perceived stress levels, especially when comparing tolerant and sensitive plants (Hussain et al., 2021).
- **Chlorophyll (mg/g FM)**: Marker of photosynthetic status
  - Salt stress directly affects photosynthesis, notably through damage to chlorophyll pigments.
  - Chlorophyll content is a good indicator of the health of the photosynthetic system, and therefore of the plant's ability to grow under stress.
  - A significant decrease indicates oxidative damage or inhibition of pigment biosynthesis (González-Espíndola et al., 2024).

2.5. Statistical Analysis

A comprehensive statistical approach was applied to evaluate the physiological responses of *Rosmarinus officinalis* to salinity stress and co-cultivation with halophyte species. Data analysis was performed using a combination of univariate and multivariate methods.

First, a two-way analysis of variance (ANOVA) was conducted for each physiological parameter (biomass, proline, chlorophyll, etc.) to assess the individual and interactive effects of the two fixed factors: NaCl concentration (0, 34, 86, 172 mM/L) and cultivation type (monoculture, co-culture with *Plantago coronopus* or *Spergularia salina*). The interaction term (NaCl × co-cultivation) was included to detect context-dependent responses.

Whenever significant main or interaction effects were detected ($p < 0.05$), a Tukey HSD post hoc test was performed to determine statistically homogeneous groups and assign significance letters. Data were checked for normality and homoscedasticity before applying the tests.

To explore the global structure of physiological variation and assess correlations among traits, a Principal Component Analysis (PCA) was performed using standardized data. A biplot was generated to visualize the projection of both samples and trait vectors, providing insight into the contribution and association of traits along the first two principal components.

In parallel, a Partial Least Squares Discriminant Analysis (PLS-DA) was applied as a supervised multivariate method to maximize discrimination among the 12 experimental groups (4 NaCl levels × 3 cultivation modes). This model allowed the extraction of VIP (Variable Importance in Projection) scores to quantify the relative contribution of each trait to group separation. Traits with VIP > 1 were considered discriminant.

To explore global patterns of physiological trait variation across treatments, a hierarchical clustering heatmap was generated using standardized data. All quantitative variables were normalized using Z-score transformation to ensure comparability across scales. Clustering was performed both on the rows (samples) and columns (variables) using Ward's linkage method and Euclidean distance as the similarity metric.

This analysis was conducted using the seaborn and scikit-learn libraries in Python. The final heatmap visualized the grouped structure of samples and traits simultaneously, supported by dendrograms illustrating the multivariate proximities. Color gradients reflected trait intensities, and treatment labels were appended to sample names to facilitate group-level interpretation.

This approach allowed the integration of all measured traits in a single visual framework, complementing the PCA and PLS-DA results by highlighting sample-level clustering and trait co-regulation patterns.

To disentangle the combined effects of salinity and co-cultivation on trait variation in *Rosmarinus officinalis*, a canonical correspondence analysis (CCA) was conducted. Prior to analysis, all quantitative physiological variables (biomass, plant height, root traits, relative water content, chlorophyll, proline, and total sugars) were standardized (Z-score transformation). Environmental variables included salinity level (NaCl in mM) as a continuous gradient and cultivation group (control, co-culture with *Plantago coronopus* or *Spergularia salina*) as a categorical variable. Co-cultivation categories were dummy-coded and concatenated with NaCl concentration to form the environmental matrix. The CCA was applied using the sklearn.cross_decomposition.CCA implementation in Python, projecting both the physiological traits and treatment conditions onto a biplot of the first two canonical axes. The trait loadings were extracted to identify the directionality and strength of trait associations with specific environmental constraints.

All statistical analyses and visualizations were conducted using Python 3.11 (libraries: statsmodels, scikit-learn, matplotlib, seaborn, pandas) and validated against standard thresholds ($p < 0.05$).

3. RESULTS

Under increasing NaCl concentrations, *Rosmarinus officinalis* exhibited a marked reduction in biomass when cultivated alone, decreasing from 25.65 g at 0 mM to 21.13 g at 172 mM. In the presence of *Plantago coronopus*, biomass values ranged from 23.60 g to 19.54 g across the same salinity gradient, indicating a moderate reduction, yet consistently higher values than the monoculture beyond 34 mM. When co-cultivated with *Spergularia salina*, biomass remained lower overall, fluctuating from 20.87 g to 18.14 g, with only a slight increase at 86 mM, suggesting a potentially less favorable or even competitive interaction (Figure 1).

Figure 1. Effect of salinity and halophyte co-cultivation on the biomass of Rosmarinus officinalis.

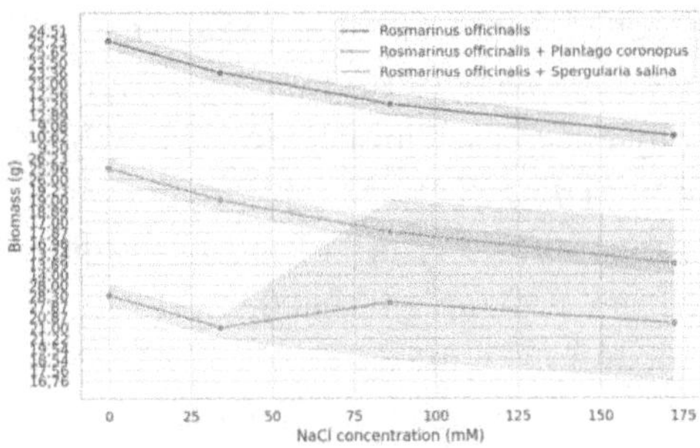

Proline content in *R. officinalis* alone increased from 2.00 mg/g DM at 0 mM to 5.34 mg/g DM at 172 mM, reflecting a classical osmotic stress response. Co-cultivation with *P. coronopus* altered this trend, as proline accumulation peaked earlier at 3.90 mg/g DM at 86 mM before slightly declining to 3.76 mg/g DM at the highest salinity, suggesting partial alleviation of stress. With S. salina, proline content remained notably lower, ranging from 2.17 mg/g DM at 0 mM to 2.25 mg/g DM at 172 mM, indicating a possibly reduced perception of osmotic stress or suppression of stress response pathways (Figure 2).

Figure 2. Modulation of proline accumulation in Rosmarinus officinalis under salinity and halophyte co-cultivation.

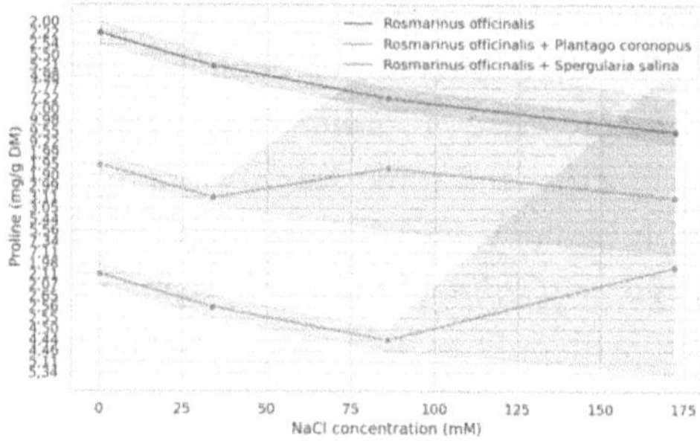

Chlorophyll content was significantly affected by salinity in all conditions. In monoculture, values dropped from 1.55 mg/g FM at 0 mM to 1.08 mg/g FM at 172 mM. The most pronounced protective effect was observed in co-culture with P. coronopus, where chlorophyll decreased initially to 1.34 mg/g FM at 86 mM but then sharply increased to 1.49 mg/g FM at 172 mM, suggesting a delayed but efficient compensatory mechanism. Conversely, in the S. salina condition, chlorophyll content declined continuously from 1.50 mg/g FM at 0 mM to 1.20 mg/g FM at 172 mM, with no indication of recovery (Figure 3).

Biomass accumulation in *Rosmarinus officinalis* was significantly influenced by both NaCl concentration and co-cultivation treatment, as revealed by two-way ANOVA ($p < 0.001$ for salinity, co-cultivation, and their interaction). In monoculture, increasing salinity led to a pronounced reduction in biomass, decreasing from 25.65 g at 0 mM to 23.50 g at 34 mM (not significant, group "a"), followed by a sharp drop to 12.98 g at 86 mM and 9.50 g at 172 mM (group "b"; Tukey HSD, $p < 0.05$). These results highlight the clear sensitivity of *R. officinalis* to salt stress when grown alone, with a threshold effect evident beyond 34 mM NaCl.

Under co-cultivation with *Plantago coronopus*, a more gradual and buffered response was observed. Biomass values declined from 26.23 g (0 mM) to 19.00 g (34 mM), both within the same statistical group "a", indicating tolerance at mild salinity. Significant reductions occurred at 86 mM (17.87 g, group "b") and further at 172 mM (13.89 g, group "c"), with clear statistical separation between each salinity level (Tukey HSD, $p < 0.05$). These results suggest that P. coronopus exerted a protective effect, delaying the onset of significant biomass loss and maintaining better growth under moderate and high salt concentrations compared to monoculture.

Conversely, co-cultivation with *Spergularia salina* did not prevent biomass reduction under salinity stress. While the initial biomass was the highest among all conditions (28.30 g at 0 mM), a notable decrease was observed as early as 34 mM (21.20 g, still group "a"), followed by a decline to 18.93 g at 86 mM (group "b") and 16.76 g at 172 mM (group “c”). The absence of statistical distinction between the control and 34 mM suggests that mild salinity was tolerated, but the rapid fall at higher concentrations indicates that S. salina did not confer a consistent facilitative benefit. In contrast to P. coronopus, this companion species appeared neutral or weakly competitive under stress.

Overall, these results demonstrate that while salinity severely limits biomass production in R. officinalis, the presence of certain halophytes can modulate this response. The facilitative effect observed with P. coronopus suggests a potential for positive plant–plant interactions under abiotic stress, whereas the response in co-culture with S. salina was less conclusive. These findings highlight the species-specific nature of co-cultivation benefits and underscore the importance of companion plant selection in saline environments.

Figure 3. Variation in chlorophyll content of Rosmarinus officinalis under salinity stress and halophyte co-cultivation

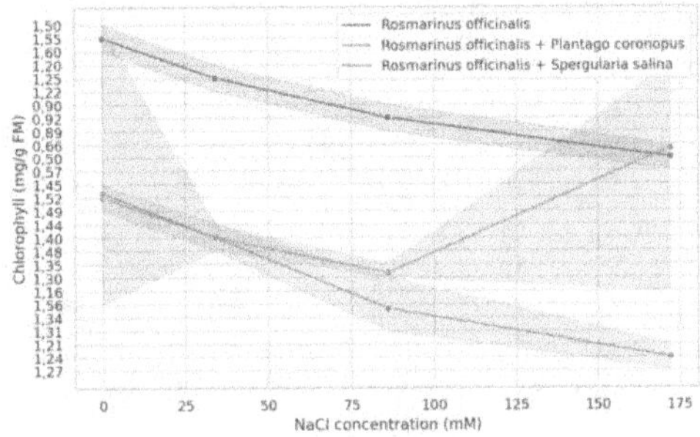

Figure 4. Effect of salinity on the biomass of Rosmarinus officinalis in monoculture.

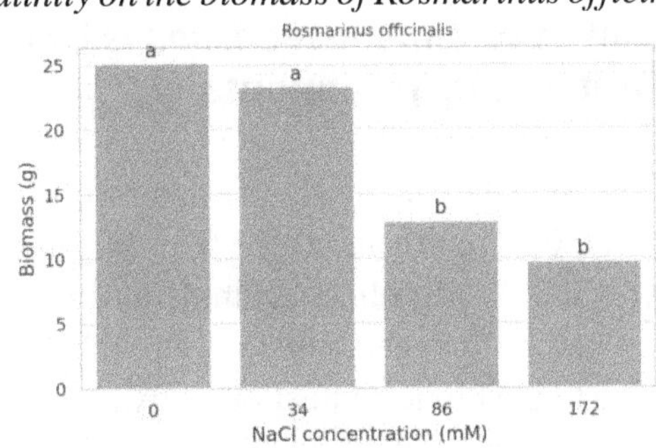

Figure 5. Effect of salinity on the biomass of Rosmarinus officinalis co-cultivated with Plantago coronopus.

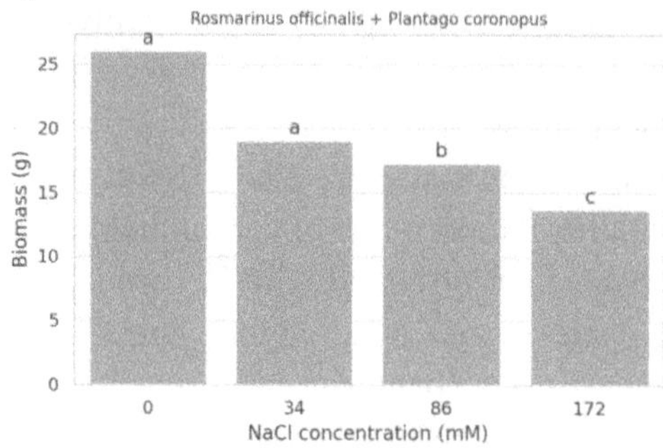

Figure 6. Effect of salinity on the biomass of Rosmarinus officinalis co-cultivated with Spergularia salina.

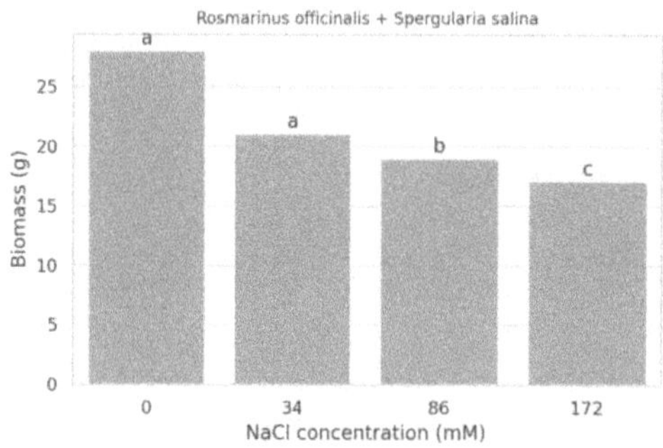

3.1. Proline Content in *Rosmarinus officinalis* Under Salinity and Co-Cultivation Treatments

The two-way ANOVA revealed highly significant effects of both NaCl concentration ($F = 1188.33$, $p < 0.001$), cultivation type ($F = 483.78$, $p < 0.001$), and their interaction ($F = 57.18$, $p < 0.001$) on proline accumulation in *Rosmarinus officinalis*. These results highlight that proline metabolism is strongly responsive not only to salinity intensity but also to the biotic context of cultivation.

In monoculture, *R. officinalis* displayed a pronounced and dose-dependent increase in proline content with rising NaCl levels, from 2.26 mg/g DM at 0 mM to 9.54 mg/g DM at 172 mM. The differences between treatments were statistically significant across all stress levels (Tukey's test, $p < 0.05$), illustrating a robust osmotic stress response (Figure 7).

When co-cultivated with *Plantago coronopus*, proline content also increased with salt stress but at a reduced magnitude. The increase remained statistically non-significant between 0 and 34 mM (1.91 to 3.05 mg/g DM), and became significant only from 86 mM onwards, reaching 7.19 mg/g DM at the highest salinity. This delay in osmotic adjustment suggests a possible facilitative interaction at moderate salinity levels (Figure 8).

In the presence of *Spergularia salina*, proline accumulation also followed a significant upward trend with salinity, albeit with the lowest levels observed among the three systems. Proline rose from 2.05 mg/g DM (0 mM) to 5.22 mg/g DM (172 mM), with significant differences emerging only at 86 and 172 mM. This pattern may reflect a more conservative osmotic strategy or a stress-attenuating effect of the neighboring halophyte (Figure 9).

Altogether, these findings demonstrate that while proline serves as a consistent stress marker, its accumulation dynamics are finely modulated by plant–plant interactions, potentially reflecting shifts in stress intensity, metabolic prioritization, or resource competition under saline conditions.

Figure 7. Effect of salinity on proline accumulation in Rosmarinus officinalis grown in monoculture.

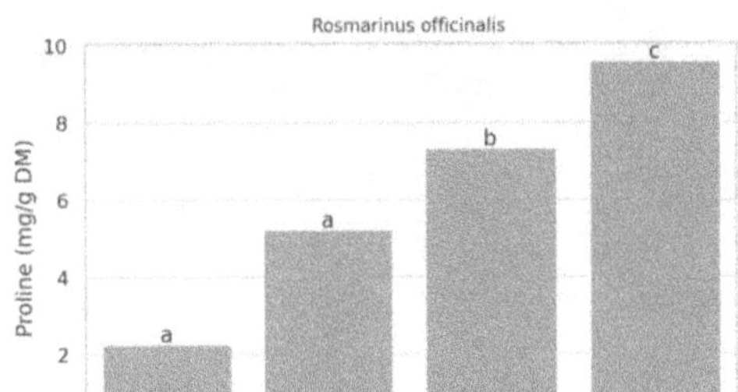

Figure 8. Effect of salinity on proline content in Rosmarinus officinalis co-cultivated with Plantago coronopus.

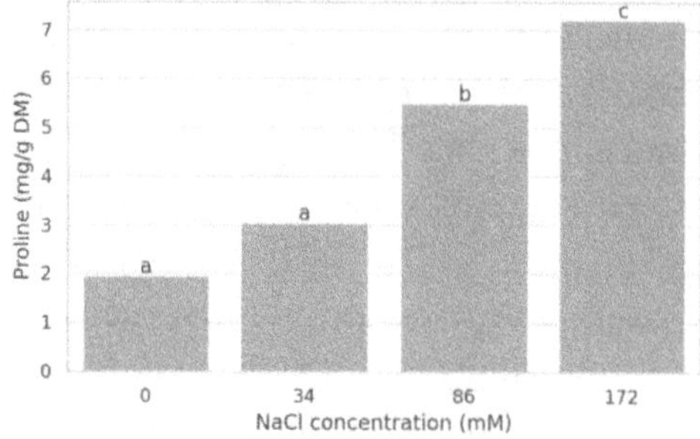

Figure 9. Proline accumulation in Rosmarinus officinalis co-cultivated with Spergularia salina under salinity stress.

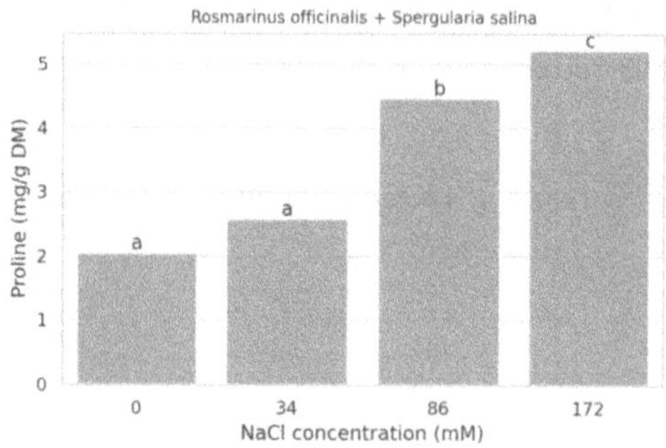

3.2. Chlorophyll Content in *Rosmarinus officinalis* Under Salinity and Co-Cultivation Conditions

The two-way ANOVA revealed that chlorophyll content in *Rosmarinus officinalis* was significantly affected by both salinity level ($F = 282.79$, $p < 0.001$), co-cultivation condition ($F = 234.82$, $p < 0.001$), and their interaction ($F = 54.32$, $p < 0.001$). These results confirm that both the abiotic factor (NaCl concentration) and the presence of associated halophytic species influence photosynthetic pigment stability.

In monoculture, chlorophyll content decreased sharply with increasing salinity. The concentration dropped from 1.54 mg/g FM at 0 mM to 1.23 mg/g FM at 34 mM (not significantly different), followed by significant reductions at 86 mM (0.90

mg/g FM) and 172 mM (0.57 mg/g FM). This marked decline reflects a typical salt-induced degradation of chlorophyll, likely associated with oxidative stress and impairment of photosynthetic apparatus integrity (Figure 10).

When co-cultivated with *Plantago coronopus*, chlorophyll levels remained relatively stable up to 34 mM (1.49 and 1.45 mg/g FM), with a mild but significant decrease from 86 mM (1.31 mg/g FM) and 172 mM (1.19 mg/g FM). The delayed onset of pigment degradation suggests a partial protective influence of P. coronopus at moderate salinity levels, potentially through microenvironmental or physiological interactions (Figure 11).

In the presence of *Spergularia salina*, chlorophyll content followed a similar trend, with higher retention under stress. The values declined from 1.52 mg/g FM at 0 mM to 1.24 mg/g FM at 172 mM, with intermediate levels at 34 mM (1.44 mg/g FM) and 86 mM (1.31 mg/g FM). Although significant reductions were observed beyond 34 mM, the pigment preservation remained more effective than in monoculture. This suggests that S. salina may exert a mitigating effect on stress-driven pigment loss, possibly by enhancing photoprotection or reducing ionic toxicity (Figure 12).

Collectively, these results demonstrate that co-cultivation, particularly with halophytes, helps attenuate salt-induced declines in chlorophyll content in R. officinalis, likely by modulating physiological responses or environmental stress load.

Figure 10. Chlorophyll content in Rosmarinus officinalis exposed to increasing NaCl concentrations under monoculture conditions.

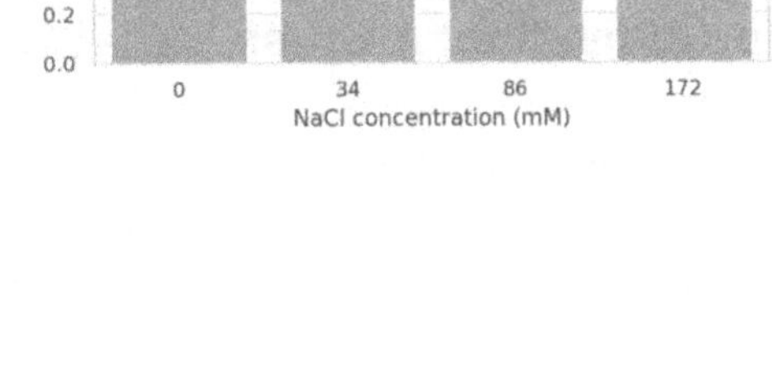

Figure 11. Chlorophyll content in Rosmarinus officinalis co-cultivated with Plantago coronopus under increasing salinity.

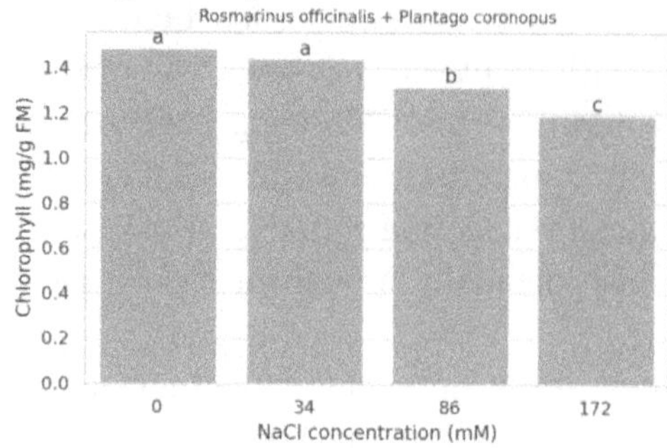

Figure 12. Chlorophyll content in Rosmarinus officinalis co-cultivated with Spergularia salina under salt stress.

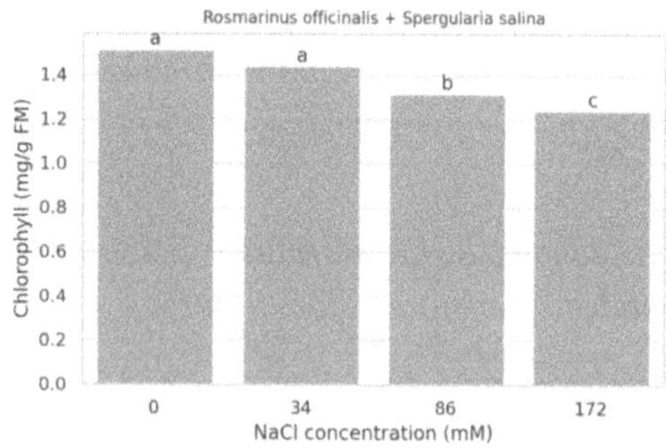

3.3. Integrated Physiological Response of *Rosmarinus officinalis* to Salinity and Co-Cultivation: A Cross-Variable Analysis

The combined analysis of biomass, proline, and chlorophyll content in *Rosmarinus officinalis* under varying NaCl concentrations and co-cultivation conditions revealed consistent yet distinct patterns of physiological response to salinity and plant–plant interaction.

Under salt stress, biomass showed a strong and progressive decline across all treatments, reflecting the general inhibitory effect of salinity on growth and resource assimilation. However, the extent of reduction was mitigated in co-cultivation settings, particularly with *Spergularia salina*, where biomass losses were less severe at moderate and high salinity. This suggests that facilitative effects or niche complementarity may buffer growth under stress.

Proline content, a classical osmoprotective compound, increased significantly with NaCl concentration, confirming its role as a biochemical marker of osmotic stress. The strongest accumulation was recorded in monoculture at 172 mM, highlighting the unmitigated stress impact. In contrast, proline levels rose more gradually in co-culture conditions, particularly with P. coronopus, suggesting a delay or attenuation of stress perception, potentially linked to altered root zone conditions or shared osmotic regulation.

Chlorophyll content exhibited a declining trend with increasing salinity, indicative of stress-induced pigment degradation and compromised photosynthetic efficiency. However, in co-cultivation systems, chlorophyll degradation was attenuated, with higher retention observed especially at moderate salinity levels. These results point to improved photoprotection or microenvironmental buffering in the presence of halophytes.

Together, these three parameters depict a coherent adaptive strategy: as salinity increases, *R. officinalis* responds by activating osmotic adjustments (proline), reducing metabolic investment (biomass), and modulating photosynthesis (chlorophyll). The presence of companion halophytes modifies these responses, sometimes reducing the magnitude of stress effects and favoring physiological resilience. This integrated view underscores the potential of strategic co-cultivation for enhancing salt stress tolerance through ecological facilitation.

3.4. Principal Component Analysis of Physiological Responses in *Rosmarinus officinalis*

Principal component analysis (PCA) was performed to explore the overall structure of physiological variation in *Rosmarinus officinalis* subjected to different salinity levels and co-cultivation conditions. The first two principal components (PC1 and PC2) accounted for 90.8% of the total variance, with PC1 alone explaining 83.6% and PC2 7.2%.

The score plot of PCA (Figure 13) revealed a clear stratification of samples along PC1 according to NaCl concentration, indicating that salinity is the dominant factor driving variation in the dataset. Samples exposed to high salinity (86 and 172 mM) were positioned on the right side of the PCA space, while control and low-salt treatments (0 and 34 mM) clustered on the left. PC2 contributed to separating co-cultivation modalities to a lesser extent, with subtle vertical shifts between monoculture and co-culture with *Plantago coronopus* or *Spergularia salina*.

The PCA biplot (Figure 14), which includes both individuals and variable vectors, provided insight into how traits contribute to group separation. Vectors representing proline, total sugars, root length, and root volume aligned strongly with the positive direction of PC1, suggesting their elevation under salt stress. In contrast, chlorophyll content, relative water content, and plant height projected in the opposite direction, highlighting their decline in response to salinity.

Trait correlations were also apparent: proline and sugars shared a similar orientation, as did chlorophyll and RWC, indicating consistent co-regulation. Biomass occupied an intermediate position, reflecting a more integrative response influenced by both stress and co-cultivation effects.

These findings demonstrate that while salinity is the principal structuring gradient, co-cultivation modulates physiological traits in R. officinalis, likely through facilitation or resource partitioning. The PCA thus offers a comprehensive overview of the multivariate responses, revealing both antagonistic and synergistic dynamics among key functional traits.

Figure 13. Principal component analysis (PCA) of physiological responses of Rosmarinus officinalis under salinity and co-cultivation conditions.

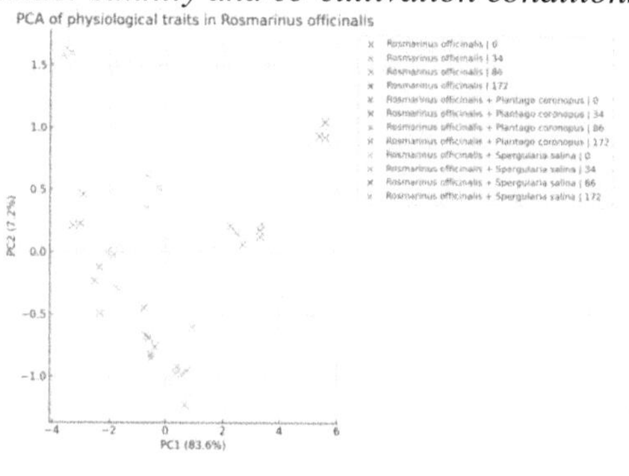

Figure 14. Biplot of PCA showing physiological traits and treatment groups in Rosmarinus officinalis under salinity and co-cultivation.

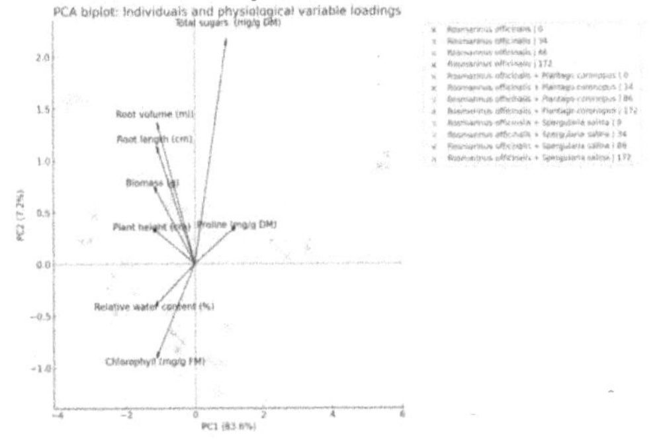

Partial least squares discriminant analysis (PLS-DA) was used to evaluate the extent to which physiological profiles could distinguish between *Rosmarinus officinalis* treatment groups defined by salinity level and co-cultivation condition. The resulting score plot revealed a clear separation among the twelve experimental groups across the first two PLS components.

PLS1 captured the main discriminatory axis, separating treatments along a salinity gradient, with highly saline conditions (86–172 mM NaCl) positioned toward the negative end of the axis. PLS2 provided orthogonal discrimination associated with co-cultivation effects, particularly distinguishing the influence of *Spergularia salina* and *Plantago coronopus* under specific NaCl levels.

Compared to the unsupervised PCA, PLS-DA enhanced class separability by incorporating class label information, enabling more precise identification of the treatments' multivariate signatures. This supervised approach confirms that physiological variables — when considered together — allow robust classification of experimental groups, reinforcing the value of combining trait data in salinity tolerance studies (Figure 15).

Figure 15. PLS-DA score plot showing discrimination of Rosmarinus officinalis treatments under salinity and co-cultivation conditions.

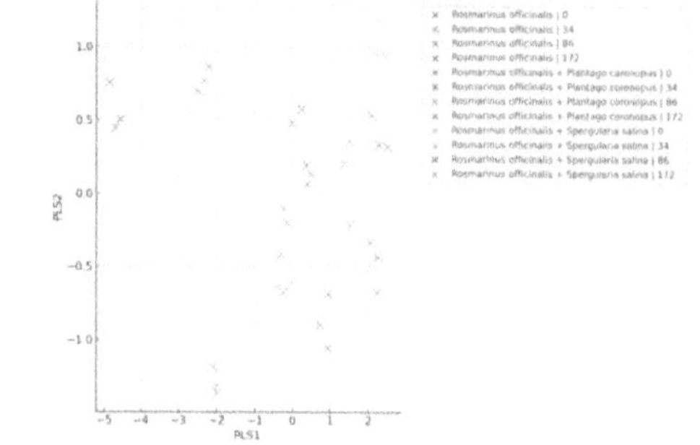

The Variable Importance in Projection (VIP) analysis identified the physiological traits that contributed most to the discrimination of treatment groups in the PLS-DA model. Among all the variables, root length (VIP = 1.51) and relative water content (VIP = 1.27) were the top contributors, followed closely by chlorophyll content and root volume. These variables had VIP scores above the commonly used significance threshold of 1.0, confirming their critical role in distinguishing among salinity levels and co-cultivation effects.

In contrast, traits such as biomass, proline, and total sugars had lower VIP scores (<1), suggesting they contributed less to group separation in the supervised model, despite their biological importance in salt response. The relatively modest VIP for proline reflects its consistent increase across conditions, but limited ability to distinguish co-culture effects specifically (Figure 16).

These results underscore the functional centrality of root and water-related traits in shaping the adaptive phenotype of *R. officinalis* under saline and interactive plant conditions. They also highlight which variables may serve as sensitive biomarkers for salt tolerance or facilitation mechanisms in intercropping systems.

Figure 16. Variable Importance in Projection (VIP) scores from PLS-DA of Rosmarinus officinalis traits under salinity and co-cultivation conditions.

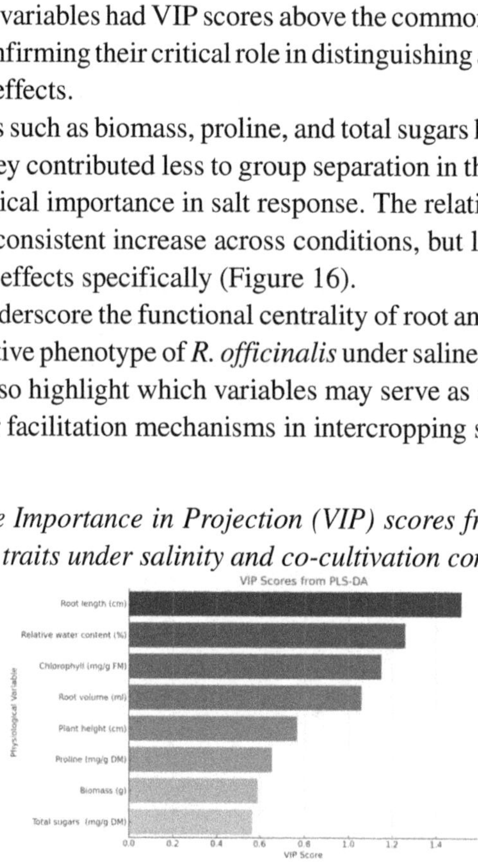

3.5. Multivariate Analysis of Physiological Traits in *Rosmarinus officinalis* Under Salinity and Co-Cultivation

To evaluate the integrated physiological response of *Rosmarinus officinalis* to salinity and co-cultivation, multivariate statistical analyses were performed, including Principal Component Analysis (PCA), Partial Least Squares Discriminant Analysis (PLS-DA), and Variable Importance in Projection (VIP) scoring.

The PCA revealed that the first two components (PC1 = 83.6%, PC2 = 7.2%) together accounted for 90.8% of the total variance. The score plot demonstrated a strong gradient along PC1, clearly separating treatments according to NaCl concentration. Co-cultivation effects were subtler and primarily reflected along PC2. The corresponding PCA biplot showed that proline, total sugars, root volume, and root length were positively correlated with PC1, indicating their enhancement under high salt stress. In contrast, chlorophyll content, relative water content, and plant height were aligned negatively with PC1, suggesting their depletion under salinity.

PLS-DA further improved group separation by integrating class information (NaCl × partner plant). The resulting score plot showed clearer discrimination among the 12 experimental groups. PLS1 reflected the salinity effect, whereas PLS2 contributed to differentiation between co-cultivation types. This analysis confirmed that the combined physiological profile was effective in distinguishing between treatments.

The VIP scores derived from the PLS-DA model highlighted the most discriminant traits. Root length (VIP = 1.51), relative water content (VIP = 1.27), chlorophyll content (VIP = 1.15), and root volume (VIP = 1.06) were identified as key drivers of treatment separation. These results emphasize the importance of water status and belowground growth parameters in determining the phenotypic response to salinity and interspecific interaction. In contrast, traits like proline and total sugars, although strongly responsive to stress, showed lower discriminative power, indicating their uniform expression across stress contexts.

Altogether, these multivariate insights suggest that while salinity remains the dominant factor shaping physiological variation in R. officinalis, co-cultivation influences trait expression patterns in a nuanced and trait-specific manner, particularly affecting root-related and water-retention mechanisms.

3.6. Heatmap With AI-Assisted Hierarchical Clustering

A hierarchical clustering heatmap was constructed based on Z-score standardized physiological variables to visualize the multivariate structure of *Rosmarinus officinalis* responses across treatments. The dendrogram revealed a strong partitioning of the experimental samples, driven primarily by salinity level, and secondarily by co-cultivation context. This structure is consistent with the separation previously observed in PCA and PLS-DA.

Two main clusters of samples emerged: the first, enriched in control and low-salt treatments (0–34 mM NaCl), was characterized by elevated levels of biomass, plant height, chlorophyll, and relative water content. These traits grouped together on the column dendrogram, indicating strong inter-correlation and a shared association with non-stress conditions. In contrast, the second major cluster contained samples exposed to higher salinity (86–172 mM NaCl), distinguished by upregulated levels of proline, total sugars, root length, and root volume, traits previously identified as discriminant in PLS-DA (VIP > 1 for root traits and RWC).

Interestingly, the co-cultivation with *Plantago coronopus* and *Spergularia salina* did not simply buffer or exacerbate salinity stress, but altered the trait expression profile in nuanced ways, as evidenced by their intermediate positioning between the extremes of the salinity axis. This confirms the PCA observation of orthogonal shifts introduced by co-culture modalities.

Overall, the heatmap and its clustering topology confirm and extend the findings from univariate and multivariate analyses: salinity remains the dominant driver of physiological reprogramming, but co-cultivation introduces a gradient of phenotypic modulation. Traits related to water status, carbon allocation, and root morphology form coherent clusters, highlighting their role in defining stress tolerance and interactive plasticity in R. officinalis.

Figure 17. Hierarchical clustering heatmap of standardized physiological traits in Rosmarinus officinalis under varying salinity and co-cultivation conditions.

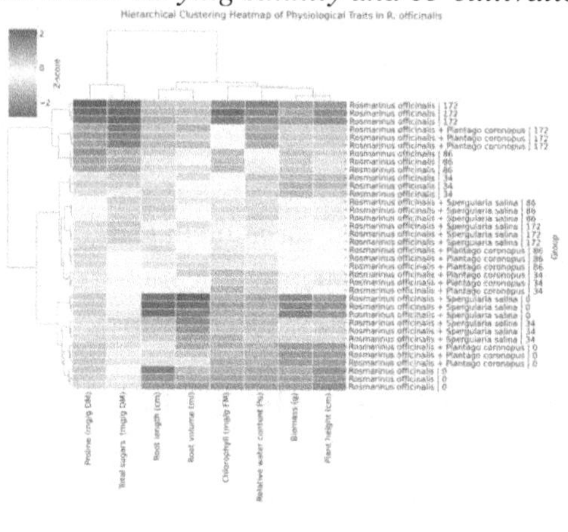

The Canonical Correspondence Analysis (CCA) biplot reveals the multivariate relationships between physiological traits of *R. officinalis* and experimental factors including NaCl concentration and co-cultivation conditions. The first canonical axis (CCA1) clearly separates high-salinity treatments from control and low-NaCl treatments, while the second axis (CCA2) captures residual variation linked to co-cultivation effects.

Traits such as proline and total sugars project strongly along the direction of high NaCl treatments, confirming their role as stress-induced osmoprotectants. In contrast, chlorophyll, relative water content, and plant height are oriented toward low-salinity treatments, highlighting their association with growth and water homeostasis. Root-related traits show distinct loading angles, suggesting a unique modulation across treatments.

The projection of sample responses (blue) and treatment profiles (orange) demonstrates treatment-specific phenotypic clustering. Treatments with high salinity and co-cultivation (e.g., with *Spergularia salina*) show unique positioning, suggesting combined environmental and biotic constraints drive distinct physiological signatures (Figure 18).

Figure 18. Canonical Correspondence Analysis (CCA) biplot: Projection of physiological traits constrained by salinity and co-cultivation treatments in Rosmarinus officinalis.

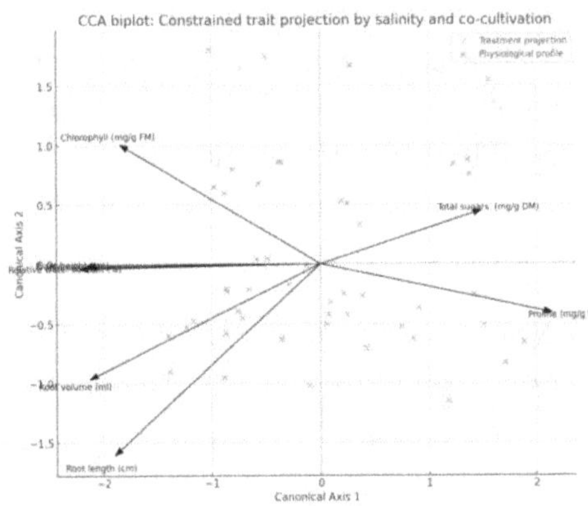

The canonical correspondence analysis (CCA) allowed for the simultaneous visualization of the distribution of physiological traits in *Rosmarinus officinalis* as constrained by salinity levels and the presence of companion halophytes (*Plantago coronopus* and *Spergularia salina*). The first two canonical axes explained most of the variance in trait distribution, with CCA1 primarily representing the salinity gradient and CCA2 capturing the contribution of co-cultivation conditions.

Traits such as biomass, plant height, and relative water content displayed strong negative loadings on CCA1, reflecting their sensitivity to high salinity stress. Among these, biomass (CCA1 = -1.05) and plant height (CCA1 = -1.07) were most strongly suppressed under elevated NaCl concentrations, particularly in the absence of facilitative interactions, confirming salt-induced growth inhibition. Interestingly, biomass also had a moderate positive loading on CCA2 (+0.30), indicating that its retention was partially mitigated under co-cultivation with *Plantago coronopus*.

Root morphological parameters such as root length and root volume showed moderate to strong negative loadings on CCA1 and high positive loadings on CCA2, with root length (CCA2 = +1.79) and root volume (CCA2 = +1.46) aligning with co-culture with *Spergularia salina* under low-salinity conditions. These findings suggest a potential facilitation effect on belowground architecture under less stressful environments.

Chlorophyll content loaded negatively on CCA1 (−0.86) and modestly positively on CCA2 (+0.38), revealing its association with plants grown under low salt stress with enhanced photosynthetic retention.

On the contrary, osmolyte accumulation traits, namely proline and total soluble sugars, had strong positive loadings on CCA1 and moderate on CCA2. Proline (CCA1 = +1.02) was most associated with severe salt exposure and the presence of either alophyte, suggesting its universal role as an osmoprotectant under abiotic stress. Soluble sugars were moderately associated with both axes, marking them as intermediaries of salinity response during progressive stress (Table 1).

These results reveal a clear partitioning of traits along abiotic and biotic gradients, supporting the hypothesis that co-cultivation with halophytes modulates stress expression and physiological performance in *Rosmarinus officinalis* in a trait-dependent manner.

Table 1. Canonical Correspondence Loadings of Physiological Traits in Rosmarinus officinalis under Salinity and Co-cultivation with Halophytes.

	CCA1_loading	CCA2_loading	Associated Conditions
Biomass (g)	-1,048881472	0,297339161	High NaCl + Plantago
Plant height (cm)	-1,066767717	0,07383208	High NaCl
Root length (cm)	-0,876718312	1,791715088	Low NaCl + Spergularia
Root volume (ml)	-1,00992668	1,458993673	Low NaCl + Spergularia
Relative water content (%)	-1,10437993	0,052792668	Controls + Spergularia
Chlorophyll (mg/g FM)	-0,855746957	0,378076612	Low NaCl
Proline (mg/g DM)	1,01597171	0,034467247	NaCl + Both alophytes
Total sugars (mg/g DM)	0,610963053	0,66756233	Intermediate stress

4. DISCUSSION

Salinity is one of the main abiotic stresses of plants, causing reductions in growth, affecting physiology and modifying metabolic functions (Arif et al., 2020). Data displayed in this study show that exposure to increasing concentrations of NaCl produced a significant reduction in biomass, height and root development in *Rosmarinus officinalis*, a glycophyte species. These observations support a hypothesis that rosemary, like other sensitive species, evolve an adaptation by minimizing growth to conserve resources due to an unfavorable salinity level.

The decline in growth, specifically in terms of foliage, has also been reported in clover, citrus and foxtail (Ziogas et al., 2021; Ma et al., 2023; Ling et al., 2025). This decline was attributed to inhibited cell division and elongation, but it can be interpreted as a controlled metabolic slowdown, where the plant is prioritizing defense (Munir et al., 2022). Root growth was inhibited as well; although, other plants did show a slight stimulation in root growth at low salinity, which is also an effect only seen in some tolerant species in which the root system acted as a specialized storage compartment for toxic ions (Na^+ and Cl^-) to limit uptake in the leaves (Joshi et al., 2022).

Relative water content (RWC) decreased with increased salinity in monoculture rosemary, indicating a water deficit created by salinity levels in the soil due to osmotic effects (Shakri et al., 2022). This again points to increased water stress in glycophytes, which can't dissociate osmotic and ionic effects of salinity, unlike halophytes, which effectively osmotic regulatory strategies, filtering salinity (Dourado et al., 2022). In treatments in association with halophytes, the drop in RWC was less than in monoculture systems, suggesting halophytes may improve water availability locally, possibly through root exudates or local changes to soil structure (Chen et al., 2022). From a physiological perspective, it is generally observed that chlorophyll concentration declines under salt stress as a result of the degradation of pigments or alteration in chloroplasts (Lu et al., 2023). Our findings confirm this trend in rosemary, although losses were more moderate in the cultivation of associated crops. Therefore, halophytes appear to have a protective effect that allows for partial maintenance of growth rates through photosynthesis (Farhat et al., 2021).

In response to salt stress, proline, a well-known osmoprotectant, accumulated biochemically and exhibited higher levels in monoculture plants. Proline exerts protective effects on cell structures, acts as a free radical scavenger, and assists with osmotic balance (Zhu et al., 2019). Nevertheless, higher amounts can potentially indicate severe cellular stress. In combination cultures, proline concentrations were lower, suggesting that plants experienced a lower level of perceived stress, possibly from a more favorable microenvironment. Total soluble sugars also play a role in osmoregulation and protection against cellular damage, and exhibited a pattern similar to that of proline, with higher concentrations overall, but were suppressed by intercropping (Mohammadi Alagoz et al., 2023).

Associated halophytes exhibit a significant ability to respond to salt through various mechanisms: exclusion of toxic ions at the root level, vacuolar compartmentalization, excretion through specialized glands, and production of antioxidant compounds (Bueno & Cordovilla, 2021). These abilities allow halophytes to survive in a saline environment and improve rhizospheric conditions for their non-tolerant neighbors (Hajiabadi et al., 2021). This positive interaction has been confirmed by various studies where crops accompanied by halophytes reduce salt toxicity, enhance nutrient availability, and stimulate the growth of sensitive species as well as having phytoremediation properties (Kerbab et al., 2021; Mouniane et al., 2025).

The CCA provides a supervised multivariate perspective that complements previous PCA and PLS-DA results. While PCA revealed global variance structure, and PLS-DA highlighted discriminatory traits, CCA uniquely identifies the direct multivariate influence of experimental conditions on plant physiology. The directionality and strength of trait vectors in the CCA biplot further validate results from the hierarchical clustering heatmap and correlation analysis, showing clear trade-offs between osmolyte accumulation and growth-related traits. CCA also supports the hypothesis that co-cultivation modulates stress responses in a treatment-specific manner, introducing phenotypic plasticity not captured by NaCl concentration alone. Halophytes intercropped with rosemary greatly mitigated the detrimental effects of salt stress (El-Khadir et al., 2024). This intercropping system inspired by agroecology represents a promising opportunity for agricultural areas impacted by salinization, improving functional biodiversity for resilient and sustainable agricultural systems (Gupta et al., 2022).

5. CONCLUSION

This study examined the effects of salt stress on *Rosmarinus officinalis* and the potential positive effects of its association with halophytic plants. The findings indicate that salt stress reduced rosemary height, biomass and relative water content, verifying that this plant can be classified as a salinity-sensitive glycophyte. However, with an association to halophytes, the stress effects were lessened due to an enhanced water regime, the regulation of proline concentrations, and sustainable photosynthetic activity. These positive interactions frame the importance of integrating halophytes into agricultural systems to enhance the salt stress tolerance of sensitive crops. Halophyte tolerance mechanisms, such as ionic exclusion, vacuolar compartmentalization, and osmoprotectant production can aid in salt stress management efforts. This agroecological approach represents an imaginative and viable strategy to encourage sustainable production in agroecosystems challenged by soil salinization and enhance food security in a changing climate. In sum, the present study opens up future opportunities for the use of halophytes as intercrops in agricultural systems to restore saline soil productivity and minimize the potential negative environmental impacts from agricultural practices.

REFERENCES

Abebe, H., & Tu, Y. (2024). Impact of salt and alkali stress on forage biomass yield, nutritive value, and animal growth performance: A comprehensive review. *Grasses*, *3*(4), 355–368. DOI: 10.3390/grasses3040026

Albahri, G., Alyamani, A. A., Badran, A., Hijazi, A., Nasser, M., Maresca, M., & Baydoun, E. (2023). Enhancing Essential Grains Yield for Sustainable Food Security and Bio-Safe Agriculture through Latest Innovative Approaches. *Agronomy (Basel)*, *13*(7), 1709. DOI: 10.3390/agronomy13071709

Andrade, J. M., Faustino, C., Garcia, C., Ladeiras, D., Reis, C. P., & Rijo, P. (2018). *Rosmarinus officinalis* L.: An update review of its phytochemistry and biological activity. *Future Science OA*, *4*(4), FSO283. Advance online publication. DOI: 10.4155/fsoa-2017-0124 PMID: 29682318

Arif, Y., Singh, P., Siddiqui, H., Bajguz, A., & Hayat, S. (2020). Salinity induced physiological and biochemical changes in plants: An omic approach towards salt stress tolerance. *Plant Physiology and Biochemistry*, *156*, 64–77. DOI: 10.1016/j.plaphy.2020.08.042 PMID: 32906023

Arnon, D. I. (1949). Copper enzymes in isolated chloroplasts. polyphenoloxidase in beta vulgaris. *Plant Physiology*, *24*(1), 1–15. DOI: 10.1104/pp.24.1.1 PMID: 16654194

Arora, N. K. (2019). Impact of climate change on agriculture production and its sustainable solutions. *Environmental Sustainability*, *2*(2), 95–96. DOI: 10.1007/s42398-019-00078-w

Aulicky, R., Vendl, T., Douda, O., Pavela, R., & Stejskal, V. (2025). Insecticidal activity of rosemary essential oil against primary and secondary storage beetles in the presence and absence of grain. *Journal of Stored Products Research*, *111*, 102498. DOI: 10.1016/j.jspr.2024.102498

Aziz, E., Batool, R., Akhtar, W., Shahzad, T., Malik, A., Shah, M. A., Iqbal, S., Rauf, A., Zengin, G., Bouyahya, A., Rebezov, M., Dutta, N., Khan, M. U., Khayrullin, M., Babaeva, M., Goncharov, A., Shariati, M. A., & Thiruvengadam, M. (2021). Rosemary species: A review of phytochemicals, bioactivities and industrial applications. *South African Journal of Botany*, *151*, 3–18. DOI: 10.1016/j.sajb.2021.09.026

Barrs, H. D. (1996). Root pressure and leaf water potential. *Science*, *152*(3726), 1266–1268. DOI: 10.1126/science.152.3726.1266 PMID: 17769543

Bates, L. S., Waldren, R. P., & Teare, I. D. (1973). Rapid determination of free proline for water studies. *Plant and Soil*, *39*(1), 205–208. DOI: 10.1007/BF00018060

Boussora, F., Triki, T., Bennani, L., Bagues, M., Ben Ali, S., Ferchichi, A., Ngaz, K., & Guasmi, F. (2024). Mineral accumulation, relative water content and gas exchange are the main physiological regulating mechanisms to cope with salt stress in barley. *Scientific Reports*, *14*(1), 14931. DOI: 10.1038/s41598-024-65967-5 PMID: 38942909

Bueno, M., & Cordovilla, M. P. (2021). Plant Growth Regulators Application Enhance Tolerance to Salinity and Benefit the Halophyte *Plantago coronopus* in Saline Agriculture. *Plants*, *10*(9), 1872. DOI: 10.3390/plants10091872 PMID: 34579404

Cerdá-Gandia, R., Agüero, Á., Arrieta, M. P., & Fenollar, O. (2024). Effect of different porous size of porous inorganic fillers on the encapsulation of rosemary essential oil for PLA-based active packaging. *Polymers*, *16*(18), 2632. DOI: 10.3390/polym16182632 PMID: 39339096

Chele, K. H., Steenkamp, P., Piater, L. A., Dubery, I. A., Huyser, J., & Tugizimana, F. (2021). A Global Metabolic Map Defines the Effects of a Si-Based Biostimulant on Tomato Plants under Normal and Saline Conditions. *Metabolites*, *11*(12), 820. DOI: 10.3390/metabo11120820 PMID: 34940578

Chen, G., Amoanimaa-Dede, H., Zeng, F., Deng, F., Xu, S., & Chen, Z. H. (2022). Stomatal regulation and adaptation to salinity in glycophytes and halophytes. *Advances in Botanical Research*, *103*, 1–42. DOI: 10.1016/bs.abr.2022.02.008

Chriqui, A., Benkhnigue, O., Moujahdi, C., El-Khadir, I., Mouniane, Y., Zidane, L., & Hmouni, D. (2024). Contribution to a floristic study of the Ouezzane region (Prerif, North Atlantic Morocco-1). *Egyptian Journal of Botany*, *64*(4), 28–48. DOI: 10.21608/ejbo.2024.226380.2439

Christoforidi, I., Kollaros, D., Manios, T., & Daliakopoulos, I. N. (2022). Drought- and salt-tolerant plants of the Mediterranean and their diverse applications: The case of Crete. *Land (Basel)*, *11*(11), 2038. DOI: 10.3390/land11112038

Dourado, P. R. M., de Souza, E. R., Santos, M. A., Lins, C. M. T., Monteiro, D. R., Paulino, M. K. S. S., & Schaffer, B. (2022). Stomatal regulation and osmotic adjustment in sorghum in response to salinity. *Agriculture*, *12*(5), 658. DOI: 10.3390/agriculture12050658

Dubois, M. K., Gils, J. K., Hanniton, P. A., & Smith, F. (1956). Use of phenol reagent for the determination of total sugar. *Analytical Chemistry*, *28*, 350–356. DOI: 10.1021/ac60111a017

El-hendawy, S. E., Hassan, W. M., Al-suhaibani, N. A., Refay, Y., & Abdella, K. A. (2017). Comparative performance of multivariable agro-physiological parameters for detecting salt tolerance of wheat cultivars under simulated saline field growing conditions. *Frontiers in Plant Science*, *8*, 435. DOI: 10.3389/fpls.2017.00435 PMID: 28424718

El-Khadir, I., Ktaoui, S., & Mouniane, Y. (2024). Improved Salt Stress Tolerance of *Salvia Officinalis* Grown in the Presence of a Halophytic Plant *Spergularia Maritima*: Analysis of Morpho-Physiological Parameters. In Mabrouki, J., & Azrour, M. (Eds.), *Integrated Solutions for Smart and Sustainable Environmental Conservation. Studies in Systems, Decision and Control* (Vol. 527). Springer. DOI: 10.1007/978-3-031-55787-3_2

Elachouri, M., Ouasti, I., Fakchich, J., & Bussmann, R. W. (2024). *Rosmarinus officinalis* L.Lamiaceae. In Bussmann, R. W., Elachouri, M., & Kikvidze, Z. (Eds.), *Ethnobotany of Northern Africa and Levant. Ethnobotany of Mountain Regions.* Springer. DOI: 10.1007/978-3-031-43105-0_179

FAO. (2019). *High level expert forum-how to feed the world in 2050. Economic and Social Development, Food and Agricultural Organization of the United Nations.* Plant Cell Tiss. Organ. Cult.

Farhat, N., Kouas, W., Braun, H.-P., & Debez, A. (2021). Stability of thylakoid protein complexes and preserving photosynthetic efficiency are crucial for the successful recovery of the halophyte Cakile maritima from high salinity. *Plant Physiology and Biochemistry*, *166*, 177–190. DOI: 10.1016/j.plaphy.2021.05.044 PMID: 34116337

Fouad, N., El-Zayat, E. M., Amr, D., El-Khishin, D. A., Abd-Elhalim, H. M., Hafez, A., Radwan, K. H., Hamwieh, A., & Tadesse, W. (2025). Characterizing wheat rhizosphere bacterial microbiome dynamics under salinity stress: Insights from 16S rRNA metagenomics for enhancing stress tolerance. *Plants*, *14*(7), 1033. DOI: 10.3390/plants14071033 PMID: 40219101

González-Espíndola, L. Á., Pedroza-Sandoval, A., Trejo-Calzada, R., Jacobo-Salcedo, M. R., García de los Santos, G., & Quezada-Rivera, J. J. (2024). Relative water content, chlorophyll index, and photosynthetic pigments on *Lotus corniculatus* L. in response to water deficit. *Plants*, *13*(7), 961. DOI: 10.3390/plants13070961 PMID: 38611490

González-Minero, F. J., Bravo-Díaz, L., & Ayala-Gómez, A. (2020). *Rosmarinus officinalis* L. (Rosemary): An ancient plant with uses in personal healthcare and cosmetics. *Cosmetics*, *7*(4), 77. DOI: 10.3390/cosmetics7040077

Gupta, A., Mishra, R., Rai, S., Bano, A., Pathak, N., Fujita, M., Kumar, M., & Hasanuzzaman, M. (2022). Mechanistic insights of plant growth promoting bacteria mediated drought and salt stress tolerance in plants for sustainable agriculture. *International Journal of Molecular Sciences, 23*(7), 3741. DOI: 10.3390/ijms23073741 PMID: 35409104

Hajiabadi, A. A., Arani, A. M., Ghasemi, S., Rad, M. H., Etesami, H., Shabazi Manshadi, S., & Dolati, A. (2021). Mining the rhizosphere of halophytic rangeland plants for halotolerant bacteria to improve growth and yield of salinity-stressed wheat. *Plant Physiology and Biochemistry, 163*, 139–153. DOI: 10.1016/j.plaphy.2021.03.059 PMID: 33845330

Hammer, M., & Junghanns, W. (2020). *Rosmarinus officinalis* L.: Rosemary. In Novak, J., & Blüthner, W. D. (Eds.), *Medicinal, Aromatic and Stimulant Plants. Handbook of Plant Breeding* (Vol. 12). Springer. DOI: 10.1007/978-3-030-38792-1_15

Hassani, A., Azapagic, A., & Shokri, N. (2021). Global predictions of primary soil salinization under changing climate in the 21st century. *Nature Communications, 12*(1), 6663. DOI: 10.1038/s41467-021-26907-3 PMID: 34795219

Hossain, A., Skalicky, M., Brestic, M., Maitra, S., Ashraful Alam, M., Syed, M. A., Hossain, J., Sarkar, S., Saha, S., Bhadra, P., Shankar, T., Bhatt, R., Kumar Chaki, A., EL Sabagh, A., & Islam, T. (2021). Consequences and mitigation strategies of abiotic stresses in wheat (*Triticum aestivum* L.) under the changing climate. *Agronomy (Basel), 11*(2), 241. DOI: 10.3390/agronomy11020241

Hussain, R., Ayyub, C. M., Shaheen, M. R., Rashid, S., Nafees, M., Ali, S., Butt, M., Ali, M., Maqsood, A., Fiaz, S., Ahmar, S., Mahmood, T., & Mora-Poblete, F. (2021). Regulation of osmotic balance and increased antioxidant activities under heat stress in *Abelmoschus esculentus* L. triggered by exogenous proline application. *Agronomy (Basel), 11*(4), 685. DOI: 10.3390/agronomy11040685

Ibrahimova, U., Zivcak, M., Gasparovic, K., Rastogi, A., Allakhverdiev, S. I., Yang, X., & Brestic, M. (2021). Electron and proton transport in wheat exposed to salt stress: Is the increase of the thylakoid membrane proton conductivity responsible for decreasing the photosynthetic activity in sensitive genotypes? *Photosynthesis Research, 150*(1-3), 195–211. DOI: 10.1007/s11120-021-00853-z PMID: 34125427

Ilakiya, T., Parameswari, E., Swarnapriya, R., Yazhini, G., Kalaiselvi, P., Davamani, V., Singh, S., Vinothini, N., Dharani, C., & Garnepudi, S.L. (2024). Unlocking the carbon sequestration potential of horticultural crops. C, 10(3), 65. DOI: 10.3390/c10030065

Joshi, S., Nath, J., Singh, A. K., Pareek, A., & Joshi, R. (2022). Ion transporters and their regulatory signal transduction mechanisms for salinity tolerance in plants. *Physiologia Plantarum*, *174*(3), e13702. DOI: 10.1111/ppl.13702 PMID: 35524987

Kaur, G., Sanwal, S. K., Kumar, A., Pundir, R. K., Yadav, M., & Sehrawat, N. (2024). Role of osmolytes dynamics in plant metabolism to cope with salinity induced osmotic stress. *Discover Agriculture.*, *2*(1), 59. DOI: 10.1007/s44279-024-00070-x

Kerbab, S., Silini, A., Chenari Bouket, A., Cherif-Silini, H., Eshelli, M., El Houda Rabhi, N., & Belbahri, L. (2021). Mitigation of NaCl Stress in Wheat by Rhizosphere Engineering Using Salt Habitat Adapted PGPR Halotolerant Bacteria. *Applied Sciences (Basel, Switzerland)*, *11*(3), 1034. DOI: 10.3390/app11031034

Khan, N., Ray, R. L., Sargani, G. R., Ihtisham, M., Khayyam, M., & Ismail, S. (2021). Current Progress and Future Prospects of Agriculture Technology: Gateway to Sustainable Agriculture. *Sustainability (Basel)*, *13*(9), 4883. DOI: 10.3390/su13094883

Li, S., Khoso, M. A., Xu, H., Zhang, C., Liu, Z., Wagan, S., Dinislam, K., & Liu, L. (2024). WRKY Transcription Factors (TFs) as Key Regulators of Plant Resilience to Environmental Stresses: Current Perspective. *Agronomy (Basel)*, *14*(10), 2421. DOI: 10.3390/agronomy14102421

Ling, Y., Wang, D., Peng, Y., Peng, D., & Li, Z. (2025). Cross-stressful adaptation to drought and high salinity is related to variable antioxidant defense, proline metabolism, and Dehydrin b expression in white clover. *Agronomy (Basel)*, *15*(1), 126. DOI: 10.3390/agronomy15010126

Lu, C., Li, L., Liu, X., Chen, M., Wan, S., & Li, G. (2023). Salt stress inhibits photosynthesis and destroys chloroplast structure by downregulating chloroplast development–related genes in Robinia pseudoacacia seedlings. *Plants*, *12*(6), 1283. DOI: 10.3390/plants12061283 PMID: 36986971

Luo, J., Zhang, S., Peng, J., Zhu, X., Lv, L., Wang, C., Li, C., Zhou, Z., & Cui, J. (2017). Effects of Soil Salinity on the Expression of Bt Toxin (Cry1Ac) and the Control Efficiency of Helicoverpa armigera in Field-Grown Transgenic Bt Cotton. *PLoS One*, *12*(1), 1–13. DOI: 10.1371/journal.pone.0170379 PMID: 28099508

Ma, Y., Liu, W., Qiao, Y., Qiao, W., Yang, H., Zhong, Y., Yang, H., Wang, H., Li, Y., Dong, B., & Liu, M. (2023). Effects of soil salinity on foxtail millet osmoregulation, grain yield, and soil water utilization under varying water conditions. *Agricultural Water Management*, *284*, 108354. DOI: 10.1016/j.agwat.2023.108354

Mansinhos, I., Gonçalves, S., & Romano, A. (2024). How climate change-related abiotic factors affect the production of industrial valuable compounds in Lamiaceae plant species: A review. *Frontiers in Plant Science, 15*, 1370810. DOI: 10.3389/fpls.2024.1370810 PMID: 39049861

Mohammadi Alagoz, S., Hadi, H., Toorchi, M., Pawłowski, T. A., Asgari Lajayer, B., Price, G. W., Farooq, M., & Astatkie, T. (2023). Morpho-physiological responses and growth indices of triticale to drought and salt stresses. *Scientific Reports, 13*(1), 8896. DOI: 10.1038/s41598-023-36119-y PMID: 37264097

Mouniane, Y., Chriqui, A., El-Khadir, I., Bensaid, A., Arabi, R., Zoubir, M. H., Mabrouki, J., & Hmouni, D. (2024). Investigation on the Phytosociology and Soil Characteristics of Carob Tree (*Ceratonia siliqua* L.) in Northeast Morocco: Implications for Biodiversity Conservation and Ecosystem Restoration. In Mabrouki, J., & Azrour, M. (Eds.), *Integrated Solutions for Smart and Sustainable Environmental Conservation. Studies in Systems, Decision and Control* (Vol. 527, pp. 133–142). Springer. DOI: 10.1007/978-3-031-55787-3_10

Mouniane, Y., Vibhute, A. D., & Kadmi, Y. (2025). Using Phytoremediation to Treat Industrial and Pharmaceutical Wastewater: Innovations and Future Prospects. In: Biotechnology Approaches to Industrial and Pharmaceutical Wastewater Treatment. IGI Global Scientific Publishing. DOI: 10.4018/979-8-3693-8487-9.ch009

Múnera-Echeverri, A., Múnera-Echeverri, J. L., & Segura-Sánchez, F. (2024). Biopesticidal potential of nanostructured lipid carriers loaded with thyme and rosemary essential oils against common ornamental flower pests. *Colloids and Interfaces, 8*(5), 55. DOI: 10.3390/colloids8050055

Munir, N., Hasnain, M., Roessner, U., & Abideen, Z. (2022). Strategies in improving plant salinity resistance and use of salinity resistant plants for economic sustainability. *Critical Reviews in Environmental Science and Technology, 52*(12), 2150–2196. DOI: 10.1080/10643389.2021.1877033

Munns, R., & Tester, M. (2008). Mechanisms of salinity tolerance. *Annual Review of Plant Biology, 59*(1), 651–681. DOI: 10.1146/annurev.arplant.59.032607.092911 PMID: 18444910

Omuto, C. T., Kome, G. K., Ramakhanna, S. J., Muzira, N. M., Ruley, J. A., Jayeoba, O. J., Raharimanana, V., Owusu Ansah, A., Khamis, N. A., Mathafeng, K. K., Elmobarak, A. A., Vargas, R. R., Koetlisi, A. K., Dembele, D., Diawara, M., Mbaikoubou, M., Maria, R. M., Adam Boukary, I., Malatji, A., & Nyamai, M. (2024). Trend of soil salinization in Africa and implications for agro-chemical use in semi-arid croplands. *The Science of the Total Environment*, *951*, 175503. DOI: 10.1016/j.scitotenv.2024.175503 PMID: 39147045

Ortiz, A., & Sansinenea, E. (2023). Genetically Modified Plants Based on Bacillus Genes and Commercial Bacillus-Based Biopesticides for Sustainable Agriculture. *Horticulturae*, *9*(9), 963. DOI: 10.3390/horticulturae9090963

Oudghiri, M., Yamani, B., Benlemlih, N., El Aammouri, S., Abid, N., Brhadda, N., Bouhassoun, S., Ziri, R., Chriqui, A., Aoujil, F. Z., El Bakkali, M., Mouniane, Y., & Ibriz, M. (2025). Effect of Heavy Metals on the Morphological and Physiological Responses of the Torro Plus Variant of *Zea mays*. *Journal of Environment and Earth Science*, *7*(4), 165–179. DOI: 10.30564/jees.v7i4.8208

Pissolato, M. D., Martins, T. S., Fajardo, Y. C. G., Souza, G. M., Machado, E. C., & Ribeiro, R. V. (2024). Stress memory in crops: What we have learned so far. *Theoretical and Experimental Plant Physiology*, *23*(3), 535–565. DOI: 10.1007/s40626-024-00315-6

Psarrou, I., Oreopoulou, A., Tsimogiannis, D., & Oreopoulou, V. (2020). Extraction Kinetics of Phenolic Antioxidants from the Hydro Distillation Residues of Rosemary and Effect of Pretreatment and Extraction Parameters. *Molecules (Basel, Switzerland)*, *25*(19), 4520. DOI: 10.3390/molecules25194520 PMID: 33023142

Raffo, A., Baiamonte, I., De Benedetti, L., Lupotto, E., Marchioni, I., Nardo, N., & Cervelli, C. (2023). Exploring volatile aroma and non-volatile bioactive compounds diversity in wild populations of rosemary (*Salvia rosmarinus* Schleid.). *Food Chemistry*, *404*, 134532. DOI: 10.1016/j.foodchem.2022.134532 PMID: 36242966

Rehman, S., Yang, J., Zhang, J., Zhang, L., Hao, X., Song, R., Chen, S., Wang, G., & Hua, L. (2025). Salt stress in wheat: A physiological and genetic perspective. *Plant Stress*, *16*, 100832. DOI: 10.1016/j.stress.2025.100832

Shabbir, R., Javed, T., Afzal, I., Sabagh, A. E., Ali, A., Vicente, O., & Chen, P. (2021). Modern Biotechnologies: Innovative and Sustainable Approaches for the Improvement of Sugarcane Tolerance to Environmental Stresses. *Agronomy (Basel)*, *11*(6), 1042. DOI: 10.3390/agronomy11061042

Shakri, T., Che-Othman, M. H., Md Isa, N., Sukiran, N. L., & Zainal, Z. (2022). Morpho-physiological and stress-related gene expression of rice varieties in response to salinity stress at early vegetative stage. *Agriculture*, *12*(5), 638. DOI: 10.3390/agriculture12050638

Shokri, N., Hassani, A., Sahimi, M. (2024). Multi-scale soil salinization dynamics from global to pore scale: A review. Reviews of Geophysics, 62(4), e2023RG000804. DOI: 10.1029/2023RG000804

Singh, A., Rajput, V. D., Lalotra, S., Agrawal, S., Ghazaryan, K., Singh, J., Minkina, T., Rajput, P., Mandzhieva, S., & Alexiou, A. (2024). Zinc oxide nanoparticles influence on plant tolerance to salinity stress: Insights into physiological, biochemical, and molecular responses. *Environmental Geochemistry and Health*, *46*(5), 148. DOI: 10.1007/s10653-024-01921-8 PMID: 38578547

Surówka, E., Grigore, M. N., Ślesak, I., Dubert, F., & Miszalski, Z. (2025). Halophytes in the Bioeconomy: New Opportunity and Challenge for Agriculture. In Arora, J., Grigore, M. N., & Joshi, A. (Eds.), *Sustainable Utilisation and Bioengineering of Halophytes. Sustainable Development and Biodiversity* (Vol. 38). Springer. DOI: 10.1007/978-981-97-9819-3_16

Tang, H., Du, L., Xia, C., & Luo, J. (2024). Bridging gaps and seeding futures : A synthesis of soil salinization and the role of plant-soil interactions under climate change. *iScience*, *27*(9), 1–10. DOI: 10.1016/j.isci.2024.110804 PMID: 39286506

Wang, X., Liu, X., Su, Y., & Shen, H. (2025). Rice Responses to Abiotic Stress: Key Proteins and Molecular Mechanisms. *International Journal of Molecular Sciences*, *26*(3), 896. DOI: 10.3390/ijms26030896 PMID: 39940666

Zhang, W. P., Surigaoge, S., Yang, H., Yu, R. P., Wu, J. P., Xing, Y., Chen, Y., & Li, L. (2024). Diversified cropping systems with complementary root growth strategies improve crop adaptation to and remediation of hostile soils. *Plant and Soil*, *502*, 7–30. DOI: 10.1007/s11104-023-06464-y

Zhu, Y. X., Gong, H. J., & Yin, J. L. (2019). Role of silicon in mediating salt tolerance in plants: A review. *Plants*, *8*(6), 147. DOI: 10.3390/plants8060147 PMID: 31159197

Ziogas, V., Tanou, G., Morianou, G., & Kourgialas, N. (2021). Drought and salinity in citriculture: Optimal practices to alleviate salinity and water stress. *Agronomy (Basel)*, *11*(7), 1283. DOI: 10.3390/agronomy11071283

Chapter 9
Understanding the Impact of Moroccan Weather on the Durability of Solar Mirror:
A Comprehensive Review

Fatima Zahra Aoujil
Ibn Zohr University, Morocco

Mohamed Bouzelmad
https://orcid.org/0000-0001-5359-7187
Laboratory of Physics, Energy, and Information Processing (Lab.PETI), Morocco

Yassin Darmane
Physics-Chemistry Department, Polydisciplinary Faculty of Ouarzazate, Morocco

Yassine Mouniane
https://orcid.org/0000-0002-6037-5301
Faculty of Sciences, Ibn Tofail University, Morocco

Mohamed El Bakkali
Faculty of Sciences, Ibn Tofail University, Morocco

ABSTRACT

The aim of this chapter is to provide an up-to-date overview of research into the durability of solar mirrors in different Moroccan environments. Taking into consid-

DOI: 10.4018/979-8-3693-9132-7.ch009

eration numerous studies carried out in different regions of Morocco, the synthesis identifies the complex combination of environmental factors, materials, and ageing mechanisms that shape the performance of solar mirrors in concentrated solar power (CSP) applications. In particular, the research highlights the critical impact of geographical location, climatic conditions, and material choices on mirror degradation. The varied nature of paint degradation, the potential of protective coatings, and the importance of rigorous testing procedures are highlighted. This review summarizes the results, highlighting the importance of Moroccan research in advancing CSP technology in the context of renewable energy development.

1. INTRODUCTION

Solar energy represents a cost-effective solution to renewable energy, and Morocco aims to change 52% of its consumption to renewables by 2030. To achieve this transition, Morocco is in the process of various solar investments and developing solar technology at various sites and in a range of climates. For solar technologies to be successful, a key component will be the strategic siting of solar plants to maximize solar resource potential. However, all of these solar developments complicate the sustainability analysis of solar power plants due to the challenges imposed by environmental conditions (i.e., temperature, irradiation, wind, dust, precipitation, humidity, and salinity). In a laboratory, all materials used in solar power plants have been subjected to accelerated testing; however, the interaction between these environmental factors under real-life conditions complicates our understanding of their role in durability and efficiency characteristic of an array of solar mirrors (Lakhouil et al., 2023).

Studies have established that regions with high levels of Direct Normal Irradiance (DNI), like the Middle East and North Africa (MENA), often experience high aerosol concentrations (Karim et al., 2014a; Karim et al., 2014b). This study seeks to understand the potential challenges solar mirrors are exposed to in this environment, including erosion from wind-borne sand particles, chemical degradation from corrosion, and surface fouling that is an unavoidable by-product in arid climates amongst others. This research aims to understand the spatial and materials attributes of solar reflectors to address wind speed, temperature, humidity amongst other climate impacts on erosion and soiling, to be able to provide a rationale for solar mirror design and maintenance in such conditions.Due to the potential impact of solar reflectors longevity on financial and operational performance of solar photovoltaic plants, this study focuses on the behaviour of the physico-chemical properties of monolithic glass mirrors and their paint coatings and protective layers (Lakhouil et al., 2023). This study assesses the stress factors of temperature, humidity and solar irradiation from the environment on the performance of paint coatings in terms of better's resistance to degradation, providing supportive data on the coatings performance over time.

Weather phenomena are closely connected to factors such as wind speed, wind direction, relative humidity, and atmospheric pressure. Dust deposition is exacerbated by high relative humidity, while low pressure is associated with rainy days (Matal et al., 2020). The weight of deposited dust could also be used to establish correlations for reflectivity losses, besides erosion of solar reflectors, as a result of sand storms, also contributes to optical losses (Boukheir et al., 2020; Lakhouil et al., 2021). The effects of erosion vary from site to site mainly due to the shapes of the particles, and this leads to different formations of microscopic cracks. The shape of the particle may have more of an effect on erosion than the hardness of the particle. Solar mirrors coated with protective coatings save water by decreasing the need to clean the solar mirrors and recover reflectivity better than uncoated mirrors. Furthermore, the anti-soiling properties of these coatings are compound under conditions of severe soiling (Lakhouil et al., 2022). The accuracy of dust level estimations is critical in estimating cleaning operations, as well as determining the economic and viability of CSP projects worldwide. The optimization of cleaning the solar field depends on the cleaning approach that minimizes potential dust deposition factors at each individual plant site. For instance, areas with high DNI potential and limited water resources indicate that additional work could be explored regarding water-efficient cleaning options. Likewise, the increase in overall solar power plant degradation potential may also increase cleaning operations which would also increase cost per KWh produced. Dust quantity and composition, along with the resulting losses, are inherently linked to the geographical and climatic attributes of each region across the globe. Therefore, mitigation approaches should be fine-tuned to account for the site-specific nature of these challenges. Recent research trends emphasize the experimental evaluation of factors influencing soiling losses in diverse global regions, moving away from the pursuit of universal standardization.

Overall, this study aims to contribute to the evolution of the solar energy field by shedding light on the nuanced relationship between Moroccan weather conditions and the durability of solar mirrors. Through an in-depth examination of degradation phenomena and optimization of material properties and configurations, the results presented here aim to guide the sustainable development of solar energy projects in Morocco's challenging but promising landscape.

2. MATERIALS AND CONFIGURATIONS

Concentrating solar collectors demand highly reflective surfaces, often using silver and aluminum due to their reflective properties within the solar spectrum. Silver generally has a significantly higher reflectance compared to aluminum across most of the spectrum, although aluminum is more suitable for UV applications. These reflective surfaces are typically metallic coatings on substrates like glass, polymers, or metals. When opaque substrates are used, the reflective layer is applied to the front surface, creating a "first-surface" reflector. When transparent substrates are employed, the coating can be applied to the rear surface, creating what is known as a "second-surface" reflector. Both metals necessitate protective coatings to inhibit corrosion, with front-surface mirrors featuring transparent top coatings, while second-surface mirrors are shielded by coatings or an additional substrate. While some mirror systems with specific refractive index gradients have been explored in laboratories, they haven't become common due to their dependence on incident angles.

Solar mirrors can be divided into two main categories, depending on the location of the reflective layer and the presence of protective coatings. Second-surface mirrors have the transparent substrate holds the reflective layer on its back, commonly glass. In such mirrors, sunlight must pass throught the transparent top layer before reaching the reflective surface. In contrast, first-surface or front-surface mirrors feature the reflective layer on top, safeguarded by an ultra-thin transparent layer, usually less than 5 µm thick, to improve durability. These mirrors are then classified according to the type of glass layer they incorporate.

Table 1. Properties of Different Solar Mirror Types

Property	Silvered Thick-Glass	Silvered Thin-Glass	Laminated Silvered Glass	Aluminum Mirrors	Silvered Polymer Film
Reflective Layer	Silver layer on glass	Silver layer on thin glass	Silver layer on laminated glass	Pure aluminum on aluminum	Silver layer on polymer film
Additional Coatings	Copper layer, protective paints	Bonded to metal, composite, or glass	Protective glass layers	Transparent coatings	Polymer layers, various substrates
Advantages	High reflectance, durability	High reflectance, lightweight	High reflectance, durability	Lightweight, flexible design	Lightweight, flexible design
Manufacturing Process	Wet chemistry process	Variable, based on structure	Thermal manufacturing process	Coil coating process	Bonding process

continued on following page

Table 1. Continued

Property	Silvered Thick-Glass	Silvered Thin-Glass	Laminated Silvered Glass	Aluminum Mirrors	Silvered Polymer Film
Considerations	Vulnerable to tarnishing and corrosion	Durability similar to thick-glass	Excellent durability expected	Lower reflectance, higher scattering	Resistance to abrasion improvements ongoing
Applications	Concentrated solar power (CSP)	Various concentrating systems	Solar applications	Low-temperature concentrators	Concentrating systems
Durability and Solutions	Durability enhanced by coatings	Relatively lightweight, potential cost reduction	Production cost is higher than thick glass	Durability may require additional glass cover	Careful bonding process required

3. GENERAL CLIMATIC CHARACTERISTICS OF MOROCCO

Morocco is situated in the northwestern corner of the African continent, with diverse geography that includes the Rif Mountains in the north, the Atlas Mountains with a peak at 4165 meters in the center, western plains and coast, eastern plateaus, and the southern desert. The country experiences a wide range of climates, from temperate to semi-arid to desert, influenced by the Atlantic Ocean to the west, the Mediterranean Sea to the north, and the Sahara Desert to the south and southeast, following the Koppen-Geiger classification (Figure 1). Most of Morocco's precipitation occurs between September and May, considered the rainy and agricultural season. Coastal areas have a Mediterranean climate with mild, wet winters and hot, dry summers, averaging around 1,200 millimeters of annual rainfall in the Rif region. Snowfall can occur in the mountainous regions from November to April. The south, in contrast, is much drier, receiving less than 100 mm of precipitation annually. Average annual temperatures exceed 22°C in the south, range from 12 to 18°C in the mountainous areas, and vary from 18 to 20°C elsewhere. In summer, daily maximum temperatures often exceed 40°C inland, while winter brings sub-zero temperatures in the mountainous zones (Driouech et al., 2010).

Figure 1. Morocco's climate according to the Koeppen-Geiger index (1981-2010)

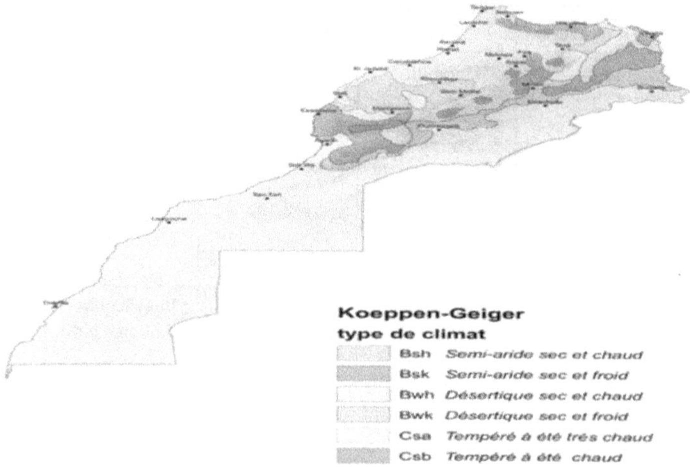

3.1. Global Irradiation

Morocco has garnered attention in numerous research studies, such as those conducted by the World Energy Council and Frontier Economics, as one of the nations boasting highly competitive renewable energy potential on a global scale. This prominence can be attributed not only to the substantial size of its renewable energy reserves but also to their consistent availability.

The geographical distribution of viable energy sites across the country allows for the even distribution and relatively steady supply of energy to the grid. In the context of solar energy, Morocco boasts an annual sunshine potential ranging from 2,500 to 3,000 hours per year, equivalent to an average of approximately 8.15 hours of sunlight per day (Figure 2). This abundant sunshine is spread across almost the entire national territory and is particularly intense on the eastern slopes of the Atlas Mountains.

Figure 2. Direct Normal Irradiance (DNI) Distribution Map of Morocco

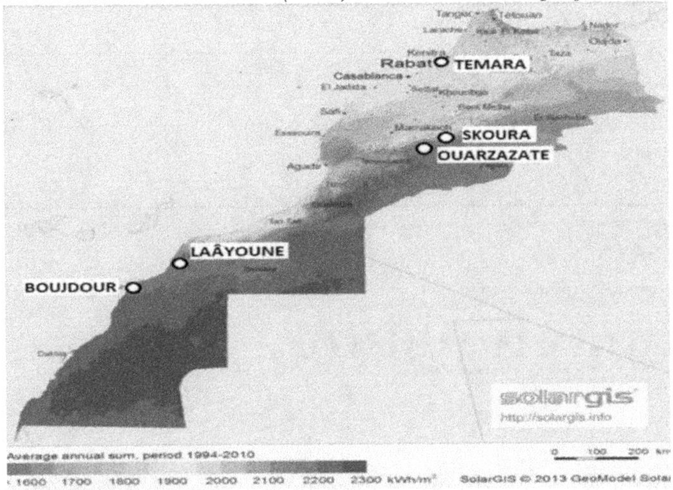

3.2. Annual Temperature

In Morocco, the average annual temperature typically stands at 19°C, with monthly averages ranging from 12°C in January to 25°C in August. However, in 2022, the annual temperature spanned from 4 to 26°C across the country. In the Atlas Mountains, situated at elevations above 1600 meters, temperatures varied between 4 and 12°C. Areas with altitudes ranging from 1000 to 1600 meters experienced temperatures between 12 and 18°C. In the remaining regions of the country, the average annual temperatures fell between 18 and 22°C, except for the southeastern slopes and the Marrakech and Laâyoune-Sakia Al Hamra areas, where temperatures ranged from 22 to 26°C. The average annual temperature surpassed the climatological norm calculated for the 1981-2010 period in Morocco. Anomalies exceeded +3.5°C in Chichaoua, Rehamna, and Taourirt, while remaining below 3°C elsewhere.

Figure 3. Temporal variation in mean annual temperature in Morocco (T_{moy}) for the period 1981-2022.

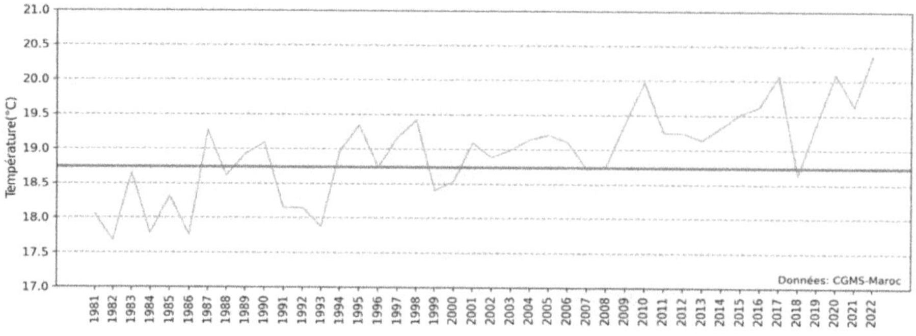

(Data source: CGMS-Maroc)

In 2022, the average maximum temperature across the entire nation reached a record level, surpassing the climatological norm of 1981-2010 by 2.1°C. This record deviation exceeded the previous highest deviations seen in 2017 and 2020, which were 2.06°C and 1.99°C, respectively. Regarding the average annual minimum temperature across the entire territory, it ranked as the second warmest since 1981, following 2010, with a deviation of approximately 1.1°C from the climatological norm.

3.3. Annual Rainfall

Average annual rainfall is 187 mm over Morocco as a whole and 377 mm in agricultural areas (Figure 1), with the three wettest months recording average rainfall of 25 mm (November), 29 mm (December) and 25 mm (February). There is practically no rainfall in June, July and August. The rainfall deficit in 2022 is the fourth in a row. In fact, the relative deviation of annual rainfall from normal is around -27%. This rainfall deficit over the last four years (2019-2022) is particularly worrying, with an average deficit of around -32%. This is the highest deficit recorded since 1981 (Driouech et al., 2010), even exceeding that in four consecutive years from 1998 to 2001, which was -22%.

Figure 4. Temporal variation in annual precipitation in Morocco for the period 1981-2022

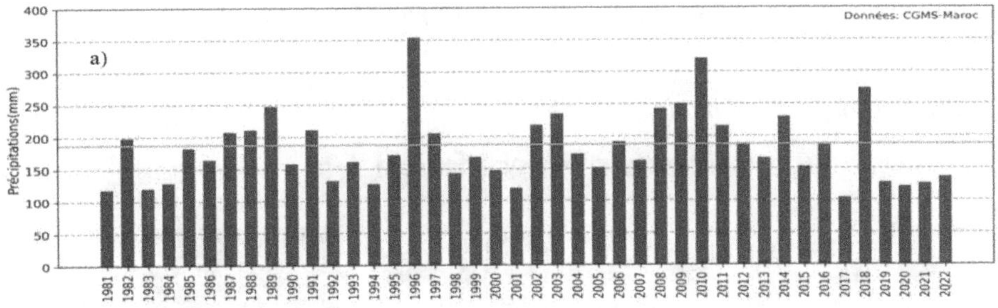

(Data source: CGMS-Maroc)

4. DEGRADATION MODES STUDIED IN THE LITERATURE

The literature describes a number of degradation modes observed in various articles on the aging and degradation of solar reflectors. Optical scattering in the top coating and filiform corrosion in the aluminum layer are notable, both having a substantial impact on specular reflectance (Boukheir et al., 2023).

Different types of reflectors have different degradation behaviors. An example is laminated reflectors without edge sealing, which degraded faster under humid heat; although lacquered reflectors have shown more resistance to aging in previous research, as mentioned by Naamane et al. (2020). The degradation of anodized aluminum reflectors is particularly important to understand, since even though they look nice, protective processes are important when exposed for long periods of time; this concern was noted in the study as well. Upgraded aluminum reflectors with no corrosion can degrade into localized corrosion and diffusion patterns, and Chbihi et al. (2022) has provided a model of degraded zones to be used for calculations of loss of reflectance.

Mirror soiling is another degradation factor and the level of soiling can be affected by temperature, humidity, and contaminants such as hematite (Guerguer et al., 2016). The same study recommended a system engineering approach to prevent mirror soiling. In more aggressive environments that contain sea salts, solar mirrors can corrode quickly, leading to edge corrosion and peeling of the protective paint. In addition to the degradation rating of protective paint, (Guerguer et al., 2021) emphasizes the importance of the combinatorial role of paint type and edge protection for corrosion rates. Additionally, (Bouaddi and Ihlal, 2015) mentions the implications of sand and soil particle interactions degrading backside paints and the implications for protecting the silver reflective back paint layer. Lastly, protective coatings are critical to the longevity of mirrors, with ongoing reports on defects spotted in initial degradation ratings related to protective coatings (Bouaddi et al., 2017).

Figure 5. Degradation modes

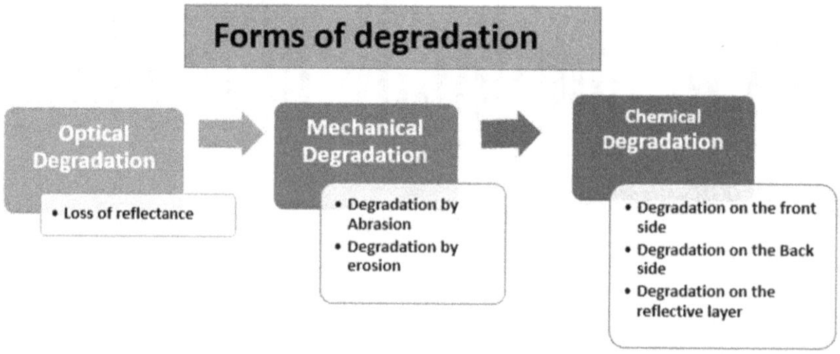

Table 2. Synthesis of Moroccan climate effects on durability of solar mirrors from previous studies

References	Device study	OET/AET	Main Results
Karim et al., 2014a	Silvered mirrors	OET: Ouarzazate 360 days AET: A horizontal sandblasting machine	The research reports on the effects of natural and accelerated aging on reflective coatings of solar mirrors. The natural low wind speed typical in the area helps minimize the impacts on mirror surfaces over time. Accelerated aging tests with real sand particles showed that wind speed, particle size, and impact angle have an effect on the wear of the surface. The importance of accounting for environmental conditions when designing mirrors for CSP applications is emphasized in the report.
Karim et al., 2014b	-	OET: Skoura and Temara city	This paper examines the effects of climatic and geological conditions on the erosion of solar mirrors at different locations. The sand particle density and the wind velocity are the primary variables effecting surface damage. For example, the Skoura desert site has large sand particles and lower wind velocity, which minimizes surface damage no matter how hard the sand grains may be. In contrast, the Temara Seaside site has relatively high humidity and higher wind speeds which subject the mirrors to greater surface damage due to increased adhesion between the soiled particles and the mirrors.
Edfouf et al., 2014	Monolithic and laminated mirror and polymeric mirrors	OET: 390 days	Erosion behavior is significantly influenced by various factors associated with the environmental conditions, particularly the properties of sand and wind. Outdoor testing has revealed that mirrors exposed in the Oceanside site exhibit more pronounced erosion compared to those in the desert site, primarily due to the higher wind speeds at the Oceanside location. However, accelerated testing results have demonstrated that desert sand is more aggressive and can lead to a higher occurrence of abrasion defects on the mirror surfaces.
Karim et al., 2015a	Silvered mirrors	OET: Rabat and sakoura 240 days AET: sandblasting chamber	This paper highlights the direct correlation between wind speed and loss of specular reflectivity. Surface erosion contributes significantly to the reduction in optical performance of mirrors. Mirrors exposed to natural weathering sites show minimal loss of reflectivity even after 240 days of exposure. For the same ageing time, the mirror at the desert site has higher roughness parameters than the mirror at the Oceanside site.
Karim et al., 2015b	Glass mirrors coated with a silver reflective film	AET: Sand blower machine	The paper discusses the effect of impact velocity and angle, as well as sand particle properties, on specular reflectance loss. The study reveals that impact speed and angle significantly influence surface roughness. Particle size, sharpness and hardness also influence reflectance loss.

continued on following page

Table 2. Continued

References	Device study	OET/AET	Main Results
Edfouf et al., 2015	Monolithic laminated glasses and polymeric mirror	OET: Rabat and Ouarzazate 1 year	After around a year of natural ageing, the study highlights the differences in the behavior of polymer and glass mirrors in coastal and desert environments. Polymer mirrors degraded in both cases, while glass mirrors underwent rapid corrosion in coastal conditions, but retained their integrity in the desert. These results underline the importance of mirror optimization and material selection for CSP applications under specific environmental conditions.
Conceição et al., 2019	TraCS mirror	OET: Green Energy Park 5 months	This article presents the initial results of a comparison of soiling measurements between Portugal and Morocco. Morocco has a significantly higher soiling rate, mainly due to dust storms in the Saharan desert. The impact of dust followed by red rain in Morocco. The study highlights the need to develop more water-efficient cleaning methods, given that both countries face or are expected to face high levels of water stress, particularly in areas where the CSP is being implemented.
Karim et al., 2016	Monolithic and polymer solar mirrors	OET: Sakoura 21 months: glass mirrors 17 months: polymer glass	The study underlines the difference in the performance of glass and polymer solar mirrors against soiling. Polymer mirrors are more sensitive to optical degradation due to their greater surface roughness. In fact, they discovered a significant correlation between the optical degradation of the mirrors caused by the deposition of dirt and the variation in wind speed and relative humidity at the exposure site.
Matal et al., 2017		AET	The aim of this research is to test the reliability and repeatability of a test bench designed to simulate real-life conditions of solar reflector erosion. The study underlines the importance of accurately reproducing external conditions during testing, and the importance of the test rig for the study of solar reflector erosion.
Guerguer et al., 2018	Paint coatings used in commercial glass mirrors	OET: Temara city 3 years	The main observations of the study include a significant transition in the colour of the paint from white to yellow, which serves as an indicator of degradation; the correlation between yellowing and UV exposure; chemical and structural changes due to UV-induced degradation in the organic binder and pigments; and physical observations of cracks, increased paint thickness and surface roughness in aged paint coatings. These results collectively underline the multifaceted nature of paint degradation and the influence of environmental conditions on this process.

continued on following page

Table 2. Continued

References	Device study	OET/AET	Main Results
Karim et al., 2017	Second surface silvered glass mirrors	OET: Ouarzazate Noor station 16 months	The main conclusions of the article are as follows: mirrors with polymeric surfaces are more prone to soiling due to their roughness than glass surfaces; climatic conditions such as precipitation and wind have a significant impact on soiling rates, with heavy precipitation and wind naturally cleaning the mirrors, while low precipitation and high humidity increase soiling. In addition, the angle of exposure of the mirrors is crucial, with horizontal exposure accumulating more soiling than normal angles, and the height and orientation of the installation in relation to the prevailing wind direction play an important role in varying soiling. The influence of exposure orientation on fouling rates is also noted, particularly in relation to wind speed.
Guerguer et al., 2019	Silvered glass mirrors	OET: First year of natural exposure	The article shows how corrosion propagates from localized defects in the paint layer and its circular shape. Corrosion has a significant impact on the reflectivity of silvered glass reflectors, particularly in marine environments.
Matal et al., 2020		AET	This study focuses on the erosion of reflector samples in concentrating solar power plants. The study demonstrates the way in which variations in air speed and impact angles influence the deterioration of the optical characteristics of reflector samples. Accurate calibration of key parameters is imperative to ensure test accuracy. In particular, when the sample is subjected to a velocity of 25 m/s, the loss of reflectance is almost twice as high at an impact angle of 90°. (11.03%) compared to 45° (5.31%).
Boukheir et al., 2020	Two types of glass solar mirror samples	OET: Midelt 1 year	The article examines the reaction of different mirror samples (M1 and M2) to erosion and degradation. M2 samples show greater erosion and a rougher surface. Mirror height and orientation also influence soiling rates and degradation.

continued on following page

Table 2. Continued

References	Device study	OET/AET	Main Results
Lakhouil et al., 2021	Silvered glass mirrors	OET: Temara city 400 days	Solar mirrors tested outdoors in a harsh coastal environment experienced rapid corrosion, mainly due to the presence of marine salts (NaCl and $MgCl_2$) which promote corrosion. Edge corrosion was the main problem, leading to detachment of the protective paint. Thick layers of paint and edge protection were important. Solar mirrors bonded to composite substrates suffered from poor resistance due to the infiltration of corrosive agents. Lead-free paints (RLA3) were less resistant than lead-containing paints (RLA2), due to the anti-corrosive properties of lead. Corrosion occurred in three stages at varying rates, influenced by factors such as paint composition, edge shape, thickness and coatings. Degradation of RLC samples results from deterioration of the adhesive, which allows moisture penetration and accelerates corrosion. The study highlighted the importance of the durability of the backing material in the evolution of corrosion, with corrosion rates ranging from 8 mm^2/day to 163 mm^2/day depending on these factors.
Lakhouil et al., 2022	Anti-soiling coated and uncoated solar mirror	**AET**	This article demonstrates the innovative combination of accelerated erosion and ageing tests for the evaluation of glass mirrors, highlighting the accelerated impact of erosion on optical performance. It underlines the significance of protective coatings, especially SiO_2, for the durability and hydrophobicity of glass mirrors, so noting the initial differences in performance between coated and uncoated samples.
Boukheir et al., 2023	Silvered-glass mirrors	**OET:** Ouarzazate 12 months to a maximum of over 31 months	The main findings of this work are as follows Outdoor exposure on a desert CSP site makes it possible to assess the behavior of materials under real-life conditions. In particular, the main wind direction (north) induces the greatest reflectance loss, and the angle of exposure affects durability and soiling rates. Surprisingly, anti-fouling coatings show only limited benefit after prolonged exposure, while hard-coated samples show reduced roughness but greater reflectance loss, indicating unexpected chemical degradation. The effectiveness of coatings varies according to environmental conditions and cleaning frequency, underlining the fact that their performance is context dependent.
Lakhouil et al., 2023	Silvered glass uncoated and hard coated	**OET:** Desertic and coastal sites	These hard-coated samples exhibit weak mechanical and chemical resistance, along with unimpressive initial reflectance. In contrast, uncoated samples, despite having high initial reflectance (95%), prove to be more durable and are suggested as better candidates for solar mirrors at the mentioned sites.

continued on following page

Table 2. Continued

References	Device study	OET/AET	Main Results
Naamane et al., 2020	Silvered glass mirrors	**OET:** 400 days Midelt and Boujdour	M2 samples showed more erosion than M1 samples, resulting in a rougher surface, particularly visible for samples exposed at a height of 2.5 m and inclined at 90° and 135°. The M1 mirrors show a significant change in yellowing, suggesting a more rapid degradation of the paint system than the M2 mirrors.
Chbihi et al., 2022	Tested mirrors M1 and M2	**OET:** Midelt (Atlas Mountain) 4 months	The study's findings illustrate the impact of wind speed and gustiness on reflectance loss due to soiling. They also highlight the role of low-volume precipitation, particularly in clay-rich soils, in promoting fouling. It is essential to understand the correlations between the various parameters. In addition, flat positions, which often accumulate fouling during sandstorms, require particular attention due to their significant role in increasing reflectance loss.
Guerguer et al., 2016	Glass and polymeric mirrors	OET: Temara city 7 months	In this article, the authors report on their experience with two methods for monitoring atmospheric salt levels in the marine environment and discuss their advantages and limitations. The study showed that both the wet candle sensor and the wash water method effectively captured atmospheric salts, including chlorides and sulfates from sea spray. It was noted that the number of salts captured is influenced by exposure time and environmental conditions, with wind speed playing an important role. The wet candle sensor was presented as a valuable tool for measuring and monitoring sea spray in the atmosphere, particularly its potential deposition on solar mirrors exposed in marine environments.
Guerguer et al., 2021	Monolithic silvered glass	**OET:** Temara and Skoura 24 months	This study examines the effect of natural ageing on the chemical structure and color changes of surface coatings under desert and marine climate of Morocco. The glass coating system used on the mirrors proved highly resistant to desert conditions, with no significant changes in color or gloss. Nevertheless, under the aggressive environmental conditions of the marine site (UV radiation, humidity, salinity), yellowing and moderate gloss degradation were observed on some mirrors. Chemical changes in the organic binder were also observed, particularly for samples exposed at the marine site. Paint yellowing was found to be influenced by factors such as humidity, salinity and precipitation, in addition to UV exposure. The results suggest that marine environments could serve as effective testing grounds for assessing the durability of glass mirror paints in a shorter timeframe.

continued on following page

Table 2. Continued

References	Device study	OET/AET	Main Results
Bouaddi and Ihlal, 2015	Silvered-glass mirrors and aluminum mirrors	OET: Agadir 6 months	In this study, the evolution of cleanliness was influenced by meteorological parameters, such as rainy events that restore cleanliness and dry, dusty summer months that lead to low levels of cleanliness. The study also assessed the normality and independence of the residuals, concluding that the residuals had a normal distribution and showed no significant dependence on each other.
Bouaddi et al., 2017	Second -surface silvered mirrors and mirrors based on an aluminum reflective layer	OET: Agadir and Tantan 6 months	In this study carried out in southwest Morocco, the fouling patterns of candidate CSP mirrors were examined at two sites, one with monthly demineralized water cleaning and the other with natural cleaning over six months. The analysis identified two common trends in cumulative mirror soiling. Rain had variable effects depending on the mirror's previous soiling level and the material used, with glass mirrors recovering reflectance better than aluminum mirrors. Mirrors cleaned once a month lost more of their cleanliness than mirrors that were already dirty. Aluminum mirrors appear suitable for use in dry desert sites where water resources are scarce. Further research into the impact of dry-cleaning techniques on these mirrors is recommended.
Merrouni et al., 2015	Aluminum reflector	OET: Oujda 3 months	This research offers initial perspectives on the influence of dust and soiling on solar mirror materials in Eastern Morocco. Notably, horizontal mirrors experienced the most substantial impact from dust, resulting in an average cleanliness reduction of 45% and 33% for glass and aluminum mirrors, respectively. In contrast, mirrors oriented at 0° and -45° angles remained unaffected by dust accumulation. Analysis via X-ray revealed that the predominant dust particles were composed primarily of Quartz (SiO_2), Calcite ($CaCO_3$), and Muscovite ($KAl_3Si_3O_{10}(OH)_2$).
Azouzoute et al., 2019	Silvered glass	OET: Green Energy Park (Bengurir) 18week	The soiling rate of aluminum and glass solar reflectors was evaluated in this study and the results demonstrate that glass mirrors suffered the most from dust in comparison to aluminum mirrors, which also had a higher density of dust deposition. The complex association between loss of reflectivity and density of dust deposition is thought to be attributed to surface properties, seasonal influences, and various climatic factors, including humidity, dew and precipitation, particularly light rain. These results suggest that aluminum mirrors would be suitable in concentrated solar power plants on arid sites for industrial processes that require lower temperatures.

continued on following page

Table 2. Continued

References	Device study	OET/AET	Main Results
Merrouni et al., 2017	Solar mirrors cut from ordinary parabolic troughs collectors	OET: Oujda 12 WEEKS	This study examined soiling effects on CSP solar mirrors in eastern Morocco and found that soiling causes significant challenges to CSP plants equipped with these mirrors in this location because, depending on the location, reflectivity can fall by as much as 30% after just one week of exposure. The study findings are considered crucial for the region's CSP plant operation and maintenance teams to better prepare for dry periods and develop suitable cleaning techniques to maintain high optical efficiency.
Azouzoute et al., 2020	Glass and Mirror samples	OET: Green Energy Park, Ben guerir 3 months	The article examines the participation of Morocco and Australia in a hybrid CSP-PV solar power plant and highlights the necessity of an assessment of fouling on the two technologies, particularly operating in the same climate conditions. This work is novel, examining a hybrid solar energy installation and directly assessing the impact of fouling on the mirror of CSP and the glass of PV. The results indicate the mirror samples are more susceptible to dust, with experienced fouling losses of approximately 35%. This was intensified by prevailing concentrations of precipitation and aerosols in the upper atmosphere.
Dahlioui et al., 2022	Silvered glass	OET: Ouarzazate 1 years	This study examined antifouling coatings for solar collectors, over the course of a year's exposure, the coated silver glass mirrors showed positive results, achieving a maximum cleanliness gain of 7 pp, which is particularly interesting for sites with high levels of soiling. Fouling rates were determined for coated and uncoated mirrors, with coated mirrors showing a 75.5% reduction in fouling. Micro-scratches and surface pitting were observed on uncoated and coated mirrors respectively, warranting further investigation into their origin.
Wiesinger et al., 2016	Second surface silver mirrors protected by low-iron glass and aluminum mirror	OET: Zagora AET	In this study, an accelerated erosion test for CSP reflective materials was developed. A strong linear correlation was established between the amount of sand used and the loss of specular reflectance. The test results closely resemble the actual outdoor effects observed in Morocco. However, frequent high winds were also observed. The study classified reflector materials according to their resistance to erosion, revealing that glass outperformed aluminum materials. In addition, aluminum reflectors with protective polymer coatings showed slightly better erosion resistance than sol-gel coatings. The angle of impact of the sand was found to have a significant effect on erosion, with perpendicular impacts causing a much greater loss of reflectance on glass mirrors than impacts at an angle of 30°.

continued on following page

Table 2. Continued

References	Device study	OET/AET	Main Results
Wiesinger et al., 2021	First surface aluminum Second surface silvered glass	OET: Zagora and Missour 24 months	The study evaluates wind speed and direction, as well as wind particle size distribution to derive single particle motion density (SPMD). Artificial facilities, using an open-loop wind tunnel, were specially constructed to carry out accelerated erosion experiments using various dust samples. The aim of these experiments was to explore the impact of particle hardness and shape on erosion results.
Sutter et al., 2018	Aluminum reflectors	OET: Arrefoud 3 years AET	The accelerated aging procedure includes a sequence of corrosion and mechanical erosion tests that reproduce the degradation mechanisms observed in outdoor conditions. Research shows a close correspondence between accelerated aging outcomes and real-world outdoor exposure, with minor mean errors of approximately 2 ppt in monochromatic specular reflectance and 1 ppt in solar hemispherical reflectance losses. These results effectively simulate the effects of three years of exposure in both 'desert' and 'coastal' environmental settings. This methodology provides an effective tool for material optimization, quality assurance and estimating the expected lifetime of newly developed coatings for CST reflectors, offering significant advances in material qualification protocols.
Wiesinger et al., 2020	Silvered glass	OET: Missour, Erfoud, Zagora, Tabernasn Chajnantor 12 months AET: Artificial sand erosion chamber	The author analyzes five prospective solar energy locations in order to assess the risk of sandstorms based on the character of the meteorological conditions, topography and soil properties that pertain to each location. The matrix of erosion potential is validated by analyzing the exposed solar reflectors at those locations where Zagora is shown to be the most subjected to sandstorms. The matrix has been shown to be useful and reliable in providing a rapid assessment of sandstorm risk at a location. The author presents a new technique of image processing analysis to determine defect density size distribution (DSDD) on the glass surface on the reflectors. Erosivity classes are assigned to outdoor locations based on the intensity of erosion. A laboratory erosion device is modified to mimic the DSDD outdoors and accurately simulate outdoor behavior; it provides a quick prediction of the performance of various material samples at outdoor exposure sites of differing erosivity classes (1, 2, 3).

continued on following page

Table 2. Continued

References	Device study	OET/AET	Main Results
Wiesinger et al., 2017	Silvered glass	OET: Zagora 1 year in	This research provides valuable insights on how height and orientation affect the erosion impact of solar reflectors. It shows that reflectors located higher, erode less, and that their orientation with respect to the prevailing wind direction influences erosion. In particular, short, high-speed wind gusts from the main wind direction caused the majority of erosion. She goes on to mention that erosion is also influenced by the conditions on site, and by the site's exposure parameters, including height, orientation, and angle of inclination.
Wette et al., 2016	Aluminum reflector	OET: Oujda, Missour, Erfoud, Zagora, Tan Tan 3 years and 7 months.	The research revealed that the type and extent of reflector degradation depends on the specific materials used and the geographical location. The study identifies five basic degradation mechanisms, namely aluminium corrosion, pitting corrosion, mechanical degradation of the surface coating and the build-up of non-removable surface deposits. Standard conventional tests often reproduce only one or two of these mechanisms and can introduce unrealistic degradation effects that have not been observed in the field. However, pioneering accelerated ageing tests using artificially soiled samples manage to reproduce three of the five mechanisms observed in real conditions, providing a more complete representation of overall degradation patterns. This research is key to improving test methodologies, product quality and site recommendations for aluminium reflector manufacturers".
Fernández-García et al., 2018	Silvered -glass reflectors	OET: Ouarzazate	The results showed that the interaction between the sand deposit and the paint layers did not lead to any degradation in the reflectance of the silver layer on the reflectors, as expected. However, it was crucial to assess the degradation of the paint layers as they protect the reflective silver layer. The study revealed that specific ambient conditions, such as extreme temperature changes and sand with specific chemical compositions, could lead to significant degradation of the paint layer.
Guerguer et al., 2017	Glass mirrors Polymeric mirrors	OET: Ouarzazate, Temara 3 years	Mirrors exposed by the sea showed deposits containing both salt and sand particles, while those located in desert environments accumulated mainly sand particles. The study also highlights the influence of environmental factors, such as wind speed and seasonal variations, on soiling rates. In addition, rain acts as a natural cleaning agent, particularly during periods of strong wind, helping to reduce soiling of these mirrors.

continued on following page

Table 2. Continued

References	Device study	OET/AET	Main Results
Raillani et al., 2019	Square mirror samples	OET: Oujda 1 Year	The aim is to establish a correlation between aerosol extinction and fouling rates. The study compares variations in AOD with measurements of square mirror reflectivity obtained from a meteorological station located at Université Mohamed First Oujda. The results indicate that larger aerosol particles play a dominant role in attenuating solar radiation, contributing in particular to its extinction. Interestingly, even during periods strongly influenced by AOD, such as the summer season, there are no reflectivity measurements. In addition, the volume distribution of suspended aerosol particles shows seasonal variations, with smaller fluctuations in winter and greater influence in summer. The sedimentation of larger aerosol particles significantly affects reflectivity during the summer season.
Abraim et al., 2022	linear Fresnel CSP	OET: semi-arid Morocco One year	The results of the study, using one year's weather and soiling data, show that optical losses due to soiling are much higher for CSP than for PV (1.21% compared with 0.24% daily average), resulting in annual energy losses of 17.76% for CSP and 1.95% for PV with monthly cleaning. Optimum cleaning every three weeks maximises efficiency for both types of power plant.

5. RESULTS AND DISCUSSION

This section presents a thorough investigation of climate impact and other factors influencing the solar mirror depreciation. It investigates the multiple degradation types considering material types, site specific conditions, and exposure times. The following paragraphs will describe these findings and detail the considerable impact of the Moroccan climate on the solar mirrors both in terms of erosion and corrosion patterns and losses to their optical properties. The findings offer significant implications for the solar energy industry in selection of mirror materials, maintenance procedures, and site selection to maximize the performance and longevity of concentrated solar power systems.

5.1. Influence of the Moroccan Climate on Degradation

Wind Speed and Direction

Studies consistently show that wind speed and direction have a significant impact on the degradation of solar mirrors. Low wind speeds in certain regions, such as Skoura, reduce the impact on mirror surfaces (Karim et al., 2014a; Karim et al., 2018). However, in coastal areas such as Temara, high wind speeds increase the risk of surface damage (Figure 6). Bouaddi et al. (2015) conducted a comprehensive analysis wherein the local linear trend emerged as the most suitable model for elucidating the prolonged evolution of cleanliness, particularly following the application of meticulous model selection criteria. Furthermore, an enhanced level of performance is observed when the model incorporates an optimal discount factor of 0.95. The direction of prevailing winds, particularly in desert regions, induces the greatest loss of reflectance; the environmental parameters prevalent in these regions pose potential challenges to the robustness of CSP mirrors, specifically, minute particles are predisposed to adhere to solar reflectors, resulting in the accumulation of a soiling layer, while larger particles exhibit a proclivity to inflict damage upon the mirrors through fracturing or scratching of the glass surface. The principal effect of wind is due to sand particles acting almost as abrasives. Based on the study of Pescheux et al. (2019), most wind speeds are from the north. As a result, samples facing the direction experience greater deterioration, and samples with a 90° inclination experience the highest reflectance loss values.

Figure 6. Impact observed on exposed solar mirrors: (a) Mirror exposed in a desert site, (b) Mirror exposed on the Oceanside site, (c) Mirror tested in a controlled environment at 90° and V=9.21 m/s.

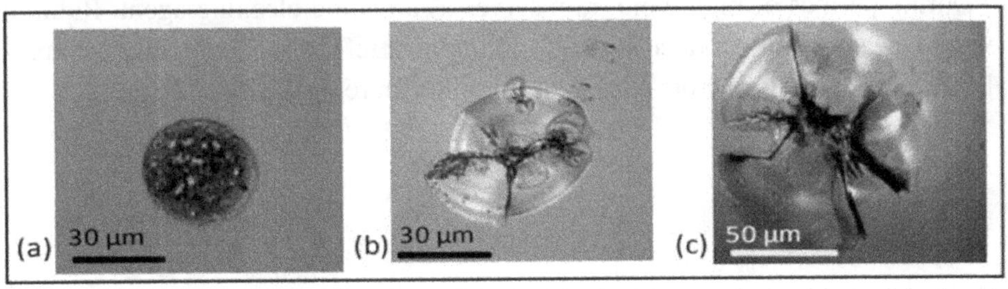

Temperature and Humidity

Several studies have pointed out that high temperatures and humidity in coastal areas, such as the Temara Seaside site, can contribute to the corrosion and chemical degradation of mirror coatings (Guerguer et al., 2021) (Figure7). In addition, exposing mirrors to extreme temperature changes can lead to degradation of the paint layers protecting the reflective silver coating. Merrouni et al. (2020) conducted a comprehensive assessment of (CSP) performance and efficiency, encompassing the evaluation of soiling levels in Morocco and Portugal. Their investigation revealed a notable discrepancy in soiling, with Morocco exhibiting a threefold higher level compared to Portugal. This disparity was attributed to prevailing weather conditions, particularly ambient temperature and (DNI). The average daily soiling rate was determined to be approximately 1.6% for Morocco and 0.06% for Portugal, underscoring the climatic influence on CSP efficiency in these distinct geographical locations.

Figure 7. Variations in color parameters (Δa and Δb) of monolithic mirrors over time of exposure to the desert site

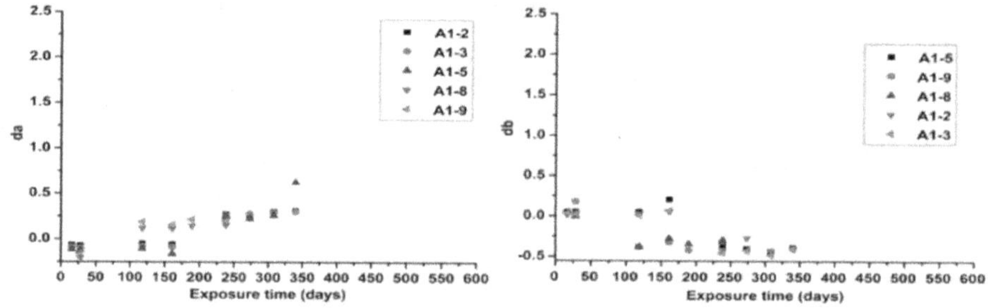

During periods of high wind, rain serves as a natural cleaning agent. Rainfall helps to clean mirrors, reduce soiling and improve reflectivity. However, in areas where rainfall is low, mirrors tend to accumulate more dirt and particles.

Figure 8. The relative specular reflectivity loss of the mirrors and the precipitation registered

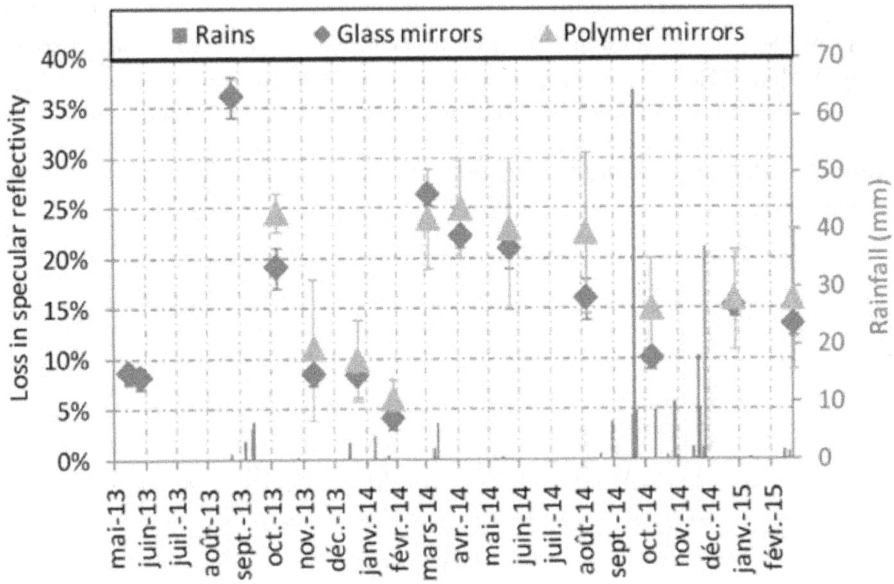

Figure 9. (a) Oceanside sand particles, (b) Desert sand particles, (c) Normalized sand particles

Salt content

The salt content of coastal environments such as Temara contains sea salts (NaCl and $MgCl_2$) which enhance corrosion (Lakhouil et al., 2021; Guerguer et al., 2016), particularly edge corrosion. This is a major problem in areas close to the coast. The execution of salt spray testing in either an accelerated or a natural environment constitutes a pivotal measure for ascertaining the durability of CSP reflectors. Nevertheless, due to elevated pollution levels and the presence of sulfuric acid (H_2SO_4), CSP power stations located in close proximity to industrial regions may encounter damage to their reflectors resulting from exposure to acid rain (Ennaceri et al., 2015).

Figure 10. XRD diagram of the dirt collected on the surface of the mirrors during the January period.

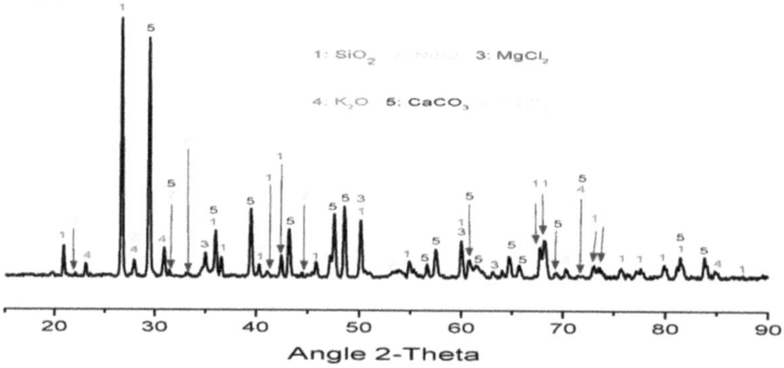

Air Aerosols and Atmospheric Conditions

Particles of air aerosols, especially the larger ones, play a dominant role in attenuating solar radiation. Seasonal variations in aerosol concentration have a direct impact on reflectivity, with a greater influence in summer when aerosols are more numerous. According to the results of Raillani et al. (2019), large particles contribute the most to the forcing of solar radiation, specifically being responsible for the extinction of this radiation. On the other hand, it is noted that throughout the monitored period, reflectivity measurement results are not presented, although still under the influence of aerosol optical thickness. This is particularly evident in the summer period when fine particles were significantly higher than in other seasons. Additionally, the volumetric distribution of suspended aerosol particles plays a significant role in reflectivity variation, with the amplitude of their variations decreasing during winter and increasing in summer. Moreover, the substantial influence of the sedimentation of large aerosol particles on reflectivity is observed during the summer period.

5.2. Forms of Degradation

A look at the forms of degradation of solar mirrors deployed in various Moroccan environments reveals a multifaceted spectrum of deterioration mechanisms. Erosion, an important mode of degradation, shows substantial variations according to geographical location. Desert sites, characterized by coarse, abrasive sand particles, contribute to more aggressive erosion (Figure9). Lakhouil et al. (2021) in their study, they suggested a solution to potential issues by recommending the application of paint systems behind reflecting layers. These paint systems serve as a barrier, preventing the diffusion of corrosive substances into the metal layer. The abrasive characteristics of sand particles in these environments lead to a higher incidence of abrasion defects on mirror surfaces. Coastal sites, on the other hand, are subject to a particular form of degradation, mainly due to the corrosive effects of sea salts such as sodium chloride and magnesium chloride. The result is rapid corrosion, with edge corrosion of particular concern, leading to the detachment of protective paint layers. Consequently, the degradation observed in coastal areas is significantly different from the abrasive wear observed in desert regions. Material considerations also have a significant impact on the degradation profile, with polymer mirrors being more susceptible to optical degradation than their glass counterparts.

Figure 11. Corrosion of solar mirror edges

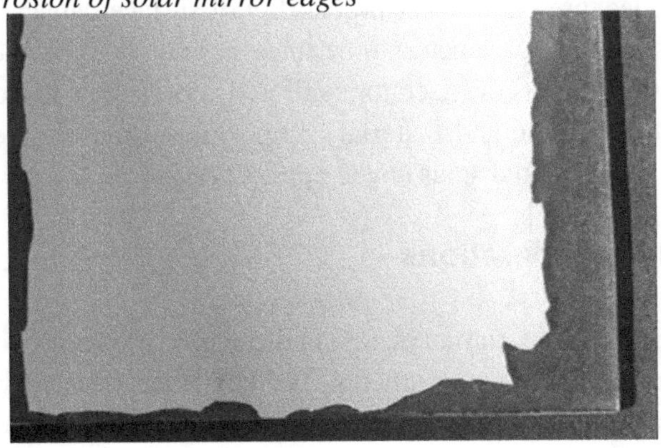

Figure 12. Scanning microscope zoom of the sand abrasion produced in Zagora, 10× magnification (a) and 100× magnification (b)

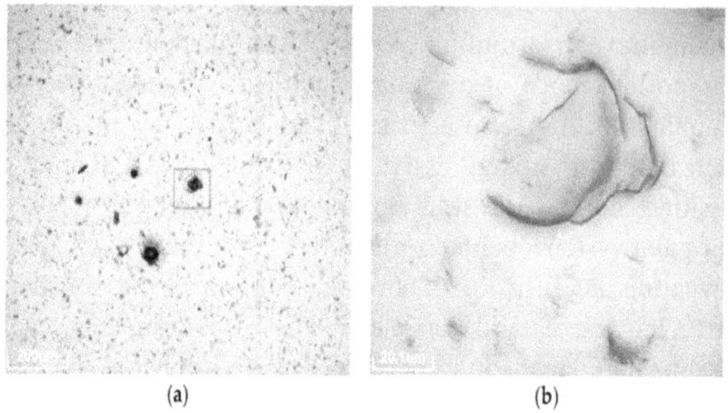

Environmental factors play a key role in the formation of degradation patterns. Variables such as humidity, precipitation, wind speed and direction exert a significant influence on the accumulation of dirt on mirror surfaces. High humidity, combined with high wind speeds, increases the likelihood of surface degradation, mainly by increasing the adhesion between soiled particles and mirrors, as observed at the Temara Seaside site. In contrast, desert sites, characterized by low wind speeds and large sand particles, present a lower probability of surface damage. The complex interplay of these climatic parameters underlines the complex, site-specific nature of degradation, and highlights the need for a tailored approach that takes into account both geographical location and material composition to ensure the durability of solar mirrors in concentrated solar power applications.

5.3. Material Considerations

Material considerations play a key role in the degradation behavior of solar mirrors deployed in Moroccan environments. The choice between polymer and glass mirrors appears to be a critical decision. Observations reveal distinct responses to the same climatic and environmental conditions. In coastal regions, glass mirrors tend to be more resistant to the corrosive impact of sea salts, retaining their integrity even as polymer mirrors undergo rapid corrosion. However, in desert regions characterized by abrasive sand particles, the situation is reversed. Polymer mirrors, with their rougher surfaces, are more susceptible to optical degradation (Neidlinger and Schissel, 1986). This intrinsic material-dependent response underlines the need to select mirrors adapted to specific environmental conditions. In addition, the type of coatings and protective layers, including innovative anti-soiling coatings, has a significant impact on reflector durability, with some coatings proving more resistant to environmental stresses. The choice of materials and coatings is not simply a matter of preference, but a determining factor in how effectively mirrors can withstand the challenges of the Moroccan climate in concentrated solar power applications.

5.4. Accelerated Ageing Tests

Accelerated ageing trials were conducted to replicate conditions found in real life accurately duplicating the external degradation styles visible in Morocco, and are critical in optimising and ensuring the quality of materials. Wiesinger et al. (2018) looked to pursue a study which sought to identify any prospective damage as a result of saltating particle salts employing the indigenous soil from the respective site; this study was conducted under the consideration that the site was expected to demonstrate eventual saltating particle salts. In this environment, calibrated to demonstrate accelerated erosion, particles of similar sizes to those sourced from Zagora and Missour were subjected to several test velocities.

Figure 13. 50x microscope images of glass reflector surface after a) 12-month exposure in Zagora and b) laboratory erosion simulation

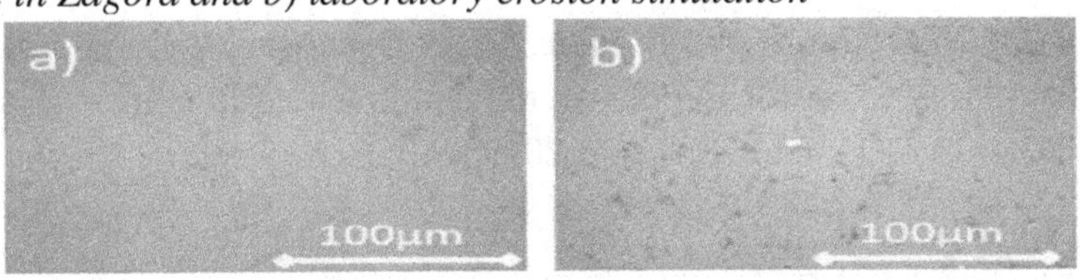

5. CONCLUSION

This review article has provided an examination of research, which focuses on the sustainability of solar mirrors throughout several areas of Morocco. The various studies collectively recognize the nuanced relationship between environmental factors, the materials used in the mirrors, and the aging processes that together influence solar mirror durability in multiple concentrated solar power (CSP) contexts. A finding recognized in the literature is the importance of geographic location and climate conditions on mirror deterioration. From the studies conducted to date, it is clear that there will be both advantages and disadvantages for mirrors located in Moroccan settings, such as in desert areas like Skoura or on the coast, as marine salts and wind particles have specific consequences for longevity. For example, low

ACKNOWLEDGMENT

We would like to express our sincere gratitude to Prof. Amol D. Vibhute of the Symbiosis Institute of Computer Studies and Research (SICSR), Symbiosis International (Deemed University), Pune, India, for his invaluable support and guidance throughout this study.

REFERENCES

Abraim, M., Salihi, M., El Alani, O., Hanrieder, N., Ghennioui, H., Ghennioui, A., El Ydrissi, M., & Azouzoute, A. (2022). Techno-economic assessment of soiling losses in CSP and PV solar power plants: A case study for the semi-arid climate of Morocco. *Energy Conversion and Management*, *270*, 116285. DOI: 10.1016/j.enconman.2022.116285

Azouzoute, A., Merrouni, A. A., Garoum, M., & Bennouna, E. G. (2019). Comparison of soiling effect of two different solar mirrors in mid-south of Morocco. *AIP Conference Proceedings*, *2126*, 190002. DOI: 10.1063/1.5117699

Azouzoute, A., Merrouni, A. A., Garoum, M., & Bennouna, E. G. (2020). Soiling loss of solar glass and mirror samples in the region with arid climate. *Energy Reports*, *6*, 693–698. DOI: 10.1016/j.egyr.2019.09.051

Bouaddi, S., & Ihlal, A. (2015). A study of the soiling patterns of CSP mirrors. 2015. *3rd International Renewable and Sustainable Energy Conference (IRSEC)*. Marrakech, Morocco. DOI: 10.1109/IRSEC.2015.7455126

Bouaddi, S., Ihlal, A., & Fernández-García, A. (2015). Soiled CSP solar reflectors modeling using dynamic linear models. *Solar Energy*, *122*, 847–863. DOI: 10.1016/j.solener.2015.09.044

Bouaddi, S., Ihlal, A., & Fernández-García, A. (2017). Comparative analysis of soiling of CSP mirror materials in arid zones. *Renewable Energy*, *101*, 437–449. DOI: 10.1016/j.renene.2016.08.067

Boukheir, S., Chbihi, A., Matal, A., Lakhouil, S., Naamane, S., Mouadine, M., & Bouhamidi, M. H. (2020). Physical and chemical reactivity of mechanically primed glass surfaces of CSP reflectors. *AIP Conference Proceedings*, *2303*, 150005. DOI: 10.1063/5.0028934

Boukheir, S., Lakhouil, S., Matal, A., Chbihi, A., Karim, M., & Naamane, S. (2023). Solar mirrors coatings contribution under arid climate and different exposure conditions. *AIP Conference Proceedings*, *2815*, 120001. DOI: 10.1063/5.0165304

Chbihi, A., Boukheir, S., Naamane, S., Mouadine, M., & Bouhamidi, M. H. (2022). Influence of soiling on CSP mirrors in a mountain climate in Morroco. *AIP Conference Proceedings*, *2445*, 080002. DOI: 10.1063/5.0085640

Conceição, R., Merrouni, A. A., Lopes, D., Alae, A., Silva, H. G., Bennouna, E. G., Collares-Pereira, M., & Ghennioui, A. (2019). A comparative study of soiling on solar mirrors in Portugal and Morocco: Preliminary results for the dry season. *AIP Conference Proceedings*, *2126*, 220001. DOI: 10.1063/1.5117760

Dahlioui, D., Wette, J., Fernández-García, A., Bouzekri, H., & Azpitarte, I. (2022). Performance assessment of the anti-soiling coating on solar mirrors soiling in the arid climate of Ouarzazate-Morocco. *Solar Energy*, *241*, 13–23. DOI: 10.1016/j.solener.2022.05.063

Driouech, F., Déqué, M., & Sánchez-Gómez, E. (2010). Weather regimes—Moroccan precipitation link in a regional climate change simulation. *Global and Planetary Change*, *72*(1-2), 1–10. DOI: 10.1016/j.gloplacha.2010.03.004

Edfouf, Z., Guerguer, M., Naamane, S., Karim, M., Zari, N., Grirate, H., Raccurt, O., Delord, C., & Couturier, R. (2014). Research on component materials used for Moroccan CSP power plants: Mirrors ageing and thermal storage. *International Renewable and Sustainable Energy Conference (IRSEC)*. Ouarzazate, Morocco. DOI: 10.1109/IRSEC.2014.7059790

Edfouf, Z., Guerguer, M., & Raccurt, O. (2015). Glass and polymeric mirrors ageing under different Moroccan weathers, an application for CSP power plants. *Energy Procedia*, *69*, 1508–1518. DOI: 10.1016/j.egypro.2015.03.101

Ennaceri, H., Benyoussef, A., Ennaoui, A., & Khaldoun, A. (2015). On the analysis of suitable ageing tests of first-surface CSP mirrors in Moroccan outdoor conditions. *3rd International Renewable and Sustainable Energy Conference (IRSEC)*. Marrakech, Morocco. DOI: 10.1109/IRSEC.2015.7455062

Fernández-García, A., Juaidi, A., Sutter, F., Martínez-Arcos, L., & Manzano-Agugliaro, F. (2018). Solar Reflector Materials Degradation Due to the Sand Deposited on the Backside Protective Paints. *Energies*, *11*(4), 808. DOI: 10.3390/en11040808

Guerguer, M., Edfouf, Z., & Raccurt, O. (2016). Comparison of two methods used to evaluate the aggressivity of a Moroccan marine site on solar mirrors. *International Renewable and Sustainable Energy Conference (IRSEC)*. Marrakech, Morocco. https://doi.org/DOI: 10.1109/IRSEC.2016.7984014

Guerguer, M., Karim, M., Naamane, S., Edfouf, Z., & Raccurt, O. (2018). Degradation of solar glass mirrors paints at marine outdoor site. *AIP Conference Proceedings*, *2033*, 230004. DOI: 10.1063/1.5067232

Guerguer, M., Karim, M., Naamane, S., Edfouf, Z., Raccurt, O., & Delord, C. (2017). Soiling deposition on solar mirrors exposed in Morocco. *AIP Conference Proceedings*, *1850*, 130005. DOI: 10.1063/1.4984499

Guerguer, M., Naamane, S., Edfouf, Z., Raccurt, O., & Bouaouine, H. (2021). Chemical degradation and color changes of paint protective coatings used in solar glass mirrors. *Coatings*, *11*(4), 476. DOI: 10.3390/coatings11040476

Guerguer, M., Naamane, S., Karim, M., & Bouaouine, H. (2019). Outdoor exposure testing of silvered-glass reflectors under marine environment. *AIP Conference Proceedings*, *2126*, 160003. DOI: 10.1063/1.5117666

Karim, M., Delord, C., Raccurt, O., & Naamane, S. (2018). Exposure conditions effect on soiling of solar glass mirrors. *AIP Conference Proceedings*, *230006*, 230006. Advance online publication. DOI: 10.1063/1.5067234

Karim, M., Naamane, S., & Delord, C. (2016). Evaluation of the behaviour of different solar mirrors against soiling phenomenon. *International Renewable and Sustainable Energy Conference (IRSEC)*. Marrakech, Morocco. DOI: 10.1109/IRSEC.2016.7984008

Karim, M., Naamane, S., Delord, C., & Bennouna, A. (2014). Surface wear damage of glass solar mirrors in Moroccan desert environment. *International Renewable and Sustainable Energy Conference (IRSEC)*. Ouarzazate, Morocco. 246-251. DOI: 10.1109/IRSEC.2014.7059770

Karim, M., Naamane, S., Delord, C., & Bennouna, A. (2015a). Study of the surface damage of glass reflectors used in concentrated solar power plants. *Energy Procedia*, *69*, 106–115. DOI: 10.1016/j.egypro.2015.03.013

Karim, M., Naamane, S., Delord, C., & Bennouna, A. (2015b). Laboratory simulation of the surface erosion of solar glass mirrors. *Solar Energy*, *118*, 520–532. DOI: 10.1016/j.solener.2015.05.044

Karim, M., Naamane, S., Hassani, I. E. A. E. L., Delord, C., Belcadi, S., Tochon, P., & Bennouna, A. (2014). Towards the prediction of CSP mirrors wear: Methodology of analysis of influencing parameters on the mirrors surface degradation: Application in two different sites in Morocco. *Solar Energy, 108*, 41–50. DOI: 10.1016/j.solener.2014.06.036

Lakhouil, S., Boukheir, S., Bouaouine, H., Baron, E. L., Disdier, A., Vuillerme, V., & Naamane, S. (2022). Erosion Effect in the Degradation of Coated and Uncoated Glass Solar Mirrors Highlighted by Coupled Accelerated Erosion/Aging Tests. *Coatings, 12*(8), 1087. DOI: 10.3390/coatings12081087

Lakhouil, S., Boukheir, S., Chbihi, A., & Naamane, S. (2021). Outdoor exposure approach in a harsh environment toward the study of corrosion aspects of primary solar reflectors. *Solar Energy, 230*, 1070–1084. DOI: 10.1016/j.solener.2021.11.013

Lakhouil, S., Boukheir, S., Naamane, S., & Bouaouine, H. (2023). Mechanical resistance assessment by real outdoor study of an anti-wear coating on glass solar mirror. *AIP Conference Proceedings, 2815*, 020007. DOI: 10.1063/5.0149405

Matal, A., Karim, M., Bannouri, O., Naamane, S., & Bouaouine, H. (2017). Setting Parameters of an Erosion Test Bench for Solar Reflectors. *International Renewable and Sustainable Energy Conference (IRSEC)*. Tangier, Morocco. DOI: 10.1109/IRSEC.2017.8477290

Matal, A., Karim, M., & Naamane, S. (2020). Development and calibration of an experimental test bench simulating solar reflectors erosion. *Solar Energy, 201*, 724–731. DOI: 10.1016/j.solener.2020.03.028

Merrouni, A. A., Conceição, R., Mouaky, A., Silva, H. G., & Ghennioui, A. (2020). CSP performance assessment including soiling measurements for Morocco and Portugal reveals similar electricity yield and cost in both countries. *Renewable Energy, 162*, 1777–1792. DOI: 10.1016/j.renene.2020.10.014

Merrouni, A. A., Mezrhab, A., Ghennioui, A., & Naimi, Z. (2017). Measurement, comparison and monitoring of solar mirror's specular reflectivity using two different Reflectometers. *Energy Procedia, 119*, 433–445. DOI: 10.1016/j.egypro.2017.07.045

Merrouni, A. A., Wolfertstetter, F., Mezrhab, A., Wilbert, S., & Pitz-Paal, R. (2015). Investigation of Soiling Effect on Different Solar Mirror Materials under Moroccan Climate. *Energy Procedia, 69*, 1948–1957. DOI: 10.1016/j.egypro.2015.03.194

Naamane, S., Chbihi, A., Boukheir, S., Mouadine, M., Tahri, K., Bouaddi, S., & Bouhamidi, M. H. (2020). Towards the choice of concentrated solar power and photovoltaic materials for Moroccan conditions. *AIP Conference Proceedings*, *150013*, 150013. Advance online publication. DOI: 10.1063/5.0029152

Neidlinger, H. H., & Schissel, P. (1986). Polymer glazings for silver mirrors. *Solar Energy Materials*, *14*(3-5), 327–339. DOI: 10.1016/0165-1633(86)90056-0

Pescheux, A. C., Le Baron, E., & Raccurt, O. (2019). Characterization of different Moroccan sands to explain their potential negative impacts on CSP solar mirrors. *Solar Energy*, *194*, 959–968. DOI: 10.1016/j.solener.2019.11.020

Raillani, B., Ouali, H. A. L., Amraqui, S., Moussaoui, M. A., Jeyar, M., & Mezrhab, A. (2019). The Impact of Aerosol Optical Depth (AOD) On Decreasing the Optical Efficiency of CSP Solar Power Plants-A case study: *Eastern Morocco. International Conference on Intelligent Systems and Advanced Computing Sciences (ISACS)*. Taza, Morocco. DOI: 10.1109/ISACS48493.2019.9068909

Sutter, F., Wette, J., Wiesinger, F., Fernández-García, A., Ziegler, S., & Dasbach, R. (2018). Lifetime prediction of aluminum solar mirrors by correlating accelerated aging and outdoor exposure experiments. *Solar Energy*, *174*, 149–163. DOI: 10.1016/j.solener.2018.09.006

Wette, J., Sutter, F., Fernández-García, A., Ziegler, S., & Dasbach, R. (2016). Comparison of Degradation on Aluminum Reflectors for Solar Collectors due to Outdoor Exposure and Accelerated Aging. *Energies*, *9*(11), 916. DOI: 10.3390/en9110916

Wiesinger, F., Sutter, F., Fernández-García, A., Reinhold, J., & Pitz-Paal, R. (2016). Sand erosion on solar reflectors: Accelerated simulation and comparison with field data. *Solar Energy Materials and Solar Cells*, *145*, 303–313. DOI: 10.1016/j.solmat.2015.10.036

Wiesinger, F., Sutter, F., Fernández-García, A., Wette, J., & Hanrieder, N. (2021). Sandstorm erosion on solar reflectors: A field study on height and orientation dependence. *Energy*, *217*, 119351. DOI: 10.1016/j.energy.2020.119351

Wiesinger, F., Sutter, F., Fernández-García, A., Wette, J., Wolfertstetter, F., Hanrieder, N., Schmücker, M., & Pitz-Paal, R. (2020). Sandstorm erosion on solar reflectors: Highly realistic modeling of artificial aging tests based on advanced site assessment. *Applied Energy*, *268*, 114925. DOI: 10.1016/j.apenergy.2020.114925

Wiesinger, F., Sutter, F., Wolfertstetter, F., Hanrieder, N., Fernández-García, A., Pitz-Paal, R., & Schmücker, M. (2018). Assessment of the erosion risk of sandstorms on solar energy technology at two sites in Morocco. *Solar Energy*, *162*, 217–228. DOI: 10.1016/j.solener.2018.01.004

Wiesinger, F., Sutter, F., Wolfertstetter, F., Hanrieder, N., Wette, J., Fernández-García, A., & Pitz-Paal, R. (2017). Sandstorm erosion simulation on solar mirrors and comparison with field data. *AIP Conference Proceedings*, *1850*, 130014. DOI: 10.1063/1.4984508

Compilation of References

Abebe, H., & Tu, Y. (2024). Impact of salt and alkali stress on forage biomass yield, nutritive value, and animal growth performance: A comprehensive review. *Grasses*, *3*(4), 355–368. DOI: 10.3390/grasses3040026

Abraim, M., Salihi, M., El Alani, O., Hanrieder, N., Ghennioui, H., Ghennioui, A., El Ydrissi, M., & Azouzoute, A. (2022). Techno-economic assessment of soiling losses in CSP and PV solar power plants: A case study for the semi-arid climate of Morocco. *Energy Conversion and Management*, *270*, 116285. DOI: 10.1016/j.enconman.2022.116285

Adelakun, N. O. (2024, Jan 1). Future Trends in Artificial Intelligence for Energy Management. RELX Group (Netherlands). DOI: 10.2139/ssrn.4703931

Adnan, M., Xiao, B., Xiao, P., Zhao, P., & Bibi, S. (2022). Heavy Metal, Waste, COVID-19, and Rapid Industrialization in This Modern Era—Fit for Sustainable Future. *Sustainability (Basel)*, *14*(8), 4746. DOI: 10.3390/su14084746

AghaKouchak, A., Huning, L. S., Chiang, F., Mazdiyasni, O., & Vahedifard, F. (2014). Global warming and changes in risk of concurrent climate extremes: Insights from the 2014 California drought. *Geophysical Research Letters*, *41*(24), 8847–8852. DOI: 10.1002/2014GL062308

Ahmed, W., Sharma, G., Singh, D., & Singh, R. (2024). Ecological impact of heavy metal pollution on *Limnodrilus cervix* -abundance, bioaccumulation, and environmental dynamics in a perennial pond of Jammu, India. *Process Safety and Environmental Protection*, *188*, 595–607. DOI: 10.1016/j.psep.2024.05.147

Akachoud, O., Bouamama, H., Facon, N., Laruelle, F., Zoubi, B., Benkebboura, A., Ghoulam, C., Qaddoury, A., & Lounès-Hadj Sahraoui, A. (2022). Mycorrhizal Inoculation Improves the Quality and Productivity of Essential Oil Distilled from Three Aromatic and Medicinal Plants: Thymus satureioides, Thymus pallidus, and *Lavandula dentata*. *Agronomy (Basel)*, *12*(9), 2223. DOI: 10.3390/agronomy12092223

Albahri, G., Alyamani, A. A., Badran, A., Hijazi, A., Nasser, M., Maresca, M., & Baydoun, E. (2023). Enhancing Essential Grains Yield for Sustainable Food Security and Bio-Safe Agriculture through Latest Innovative Approaches. *Agronomy (Basel)*, *13*(7), 1709. DOI: 10.3390/agronomy13071709

Alemany, S., Beltran, J., Perez, A., & Ganzfried, S. (2019). Predicting hurricane trajectories using a recurrent neural network. Proceedings of the AAAI conference on artificial intelligence, Ampatzidis, Y., Partel, V., Meyering, B., & Albrecht, U. (2019). Citrus rootstock evaluation utilizing UAV-based remote sensing and artificial intelligence. *Computers and Electronics in Agriculture*, *164*, 104900.

Ali, H., Khan, E., & Sajad, M. A. (2022). Phytotoxicity of heavy metals: Impacts on plants and possible remediation strategies. *Environmental Science and Pollution Research International*, *29*(3), 1803–1817.

Ali, J., Ali, F., Ahmad, I., Rafique, M., Munis, M. F. H., Hassan, S. W., Sultan, T., Iftikhar, M., & Chaudhary, H. J. (2021). Mechanistic elucidation of germination potential and growth of *Sesbania sesban* seedlings with *Bacillus anthracis* PM21 under heavy metals stress: An *in vitro* study. *Ecotoxicology and Environmental Safety*, *208*, 111769. DOI: 10.1016/j.ecoenv.2020.111769 PMID: 33396087

Ali, S. S., & Choi, B. J. (2020, Jun 22). State-of-the-Art Artificial Intelligence Techniques for Distributed Smart Grids: A Review. *Electronics (Basel)*, *9*(6), 1030–1030. DOI: 10.3390/electronics9061030

Ali, S., Murtaza, G., Hedvicakova, M., Jiang, J., & Naeem, M. (2022). Intellectual capital and financial performance: A comparative study. *Frontiers in Psychology*, *13*, 967820. DOI: 10.3389/fpsyg.2022.967820 PMID: 36081720

AlMarri, M., Al-Ali, M., Alzarooni, M., AlTeneiji, A., Al-Ali, K., & Bahroun, Z. (2025). Enterprise Resource Planning Systems for Health, Safety, and Environment Management: Analyzing Critical Success Factors. *Sustainability (Basel)*, *17*(7), 2947. DOI: 10.3390/su17072947

Ameh, B. (2024). Technology-integrated sustainable supply chains: Balancing domestic policy goals, global stability, and economic growth. *Int J Sci Res Arch*, *13*(2), 1811–1828. DOI: 10.30574/ijsra.2024.13.2.2369

Andeobu, L., Wibowo, S., & Grandhi, S. (2021). An assessment of e-waste generation and environmental management of selected countries in Africa, Europe and North America: A systematic review. *The Science of the Total Environment*, *792*, 148078. DOI: 10.1016/j.scitotenv.2021.148078 PMID: 34147806

Andrade, J. M., Faustino, C., Garcia, C., Ladeiras, D., Reis, C. P., & Rijo, P. (2018). *Rosmarinus officinalis* L.: An update review of its phytochemistry and biological activity. *Future Science OA*, *4*(4), FSO283. Advance online publication. DOI: 10.4155/fsoa-2017-0124 PMID: 29682318

Angon, B. P., Islam, M. S., Kc, S., Das, A., Anjum, N., Poudel, A., & Suchi, S. A. (2024). Sources, effects and present perspectives of heavy metals contamination: Soil, plants and human food chain. *Heliyon*, *10*(7), e28357. DOI: 10.1016/j.heliyon.2024.e28357 PMID: 38590838

Anser, M. K., Naeem, M., Ali, S., Huizhen, W., & Farooq, S. (2024). From knowledge to profit: Business reputation as a mediator in the impact of green intellectual capital on business performance. *Journal of Intellectual Capital*, *25*(5/6), 1133–1153. DOI: 10.1108/JIC-01-2024-0025

Arif, Y., Singh, P., Siddiqui, H., Bajguz, A., & Hayat, S. (2020). Salinity induced physiological and biochemical changes in plants: An omic approach towards salt stress tolerance. *Plant Physiology and Biochemistry*, *156*, 64–77. DOI: 10.1016/j.plaphy.2020.08.042 PMID: 32906023

Arnbjerg-Nielsen, K., Willems, P., Olsson, J., Beecham, S., Pathirana, A., Bülow Gregersen, I., Madsen, H., & Nguyen, V.-T.-V. (2013). Impacts of Climate Change on Rainfall Extremes and Urban Drainage Systems: A Review. *Water Science and Technology*, *68*(1), 16–28. DOI: 10.2166/wst.2013.251 PMID: 23823535

Arnon, D. I. (1949). Copper Enzymes in Isolated Chloroplasts. Polyphenoloxidase in Beta Vulgaris. *Plant Physiology*, *24*(1), 1–15. DOI: 10.1104/pp.24.1.1 PMID: 16654194

Arocena, P., Orcos, R., & Zouaghi, F. (2021). The impact of ISO 14001 on firm environmental and economic performance: The moderating role of size and environmental awareness. *Business Strategy and the Environment*, *30*(2), 955–967. DOI: 10.1002/bse.2663

Arora, N. K. (2019). Impact of climate change on agriculture production and its sustainable solutions. *Environmental Sustainability*, *2*(2), 95–96. DOI: 10.1007/s42398-019-00078-w

Arora, N. K., Fatima, T., Mishra, I., Verma, M., Mishra, J., & Mishra, V. (2018). Environmental sustainability: Challenges and viable solutions. *Environmental Sustainability*, *1*(4), 309–340. DOI: 10.1007/s42398-018-00038-w

Aslanidis, P. S. C., & Golia, E. E. (2022). Urban Sustainability at Risk Due to Soil Pollution by Heavy Metals—Case Study: Volos, Greece. *Land (Basel)*, *11*(7), 1016. DOI: 10.3390/land11071016

Atrous, G., Piqueras, A., Diaz-Vivancos, P., Hernández-Cánovas, A., Hamed, K. B., Hernández, J. A., & Barba-Espín, G. (2025). In Vitro-Selected Clones of the Halophyte Arthrocaulon macrostachyum Display Enhanced Salinity Stress Tolerance. *Plants*, *14*(8), 1164. DOI: 10.3390/plants14081164 PMID: 40284052

Aulicky, R., Vendl, T., Douda, O., Pavela, R., & Stejskal, V. (2025). Insecticidal activity of rosemary essential oil against primary and secondary storage beetles in the presence and absence of grain. *Journal of Stored Products Research*, *111*, 102498. DOI: 10.1016/j.jspr.2024.102498

Awewomom, J., Dzeble, F., Takyi, Y. D., Ashie, W. B., Ettey, E. N. Y. O., Afua, P. E., Sackey, L. N. A., Opoku, F., & Akoto, O. (2024). Addressing global environmental pollution using environmental control techniques: A focus on environmental policy and preventive environmental management. *Discover Environment*, *2*(1), 8. DOI: 10.1007/s44274-024-00033-5

Aziz, E., Batool, R., Akhtar, W., Shahzad, T., Malik, A., Shah, M. A., Iqbal, S., Rauf, A., Zengin, G., Bouyahya, A., Rebezov, M., Dutta, N., Khan, M. U., Khayrullin, M., Babaeva, M., Goncharov, A., Shariati, M. A., & Thiruvengadam, M. (2021). Rosemary species: A review of phytochemicals, bioactivities and industrial applications. *South African Journal of Botany*, *151*, 3–18. DOI: 10.1016/j.sajb.2021.09.026

Azouzoute, A., Merrouni, A. A., Garoum, M., & Bennouna, E. G. (2019). Comparison of soiling effect of two different solar mirrors in mid-south of Morocco. *AIP Conference Proceedings*, *2126*, 190002. DOI: 10.1063/1.5117699

Azouzoute, A., Merrouni, A. A., Garoum, M., & Bennouna, E. G. (2020). Soiling loss of solar glass and mirror samples in the region with arid climate. *Energy Reports*, *6*, 693–698. DOI: 10.1016/j.egyr.2019.09.051

Bai, X., Bol, R., Chen, H., Cui, Q., Qiu, T., Zhao, S., & Fang, L. (2024). A meta-analysis on crop growth and heavy metals accumulation with PGPB inoculation in contaminated soils. *Journal of Hazardous Materials*, *471*, 134370. DOI: 10.1016/j.jhazmat.2024.134370 PMID: 38688214

Balogun, A.-L., Marks, D., Sharma, R., Shekhar, H., Balmes, C., Maheng, D., Arshad, A., & Salehi, P. (2019). Assessing the potentials of digitalization as a tool for climate change adaptation and sustainable development in urban centres. *Sustainable Cities and Society*, *58*, 1–12.

Barbosa, G. S., Drach, P. R., & Corbella, O. D. (2014). A conceptual review of the terms sustainable development and sustainability. *Journal of Social Sciences*, *3*(2), 1.

Barnes, E A., Hurrell, J W., Ebert-Uphoff, I., Anderson, C W., & Anderson, D. (2019, Nov 19). Viewing Forced Climate Patterns Through an AI Lens. American Geophysical Union, 46(22), 13389-13398. DOI: 10.1029/2019GL084944

Barney, J. B., Ketchen, D. J.Jr, & Wright, M. (2011). The future of resource-based theory: Revitalization or decline? *Journal of Management*, *37*(5), 1299–1315. DOI: 10.1177/0149206310391805

Barrs, H. D. (1996). Root pressure and leaf water potential. *Science*, *152*(3726), 1266–1268. DOI: 10.1126/science.152.3726.1266 PMID: 17769543

Batarseh, F. A., Donti, P. L., Drgoňa, J., Fletcher, K. M., Hanania, P., Hatton, M., Keshav, S., Knowles, B., Kotsch, R., McGinnis, S., Mitra, P., Philp, A., Spohrer, J., Stein, F., Tare, M., Volkov, S., & Wen, G. (2022, Jan 1). *Proceedings of AAAI 2022 Fall Symposium: The Role of AI in Responding to Climate Challenges*. Cornell University. DOI: 10.48550/arxiv.2212.13631

Bates, B. C.(2008). *Climate Change and Water*. Intergovernmental Panel on Climate Change.

Bates, L. S., Waldren, R. P., & Teare, I. D. (1973). Rapid determination of free proline for water studies. *Plant and Soil*, *39*(1), 205–208. DOI: 10.1007/BF00018060

Bazihizina, N., Papenbrock, J., Aronsson, H., Ben Hamed, K., Elmaz, Ö., Dafku, Z., Custódio, L., Rodrigues, M. J., Atzori, G., & Negacz, K. (2024). The Sustainable Use of Halophytes in Salt-Affected Land: State-of-the-Art and Next Steps in a Saltier World. *Plants*, *13*(16), 2322. DOI: 10.3390/plants13162322 PMID: 39204758

Bhatia, S. (2017). Artificial intelligence for better climate governance. *Journal of Artificial Intelligence Research & Advances*, *4*(3), 37–43.

Biesbroek, R., Badloe, S., & Athanasiadis, I. N. (2020). Machine learning for research on climate change adaptation policy integration: An exploratory UK case study. *Regional Environmental Change*, *20*(3), 85. DOI: 10.1007/s10113-020-01677-8

Bolton, T., & Zanna, L. (2019). Applications of deep learning to ocean data inference and subgrid parameterization. *Journal of Advances in Modeling Earth Systems*, *11*(1), 376–399. DOI: 10.1029/2018MS001472

Booth, M. (2018). Climate change and the neglected tropical diseases. *Advances in Parasitology*, *100*, 39–126. DOI: 10.1016/bs.apar.2018.02.001 PMID: 29753342

Boros-Lajszner, E., Wyszkowska, J., Borowik, A., & Kucharski, J. (2021). The Response of the Soil Microbiome to Contamination with Cadmium, Cobalt and Nickel in Soil Sown with Brassica napus. *Minerals (Basel), 11*(5), 498. DOI: 10.3390/min11050498

Bouaddi, S., & Ihlal, A. (2015). A study of the soiling patterns of CSP mirrors. 2015. *3rd International Renewable and Sustainable Energy Conference (IRSEC)*. Marrakech, Morocco. DOI: 10.1109/IRSEC.2015.7455126

Bouaddi, S., Ihlal, A., & Fernández-García, A. (2015). Soiled CSP solar reflectors modeling using dynamic linear models. *Solar Energy, 122*, 847–863. DOI: 10.1016/j.solener.2015.09.044

Bouaddi, S., Ihlal, A., & Fernández-García, A. (2017). Comparative analysis of soiling of CSP mirror materials in arid zones. *Renewable Energy, 101*, 437–449. DOI: 10.1016/j.renene.2016.08.067

Boukheir, S., Chbihi, A., Matal, A., Lakhouil, S., Naamane, S., Mouadine, M., & Bouhamidi, M. H. (2020). Physical and chemical reactivity of mechanically primed glass surfaces of CSP reflectors. *AIP Conference Proceedings, 2303*, 150005. DOI: 10.1063/5.0028934

Boukheir, S., Lakhouil, S., Matal, A., Chbihi, A., Karim, M., & Naamane, S. (2023). Solar mirrors coatings contribution under arid climate and different exposure conditions. *AIP Conference Proceedings, 2815*, 120001. DOI: 10.1063/5.0165304

Boussora, F., Triki, T., Bennani, L., Bagues, M., Ben Ali, S., Ferchichi, A., Ngaz, K., & Guasmi, F. (2024). Mineral accumulation, relative water content and gas exchange are the main physiological regulating mechanisms to cope with salt stress in barley. *Scientific Reports, 14*(1), 14931. DOI: 10.1038/s41598-024-65967-5 PMID: 38942909

Breiman, L. (2001). Random Forests. *Machine Learning, 45*(1), 5–32. DOI: 10.1023/A:1010933404324

Buckland, C., Bailey, R., & Thomas, D. (2019). Using artificial neural networks to predict future dryland responses to human and climate disturbances. *Scientific Reports, 9*(1), 3855. DOI: 10.1038/s41598-019-40429-5 PMID: 30846833

Bueno, M., & Cordovilla, M. P. (2021). Plant Growth Regulators Application Enhance Tolerance to Salinity and Benefit the Halophyte *Plantago coronopus* in Saline Agriculture. *Plants, 10*(9), 1872. DOI: 10.3390/plants10091872 PMID: 34579404

Buhmann, A. K., Waller, U., Wecker, B., & Papenbrock, J. (2015). Optimization of culturing conditions and selection of species for the use of halophytes as biofilter for nutrient-rich saline water. *Agricultural Water Management*, *149*, 102–114. DOI: 10.1016/j.agwat.2014.11.001

Callaghan, M., Schleussner, C.-F., Nath, S., Lejeune, Q., Knutson, T. R., Reichstein, M., Hansen, G., Theokritoff, E., Andrijevic, M., Brecha, R. J., Hegarty, M., Jones, C., Lee, K., Lucas, A., van Maanen, N., Menke, I., Pfleiderer, P., Yesil, B., & Minx, J. C. (2021). Machine-learning-based evidence and attribution mapping of 100,000 climate impact studies. *Nature Climate Change*, *11*(11), 966–972. DOI: 10.1038/s41558-021-01168-6

Caudal, F., Tapissier-Bontemps, N., & Edrada-Ebel, R. A. (2022). Impact of Co-Culture on the Metabolism of Marine Microorganisms. *Marine Drugs*, *20*(2), 153. DOI: 10.3390/md20020153 PMID: 35200682

Cerdá-Gandia, R., Agüero, Á., Arrieta, M. P., & Fenollar, O. (2024). Effect of different porous size of porous inorganic fillers on the encapsulation of rosemary essential oil for PLA-based active packaging. *Polymers*, *16*(18), 2632. DOI: 10.3390/polym16182632 PMID: 39339096

Chakraborty, S. (2024, May 22). Towards A Comprehensive Assessment of AI's Environmental Impact. Cornell University. https://doi.org//arxiv.2405.14004DOI: 10.48550

Challinor, A. J., Adger, W. N., Benton, T. G., Conway, D., Joshi, M., & Frame, D. (2018). Transmission of climate risks across sectors and borders. *Philosophical Transactions. Series A, Mathematical, Physical, and Engineering Sciences*, *376*(2121), 20170301. DOI: 10.1098/rsta.2017.0301 PMID: 29712795

Challoumis, C. (2024, October). Building a sustainable economy-how ai can optimize resource allocation. In *XVI International Scientific Conference* (pp. 190-224).

Chang, H. (2016). Exploring the Relationship between Land Use Change and Flood Damage through an Agent-Based Model. *The Science of the Total Environment*, *550*, 711–721.

Chatterjee, S., Chaudhuri, R., Kamble, S., Gupta, S., & Sivarajah, U. (2023). Adoption of artificial intelligence and cutting-edge technologies for production system sustainability: A moderator-mediation analysis. *Information Systems Frontiers*, *25*(5), 1779–1794. DOI: 10.1007/s10796-022-10317-x

Chbihi, A., Boukheir, S., Naamane, S., Mouadine, M., & Bouhamidi, M. H. (2022). Influence of soiling on CSP mirrors in a mountain climate in Morroco. *AIP Conference Proceedings, 2445*, 080002. DOI: 10.1063/5.0085640

Chele, K. H., Steenkamp, P., Piater, L. A., Dubery, I. A., Huyser, J., & Tugizimana, F. (2021). A Global Metabolic Map Defines the Effects of a Si-Based Biostimulant on Tomato Plants under Normal and Saline Conditions. *Metabolites, 11*(12), 820. DOI: 10.3390/metabo11120820 PMID: 34940578

Chen, L., Chen, Z., Zhang, Y., Liu, Y., Osman, A I., Farghali, M., Hua, J., Al-Fatesh, A S., Ihara, I., Rooney, D W., & Yap, P. (2023, Jun 13). Artificial intelligence-based solutions for climate change: a review. Springer Science+Business Media, 21(5), 2525-2557. DOI: 10.1007/s10311-023-01617-y

Chen, G., Amoanimaa-Dede, H., Zeng, F., Deng, F., Xu, S., & Chen, Z. H. (2022). Stomatal regulation and adaptation to salinity in glycophytes and halophytes. *Advances in Botanical Research, 103*, 1–42. DOI: 10.1016/bs.abr.2022.02.008

Chen, R., Su, Y., & Tran, L. (2024). Small Farmer's Perceptions of Climate Change and Adoption of Climate-Smart Practices: Evidence from Missouri, USA. *Sustainability (Basel), 16*(21), 9525. DOI: 10.3390/su16219525

Chen, T., & Guestrin, C. (2016). XGBoost : A Scalable Tree Boosting System. *Proceedings of the 22nd ACM SIGKDD International Conference on Knowledge Discovery and Data Mining*. 785-794. DOI: 10.1145/2939672.2939785

Chernozhukov, V., Demirer, M., Duflo, E., & Fernandez-Val, I. (2018). *Generic machine learning inference on heterogeneous treatment effects in randomized experiments, with an application to immunization in India.*

Chopra, S. S., Senadheera, S. S., Dissanayake, P. D., Withana, P. A., Chib, R., Rhee, J. H., & Ok, Y. S. (2024). Navigating the Challenges of Environmental, Social, and Governance (ESG) Reporting: The Path to Broader Sustainable Development. *Sustainability (Basel), 2024*(16), 606. DOI: 10.3390/su16020606

Choy, S., Jiang, S., Liao, S., & Wang, E. (2024). Public environmental enforcement and private lender monitoring: Evidence from environmental covenants. *Journal of Accounting and Economics, 77*(2-3), 101621. DOI: 10.1016/j.jacceco.2023.101621

Chriqui, A., Benkhnigue, O., Moujahdi, C., El-Khadir, I., Mouniane, Y., Zidane, L., & Hmouni, D. (2024). Contribution to a floristic study of the Ouezzane region (Prerif, North Atlantic Morocco-1). *Egyptian Journal of Botany, 64*(4), 28–48. DOI: 10.21608/ejbo.2024.226380.2439

Christoforidi, I., Kollaros, D., Manios, T., & Daliakopoulos, I. N. (2022). Drought- and salt-tolerant plants of the Mediterranean and their diverse applications: The case of Crete. *Land (Basel), 11*(11), 2038. DOI: 10.3390/land11112038

Coeckelbergh, M. (2020, Oct 5). AI for climate: Freedom, justice, and other ethical and political challenges. *AI and Ethics, 1*(1), 67–72. DOI: 10.1007/s43681-020-00007-2

Conceição, R., Merrouni, A. A., Lopes, D., Alae, A., Silva, H. G., Bennouna, E. G., Collares-Pereira, M., & Ghennioui, A. (2019). A comparative study of soiling on solar mirrors in Portugal and Morocco: Preliminary results for the dry season. *AIP Conference Proceedings, 2126*, 220001. DOI: 10.1063/1.5117760

Cowls, J., Tsamados, A., Taddeo, M., & Floridi, L. (2021, Oct 18). The AI gambit: Leveraging artificial intelligence to combat climate change—opportunities, challenges, and recommendations. *SSRN, 38*(1), 283–307. DOI: 10.2139/ssrn.3804983 PMID: 34690449

Crane-Droesch, A. (2018). Machine learning methods for crop yield prediction and climate change impact assessment in agriculture. *Environmental Research Letters, 13*(11), 114003. DOI: 10.1088/1748-9326/aae159

Daba, A. W., & Qureshi, A. S. (2021). Review of Soil Salinity and Sodicity Challenges to Crop Production in the Lowland Irrigated Areas of Ethiopia and Its Management Strategies. *Land (Basel), 10*(12), 1377. DOI: 10.3390/land10121377

Dahlioui, D., Wette, J., Fernández-García, A., Bouzekri, H., & Azpitarte, I. (2022). Performance assessment of the anti-soiling coating on solar mirrors soiling in the arid climate of Ouarzazate-Morocco. *Solar Energy, 241*, 13–23. DOI: 10.1016/j.solener.2022.05.063

Dankers, R., & Feyen, L. (2008). Climate Change Impact on Floods in the United Kingdom Due to Changes in Extremes. *Climate Research, 32*(1), 163–172.

Deguine, J. P., Aubertot, J. N., Flor, R. J., Lescourret, F., Wyckhuys, K. A. G., & Ratnadass, A. (2021). Integrated pest management: Good intentions, hard realities. A review. *Agronomy for Sustainable Development, 41*(3), 38. DOI: 10.1007/s13593-021-00689-w

Deines, J. M., Patel, R., Liang, S.-Z., Dado, W., & Lobell, D. B. (2021). A million kernels of truth: Insights into scalable satellite maize yield mapping and yield gap analysis from an extensive ground dataset in the US Corn Belt. *Remote Sensing of Environment, 253*, 112174. DOI: 10.1016/j.rse.2020.112174

Delaney, P. G., De Vos, S., Eisner, Z. J., Friesen, J., Hingi, M., Mirza, U. J., Kharel, R., Moussally, J., Smith, N., Slingers, M., Sun, J., & Thullah, A. H. (2025). Challenges, opportunities, and priorities for tier-1 emergency medical services (EMS) development in low- and middle-income countries: A modified Delphi-based consensus study among the global prehospital consortium. *Injury*, *56*(1), 111522. DOI: 10.1016/j.injury.2024.111522 PMID: 38599953

Dessai, S., & Hulme, M. (2007). Assessing the robustness of adaptation decisions to climate change uncertainties: A case study on water resources management in the East of England. *Global Environmental Change*, *17*(1), 59–72. DOI: 10.1016/j.gloenvcha.2006.11.005

Djordjevic´, S., Prodanovic´, D., & Maksimovic´, C. (2005). ˇ.; Maksimovic´, Cˇ.; Ivetic´, M.; Savic´D. SIPSON—Simulation of Interaction between Pipe flow and Surface Over-land flow in Networks. *Water Science and Technology*, *52*, 275–283. DOI: 10.2166/wst.2005.0143 PMID: 16248205

Donnelly, B F A G. (2022, Aug 26). Here's how AI could transform the US power grid

Doorn, N. (2021). Artificial intelligence in the water domain: Opportunities for responsible use. *The Science of the Total Environment*, *755*, 142561. DOI: 10.1016/j.scitotenv.2020.142561 PMID: 33039891

Dourado, P. R. M., de Souza, E. R., Santos, M. A., Lins, C. M. T., Monteiro, D. R., Paulino, M. K. S. S., & Schaffer, B. (2022). Stomatal regulation and osmotic adjustment in sorghum in response to salinity. *Agriculture*, *12*(5), 658. DOI: 10.3390/agriculture12050658

Dresser, L. P., & Kohn, M. A. (2023, Aug 1). Artificial Intelligence and the Evaluation and Treatment of Stroke. *Delaware Journal of Public Health*, *9*(3), 82–84. DOI: 10.32481/djph.2023.08.014 PMID: 37701474

Driouech, F., Déqué, M., & Sánchez-Gómez, E. (2010). Weather regimes—Moroccan precipitation link in a regional climate change simulation. *Global and Planetary Change*, *72*(1-2), 1–10. DOI: 10.1016/j.gloplacha.2010.03.004

Duarte, B., & Caçador, I. (2021). Iberian Halophytes as Agroecological Solutions for Degraded Lands and Biosaline Agriculture. *Sustainability (Basel)*, *13*(2), 1005. DOI: 10.3390/su13021005

Duarte, B., Sleimi, N., & Caçador, I. (2014). Biophysical and biochemical constraints imposed by salt stress: Learning from halophytes. *Frontiers in Plant Science*, *5*, 746. DOI: 10.3389/fpls.2014.00746 PMID: 25566311

Dubois, M. K., Gils, J. K., & Hanniton, P. A. (1956). Use of phenol reagent for the determination of total sugar. *Analytical Chemistry*, *28*, 350–356. DOI: 10.1021/ac60111a017

Dutta, D. (2013). A 2D Flood Inundation Model for Brisbane River: Model Development and Sensitivity Analysis. *Environmental Modelling & Software*, *43*, 55–66.

Edfouf, Z., Guerguer, M., Naamane, S., Karim, M., Zari, N., Grirate, H., Raccurt, O., Delord, C., & Couturier, R. (2014). Research on component materials used for Moroccan CSP power plants: Mirrors ageing and thermal storage. *International Renewable and Sustainable Energy Conference (IRSEC)*. Ouarzazate, Morocco. DOI: 10.1109/IRSEC.2014.7059790

Edfouf, Z., Guerguer, M., & Raccurt, O. (2015). Glass and polymeric mirrors ageing under different Moroccan weathers, an application for CSP power plants. *Energy Procedia*, *69*, 1508–1518. DOI: 10.1016/j.egypro.2015.03.101

Elachouri, M., Ouasti, I., Fakchich, J., & Bussmann, R. W. (2024). *Rosmarinus officinalis* L.Lamiaceae. In Bussmann, R. W., Elachouri, M., & Kikvidze, Z. (Eds.), *Ethnobotany of Northern Africa and Levant. Ethnobotany of Mountain Regions*. Springer. DOI: 10.1007/978-3-031-43105-0_179

Elbouzidi, A., Taibi, M., El Hachlafi, N., Haddou, M., Jeddi, M., Baraich, A., Aouraghe, A., Bellaouchi, R., Mothana, R. A., Hawwal, M. F., Mesnard, F., Hano, C., Asehraou, A., Chaabane, K., El Guerrouj, B., & Addi, M. (2024). Formulation of a Three-Component Essential Oil Mixture from *Lavandula dentata, Rosmarinus officinalis*, and Myrtus communis for Improved Antioxidant Activity. *Pharmaceuticals (Basel, Switzerland)*, *17*(8), 1071. DOI: 10.3390/ph17081071 PMID: 39204175

El-hendawy, S. E., Hassan, W. M., Al-suhaibani, N. A., Refay, Y., & Abdella, K. A. (2017). Comparative performance of multivariable agro-physiological parameters for detecting salt tolerance of wheat cultivars under simulated saline field growing conditions. *Frontiers in Plant Science*, *8*, 435. DOI: 10.3389/fpls.2017.00435 PMID: 28424718

El-Khadir, I., Ktaoui, S., & Mouniane, Y. (2024). Improved Salt Stress Tolerance of *Salvia Officinalis* Grown in the Presence of a Halophytic Plant *Spergularia Maritima*: Analysis of Morpho-Physiological Parameters. In Mabrouki, J., & Azrour, M. (Eds.), *Integrated Solutions for Smart and Sustainable Environmental Conservation. Studies in Systems, Decision and Control* (Vol. 527). Springer., DOI: 10.1007/978-3-031-55787-3_2

Emamverdian, A., Ding, Y., Mokhberdoran, F., & Xie, Y. (2015). Heavy Metal Stress and Some Mechanisms of Plant Defense Response. The Scientific World Journal. *TheScientificWorldJournal, 756120*(1), 756120. Advance online publication. DOI: 10.1155/2015/756120 PMID: 25688377

Ennaceri, H., Benyoussef, A., Ennaoui, A., & Khaldoun, A. (2015). On the analysis of suitable ageing tests of first-surface CSP mirrors in Moroccan outdoor conditions. *3rd International Renewable and Sustainable Energy Conference (IRSEC)*. Marrakech, Morocco. DOI: 10.1109/IRSEC.2015.7455062

Environmental Protection Agency (EPA). (2021). Environmental compliance and enforcement annual report. https://www.epa.gov/

Ester, M., Kriegel, H.P., Sander, J., Xu, X. (1996). A density-based algorithm for discovering clusters in large spatial databases with noise. kdd, 96(34), 226-231.

Etesami, H., & Noori, F. (2019). Soil Salinity as a Challenge for Sustainable Agriculture and Bacterial-Mediated Alleviation of Salinity Stress in Crop Plants. In Kumar, M., Etesami, H., & Kumar, V. (Eds.), *Saline Soil-based Agriculture by Halotolerant Microorganisms*. Springer. DOI: 10.1007/978-981-13-8335-9_1

Evans, S., Vladimirova, D., Holgado, M., Van Fossen, K., Yang, M., Silva, E. A., & Barlow, C. Y. (2017). Business model innovation for sustainability: Towards a unified perspective for creation of sustainable business models. *Business Strategy and the Environment, 26*(5), 597–608. DOI: 10.1002/bse.1939

FAO. (2019). *High level expert forum-how to feed the world in 2050. Economic and Social Development, Food and Agricultural Organization of the United Nations*. Plant Cell Tiss. Organ. Cult.

Farayola, O. A., Abdul, A. A., Irabor, B. O., & Okeleke, E. C.Oluwatoyin Ajoke FarayolaAdekunle Abiola AbdulBlessing Otohan IraborEvelyn Chinedu Okeleke. (2023). Innovative business models driven by ai technologies: A review. *Computer Science & IT Research Journal, 4*(2), 85–110. DOI: 10.51594/csitrj.v4i2.608

Farhat, N., Kouas, W., Braun, H. P., & Debez, A. (2021). Stability of thylakoid protein complexes and preserving photosynthetic efficiency are crucial for the successful recovery of the halophyte Cakile maritima from high salinity. *Plant Physiology and Biochemistry, 166*, 177–190. DOI: 10.1016/j.plaphy.2021.05.044 PMID: 34116337

Farooq, M., & Ahmad, N. (2023). Nexus between board characteristics, firm performance and intellectual capital: An emerging market evidence. *Corporate Governance (Bradford), 23*(6), 1269–1297. DOI: 10.1108/CG-08-2022-0355

Farooq, M., Noor, A., & Naeem, M. (2023). Does family ownership moderate the relationship between board characteristics and corporate social responsibility? Evidence from an emerging market. *Asian Journal of Business Ethics, 12*(1), 71–99. DOI: 10.1007/s13520-022-00164-z

Fateh, S., Shah, I. A., Sial, Q., & Jhanjhi, N. Z. (2025). Deep Learning Applications for Healthcare Risk Assessment. In *Generative AI Techniques for Sustainability in Healthcare Security* (pp. 59–76). IGI Global Scientific Publishing.

Faure, M., & Svatikova, K. (2012). Environmental law enforcement: Comparative analysis. *Environmental Law Review, 14*(1), 1–22.

Fernández-García, A., Juaidi, A., Sutter, F., Martínez-Arcos, L., & Manzano-Agugliaro, F. (2018). Solar Reflector Materials Degradation Due to the Sand Deposited on the Backside Protective Paints. *Energies, 11*(4), 808. DOI: 10.3390/en11040808

Flowers, T. J. (1985). Physiology of halophytes. In Pasternak, D., & San Pietro, A. (Eds.), *Biosalinity in Action: Bioproduction with Saline Water. Developments in Plant and Soil Sciences* (Vol. 17). Springer., DOI: 10.1007/978-94-009-5111-2_3

Fouad, N., El-Zayat, E. M., Amr, D., El-Khishin, D. A., Abd-Elhalim, H. M., Hafez, A., Radwan, K. H., Hamwieh, A., & Tadesse, W. (2025). Characterizing wheat rhizosphere bacterial microbiome dynamics under salinity stress: Insights from 16S rRNA metagenomics for enhancing stress tolerance. *Plants, 14*(7), 1033. DOI: 10.3390/plants14071033 PMID: 40219101

Fowler, H. J., Blenkinsop, S., & Tebaldi, C. (2007). Linking Climate Change Modelling to Impacts Studies: Recent Advances in Downscaling Techniques for Hydrological Modelling. *International Journal of Climatology, 27*(12), 1547–1578. DOI: 10.1002/joc.1556

Gao, C., Gao, K., Yang, H., Ju, T., Zhu, J., Tang, Z., Zhao, L., & Chen, Q. (2022). Genome-wide analysis of metallothionein gene family in maize to reveal its role in development and stress resistance to heavy metal. *Biological Research, 55*(1), 1. DOI: 10.1186/s40659-021-00368-w PMID: 35012672

Gashi, B., Buqaj, L., Vataj, R., & Tuna, M. (2024). Chlorophyll biosynthesis suppression, oxidative level and cell cycle arrest caused by Ni, Cr and Pb stress in maize exposed to treated soil from the Ferronikel smelter in Drenas, Kosovo. *Plant Stress, 11*, 100379. DOI: 10.1016/j.stress.2024.100379

Gesami, B. K., & Nunoo, J. (2024, Jul 18). Artificial intelligence in marine ecosystem management: Addressing climate threats to Kenya's blue economy. *Frontiers in Marine Science, 11*, 1404104. DOI: 10.3389/fmars.2024.1404104

Ghate, S., Kolhe, M., & Narkhede, V. (2020). Climate change induced flood risk assessment and adaptation in urban areas. *The Science of the Total Environment, 703*, 135558.

Godoy, F., Olivos-Hernández, K., Stange, C., & Handford, M. (2021). Abiotic stress in crop species: Improving tolerance by applying plant metabolites. *Plants, 10*(2), 186. DOI: 10.3390/plants10020186 PMID: 33498148

González-Espíndola, L. Á., Pedroza-Sandoval, A., Trejo-Calzada, R., Jacobo-Salcedo, M. R., García de los Santos, G., & Quezada-Rivera, J. J. (2024). Relative water content, chlorophyll index, and photosynthetic pigments on *Lotus corniculatus* L. in response to water deficit. *Plants, 13*(7), 961. DOI: 10.3390/plants13070961 PMID: 38611490

González-Minero, F. J., Bravo-Díaz, L., & Ayala-Gómez, A. (2020). *Rosmarinus officinalis* L. (Rosemary): An ancient plant with uses in personal healthcare and cosmetics. *Cosmetics, 7*(4), 77. DOI: 10.3390/cosmetics7040077

Government of Pakistan, Ministry of Water Resources. (2018). Annual Flood Report 2018. Office of the Chief Engineering Advisor & Chairman Federal Flood Commission, Islamabad. https://ffc.gov.pk/wp-content/uploads/2020/06/Annual-Flood-Report2018.pdf

Gregurec, I., Tomičić Furjan, M., & Tomičić-Pupek, K. (2021). The impact of COVID-19 on sustainable business models in SMEs. *Sustainability (Basel), 13*(3), 1098. DOI: 10.3390/su13031098

Guerguer, M., Edfouf, Z., & Raccurt, O. (2016). Comparison of two methods used to evaluate the aggressivity of a Moroccan marine site on solar mirrors. *International Renewable and Sustainable Energy Conference (IRSEC)*. Marrakech, Morocco. https://doi.org/DOI: 10.1109/IRSEC.2016.7984014

Guerguer, M., Karim, M., Naamane, S., Edfouf, Z., & Raccurt, O. (2018). Degradation of solar glass mirrors paints at marine outdoor site. *AIP Conference Proceedings, 2033*, 230004. DOI: 10.1063/1.5067232

Guerguer, M., Karim, M., Naamane, S., Edfouf, Z., Raccurt, O., & Delord, C. (2017). Soiling deposition on solar mirrors exposed in Morocco. *AIP Conference Proceedings, 1850*, 130005. DOI: 10.1063/1.4984499

Guerguer, M., Naamane, S., Edfouf, Z., Raccurt, O., & Bouaouine, H. (2021). Chemical degradation and color changes of paint protective coatings used in solar glass mirrors. *Coatings, 11*(4), 476. DOI: 10.3390/coatings11040476

Guerguer, M., Naamane, S., Karim, M., & Bouaouine, H. (2019). Outdoor exposure testing of silvered-glass reflectors under marine environment. *AIP Conference Proceedings*, *2126*, 160003. DOI: 10.1063/1.5117666

Gupta, A., Mishra, R., Rai, S., Bano, A., Pathak, N., Fujita, M., Kumar, M., & Hasanuzzaman, M. (2022). Mechanistic insights of plant growth promoting bacteria mediated drought and salt stress tolerance in plants for sustainable agriculture. *International Journal of Molecular Sciences*, *23*(7), 3741. DOI: 10.3390/ijms23073741 PMID: 35409104

Habán, M., Korczyk-Szabó, J., Čerteková, S., & Ražná, K. (2023). Lavandula Species, Their Bioactive Phytochemicals, and Their Biosynthetic Regulation. *International Journal of Molecular Sciences*, *24*(10), 8831. DOI: 10.3390/ijms24108831 PMID: 37240177

Hafeez, A., Rasheed, R., Ashraf, M. A., Qureshi, F. F., Hussain, I., & Iqbal, M. (2023). Effect of heavy metals on growth, physiological and biochemical responses of plants. In *Azamal Husen. Plants and Their Interaction to Environmental Pollution*. Elsevier., DOI: 10.1016/B978-0-323-99978-6.00006-6

Hajiabadi, A. A., Arani, A. M., Ghasemi, S., Rad, M. H., Etesami, H., Shabazi Manshadi, S., & Dolati, A. (2021). Mining the rhizosphere of halophytic rangeland plants for halotolerant bacteria to improve growth and yield of salinity-stressed wheat. *Plant Physiology and Biochemistry*, *163*, 139–153. DOI: 10.1016/j.plaphy.2021.03.059 PMID: 33845330

Hakkou, R., Benzaazoua, M., & Bussière, B. (2020). Environmental impacts of mine tailings in Morocco: A review. *Environmental Geochemistry and Health*, *42*(2), 497–516. PMID: 33141377

Hammer, M., & Junghanns, W. (2020). *Rosmarinus officinalis* L.: Rosemary. In Novak, J., & Blüthner, W. D. (Eds.), *Medicinal, Aromatic and Stimulant Plants. Handbook of Plant Breeding* (Vol. 12). Springer. DOI: 10.1007/978-3-030-38792-1_15

Harrison, P. A., Dunford, R. W., Holman, I. P., & Rounsevell, M. D. (2016). Climate change impact modelling needs to include cross-sectoral interactions. *Nature Climate Change*, *6*(9), 885–890. DOI: 10.1038/nclimate3039

Hasanuzzaman, M., Parvin, K., Bardhan, K., Nahar, K., Anee, T. I., Masud, A. A. C., & Fotopoulos, V. (2021). Biostimulants for the Regulation of Reactive Oxygen Species Metabolism in Plants under Abiotic Stress. *Cells*, *10*(10), 2537. DOI: 10.3390/cells10102537 PMID: 34685517

Hasegawa, A., Oyanagi, T., Minagawa, R., Fujii, Y., & Sasamoto, H. (2014). An inverse relationship between allelopathic activity and salt tolerance in suspension cultures of three mangrove species, Sonneratia alba, S. caseolaris and S. ovata: Development of a bioassay method for allelopathy, the protoplast co-culture method. *Journal of Plant Research, 127*(6), 755–761. DOI: 10.1007/s10265-014-0651-1 PMID: 25062702

Hassani, A., Azapagic, A., & Shokri, N. (2021). Global predictions of primary soil salinization under changing climate in the 21st century. *Nature Communications, 12*(1), 6663. DOI: 10.1038/s41467-021-26907-3 PMID: 34795219

Hawkins, E., & Sutton, R. (2011). The Potential to Narrow Uncertainty in Regional Climate Predictions. *Bulletin of the American Meteorological Society, 90*(8), 1095–1107. DOI: 10.1175/2009BAMS2607.1

He, W., Hong, D., Scarpa, G., Uezato, T., & Yokoya, N. (2021). Multisource Remote Sensing Image Fusion. *Deep Learning for the Earth Sciences: A Comprehensive Approach to Remote Sensing, Climate Science, and Geosciences*, 136-149.

Helmrich, A. M., & Chester, M. V. (2022). Reconciling complexity and deep uncertainty in infrastructure design for climate adaptation. *Sustainable and Resilient Infrastructure, 7*(2), 83–99. DOI: 10.1080/23789689.2019.1708179

He, Y., Yan, X., Fan, A., Yin, X., Meng, Q., Zhang, L., & Liu, P. (2024). Assessing the influence of actual LNG emission factors within the EU emissions trading system. *Transport Policy, 159*, 345–358. DOI: 10.1016/j.tranpol.2024.10.030

Heymann, F., Quest, H., Garcia, T P., Ballif, C., & Galus, M D. (2023, Nov 19). Reviewing 40 years of artificial intelligence applied to power systems – A taxonomic perspective. Elsevier BV, 15, 100322-100322. DOI: 10.1016/j.egyai.2023.100322

Hirabayashi, Y., Mahendran, R., Koirala, S., Konoshima, L., Yamazaki, D., Watanabe, S., Kim, H., & Kanae, S. (2013). Global Flood Risk under Climate Change. *Nature Climate Change, 3*(9), 816–821. DOI: 10.1038/nclimate1911

Holman, I. P., Brown, C., Carter, T. R., Harrison, P. A., & Rounsevell, M. (2019). Improving the representation of adaptation in climate change impact models. *Regional Environmental Change, 19*(3), 711–721. DOI: 10.1007/s10113-018-1328-4 PMID: 30956567

Hopmans, J. W., Qureshi, A. S., Kisekka, I., Munns, R., Grattan, S. R., Rengasamy, P., Ben-Gal, A., Assouline, S., Javaux, M., Minhas, P. S., Raats, P. A., Skaggs, T. H., Wang, G., De Jong van Lier, Q., Jiao, H., Lavado, R. S., Lazarovitch, N., Li, B., & Taleisnik, E. (2021). Critical knowledge gaps and research priorities in global soil salinity. *Advances in Agronomy*, *169*, 1–191. DOI: 10.1016/bs.agron.2021.03.001

Hossain, A., Skalicky, M., Brestic, M., Maitra, S., Ashraful Alam, M., Syed, M. A., Hossain, J., Sarkar, S., Saha, S., Bhadra, P., Shankar, T., Bhatt, R., Kumar Chaki, A., EL Sabagh, A., & Islam, T. (2021). Consequences and mitigation strategies of abiotic stresses in wheat (*Triticum aestivum* L.) under the changing climate. *Agronomy (Basel)*, *11*(2), 241. DOI: 10.3390/agronomy11020241

Hotelling, H. (1933). Analysis of a complex of statistical variables into principal components. *Journal of Educational Psychology*, *24*(6), 417–441. DOI: 10.1037/h0071325

Huang, G., Luo, H., Lu, X., Yang, C., Wang, Z., Huang, T., & Ma, J. (2020). Urban flood risk analysis and zoning methods. *Water Resources. Prot.*, *36*, 1–17.

Huntingford, C., Jeffers, E. S., Bonsall, M. B., Christensen, H. M., Lees, T., & Yang, H. (2019). Machine learning and artificial intelligence to aid climate change research and preparedness. *Environmental Research Letters*, *14*(12), 124007. DOI: 10.1088/1748-9326/ab4e55

Hussain, R., Ayyub, C. M., Shaheen, M. R., Rashid, S., Nafees, M., Ali, S., Butt, M., Ali, M., Maqsood, A., Fiaz, S., Ahmar, S., Mahmood, T., & Mora-Poblete, F. (2021). Regulation of osmotic balance and increased antioxidant activities under heat stress in *Abelmoschus esculentus* L. triggered by exogenous proline application. *Agronomy (Basel)*, *11*(4), 685. DOI: 10.3390/agronomy11040685

Ibrahimova, U., Zivcak, M., Gasparovic, K., Rastogi, A., Allakhverdiev, S. I., Yang, X., & Brestic, M. (2021). Electron and proton transport in wheat exposed to salt stress: Is the increase of the thylakoid membrane proton conductivity responsible for decreasing the photosynthetic activity in sensitive genotypes? *Photosynthesis Research*, *150*(1-3), 195–211. DOI: 10.1007/s11120-021-00853-z PMID: 34125427

Ikpe, E. O. (2024). Adoption and implementation of artificial intelligence in small businesses in selected developing countries. *Journal of Health. Applied Sciences and Management*, *8*(1), 26–34.

Ilakiya, T., Parameswari, E., Swarnapriya, R., Yazhini, G., Kalaiselvi, P., Davamani, V., Singh, S., Vinothini, N., Dharani, C., & Garnepudi, S.L. (2024). Unlocking the carbon sequestration potential of horticultural crops. C, 10(3), 65. DOI: 10.3390/c10030065

IPCC. (2014). Synthesis Report. Contribution of Working Groups I, II and III to the Fifth Assessment Report of the Intergovernmental Panel on Climate Change. *Climatic Change*, 2014.

Irfan, M., Mudassir, M., Khan, M. J., Dawar, K. M., Muhammad, D., Mian, I. A., Ali, W., Fahad, S., Saud, S., Hayat, Z., Nawaz, T., Khan, S. A., Alam, S., Ali, B., Banout, J., Ahmed, S., Mubeen, S., Danish, S., Datta, R., & Dewil, R. (2021). Heavy metals immobilization and improvement in maize (*Zea mays* L.) growth amended with biochar and compost. *Scientific Reports*, *11*(1), 18416. DOI: 10.1038/s41598-021-97525-8 PMID: 34531439

Iyelolu, T. V., Agu, E. E., Idemudia, C., & Ijomah, T. I. (2024). Driving SME innovation with AI solutions: overcoming adoption barriers and future growth opportunities. *International Journal of Science and Technology Research Archive, 7*(1), 036-054.

Jaiganesh, N., Kuo, P., Woudstra, T., Kumar, R. A., & Aravind, P. (2022, Jun 27). Negative Emission Power Plants: Thermodynamic Modeling and Evaluation of a Biomass-Based Integrated Gasification Solid Oxide Fuel Cell/Gas Turbine System for Power. *Frontiers in Energy Research*, *10*, 803756. DOI: 10.3389/fenrg.2022.803756

Jhanjhi, N. Z., Shah, I. A., & Brohi, S. N. (2025). Hardware Vulnerabilities: Taxonomy and Business Security Models. In *Generative AI for Web Engineering Models* (pp. 129-146). IGI Global.

Jin, D., Ocone, R., Jiao, K., & Xuan, J. (2020, Apr 20). Energy and AI. Elsevier BV, 1, 100002-100002. DOI: 10.1016/j.egyai.2020.100002

Jones, A., Kuehnert, J., Fraccaro, P., Meuriot, O., Ishikawa, T., Edwards, B., Stoyanov, N., Remy, S. L., Weldemariam, K., & Assefa, S. (2023, Jun 8). AI for climate impacts: Applications in flood risk. *NPJ Climate and Atmospheric Science*, *6*(1), 63. DOI: 10.1038/s41612-023-00388-1

Joshi, S., Nath, J., Singh, A. K., Pareek, A., & Joshi, R. (2022). Ion transporters and their regulatory signal transduction mechanisms for salinity tolerance in plants. *Physiologia Plantarum*, *174*(3), e13702. DOI: 10.1111/ppl.13702 PMID: 35524987

Joswig, T., & Kurz, W. (2025). Regulatory and Compliance Requirements for SMEs Operating AI Systems through Data Centers in the EU, with a Focus on Data Protection Challenges in Germany. *Journal of Next-Generation Research 5.0*.

Jurado, C., Díaz-Vivancos, P., Gregorio, B. E., Acosta-Motos, J. R., & Hernández, J. A. (2024). Effect of halophyte-based management in physiological and biochemical responses of tomato plants under moderately saline greenhouse conditions. *Plant Physiology and Biochemistry, 206*, 108228. DOI: 10.1016/j.plaphy.2023.108228 PMID: 38043255

Kadam, P., & Sen, D. (2012). Flood inundation simulation in Ajoy River using MIKE-FLOOD. *ISH Journal of Hydraulic Engineering, 18*(2), 129–141. DOI: 10.1080/09715010.2012.695449

Kalisdha, A. (2024, Feb 1). The Impact of Artificial Intelligence and Machine Learning in Library and Information Science

Karim, M., Delord, C., Raccurt, O., & Naamane, S. (2018). Exposure conditions effect on soiling of solar glass mirrors. *AIP Conference Proceedings, 230006*, 230006. Advance online publication. DOI: 10.1063/1.5067234

Karim, M., Naamane, S., & Delord, C. (2016). Evaluation of the behaviour of different solar mirrors against soiling phenomenon. *International Renewable and Sustainable Energy Conference (IRSEC)*. Marrakech, Morocco. DOI: 10.1109/IRSEC.2016.7984008

Karim, M., Naamane, S., Delord, C., & Bennouna, A. (2014). Surface wear damage of glass solar mirrors in Moroccan desert environment. *International Renewable and Sustainable Energy Conference (IRSEC)*. Ouarzazate, Morocco. 246-251. DOI: 10.1109/IRSEC.2014.7059770

Karim, M., Naamane, S., Delord, C., & Bennouna, A. (2015a). Study of the surface damage of glass reflectors used in concentrated solar power plants. *Energy Procedia, 69*, 106–115. DOI: 10.1016/j.egypro.2015.03.013

Karim, M., Naamane, S., Delord, C., & Bennouna, A. (2015b). Laboratory simulation of the surface erosion of solar glass mirrors. *Solar Energy, 118*, 520–532. DOI: 10.1016/j.solener.2015.05.044

Karim, M., Naamane, S., Hassani, I. E. A. E. L., Delord, C., Belcadi, S., Tochon, P., & Bennouna, A. (2014). Towards the prediction of CSP mirrors wear: Methodology of analysis of influencing parameters on the mirrors surface degradation: Application in two different sites in Morocco. *Solar Energy, 108*, 41–50. DOI: 10.1016/j.solener.2014.06.036

Karunathilake, E. M. B. M., Le, A. T., Heo, S., Chung, Y. S., & Mansoor, S. (2023, Aug 11). The Path to Smart Farming: Innovations and Opportunities in Precision Agriculture. *Agriculture, 13*(8), 1593–1593. DOI: 10.3390/agriculture13081593

Kaur, G., Sanwal, S. K., Kumar, A., Pundir, R. K., Yadav, M., & Sehrawat, N. (2024). Role of osmolytes dynamics in plant metabolism to cope with salinity induced osmotic stress. *Discover Agriculture.*, *2*(1), 59. DOI: 10.1007/s44279-024-00070-x

Kerbab, S., Silini, A., Chenari Bouket, A., Cherif-Silini, H., Eshelli, M., El Houda Rabhi, N., & Belbahri, L. (2021). Mitigation of NaCl Stress in Wheat by Rhizosphere Engineering Using Salt Habitat Adapted PGPR Halotolerant Bacteria. *Applied Sciences (Basel, Switzerland)*, *11*(3), 1034. DOI: 10.3390/app11031034

Khamidov, M., Ishchanov, J., Hamidov, A., Donmez, C., & Djumaboev, K. (2022). Assessment of Soil Salinity Changes under the Climate Change in the Khorezm Region, Uzbekistan. *International Journal of Environmental Research and Public Health*, *19*(14), 8794. DOI: 10.3390/ijerph19148794 PMID: 35886646

Khanam, Z., Sultana, F. M., & Mushtaq, F. (2023). Environmental Pollution Control Measures and Strategies: An Overview of Recent Developments. In Mushtaq, F., Farooq, M., Mukherjee, A. B., & Ghosh Nee Lala, M. (Eds.), *Geospatial Analytics for Environmental Pollution Modeling*. Springer., DOI: 10.1007/978-3-031-45300-7_15

Khandaker, T., Hossain, M. S., Dhar, P. K., Rahman, M. S., Hossain, M. A., & Ahmed, M. B. (2020, May 30). Efficacies of Carbon-Based Adsorbents for Carbon Dioxide Capture. *Processes (Basel, Switzerland)*, *8*(6), 654–654. DOI: 10.3390/pr8060654

Khan, M. N., Ullah, R., & Abbas, T. (2023). Heavy metal stress and its impact on crop productivity: A global perspective. *Journal of Soil Science and Plant Nutrition*, *23*(1), 45–60.

Khan, N., Ray, R. L., Sargani, G. R., Ihtisham, M., Khayyam, M., & Ismail, S. (2021). Current Progress and Future Prospects of Agriculture Technology: Gateway to Sustainable Agriculture. *Sustainability (Basel)*, *13*(9), 4883. DOI: 10.3390/su13094883

Khondoker, M., Mandal, S., Gurav, R., & Hwang, S. (2023). Freshwater Shortage, Salinity Increase, and Global Food Production: A Need for Sustainable Irrigation Water Desalination—A Scoping Review. *Earth (Basel, Switzerland)*, *4*(2), 223–240. DOI: 10.3390/earth4020012

Kim, J. (2023, Nov 22). Four ways AI is making the power grid faster and more resilient

Kiss, A., & Shelton, D. (2021). *International environmental law*. Oxford University Press.

Kosakivska, I. V., Babenko, L. M., Romanenko, K. O., Korotka, I. Y., & Potters, G. (2021). Molecular mechanisms of plant adaptive responses to heavy metals stress. *Cell Biology International*, *45*(2), 258–272. DOI: 10.1002/cbin.11503 PMID: 33200493

Kuglitsch, M. M., Cox, J., Luterbacher, J., Jamoussi, B., Xoplaki, E., Thummarukudy, M., Radwan, G. S., Yasukawa, S., McClain, S. N., Albayrak, A., Oehmen, D., & Ward, T. (2024, Oct 1). AI to the rescue: How to enhance disaster early warnings with tech tools. *Nature*, *634*(8032), 27–29. DOI: 10.1038/d41586-024-03149-z PMID: 39354079

Kulkov, I., Kulkova, J., Rohrbeck, R., Menvielle, L., Kaartemo, V., & Makkonen, H. (2024). Artificial intelligence-driven sustainable development: Examining organizational, technical, and processing approaches to achieving global goals. *Sustainable Development (Bradford)*, *32*(3), 2253–2267. DOI: 10.1002/sd.2773

Kumar, P., Debele, S., Khalili, S., Halios, C., Sahani, J., Aghamohammadi, N., Andrade, M D F., Athanassiadou, M., Bhui, K., Calvillo, N., Cao, S., Coulon, F., Edmondson, J L., Fletcher, D., Freitas, E D D., Guo, H., Hort, M., Katti, M., Kjeldsen, T., . . . Jones, L. (2024, Mar 1). Urban heat mitigation by green and blue infrastructure: Drivers, effectiveness, and future needs. Elsevier BV, 5(2), 100588-100588. DOI: 10.1016/j.xinn.2024.100588

Kumar, M., Singh, S., Jain, A., Yadav, S., Dubey, A., & Trivedi, S. P. (2024). A review on heavy metal-induced toxicity in fishes: Bioaccumulation, antioxidant defense system, histopathological manifestations, and transcriptional profiling of genes. *Journal of Trace Elements in Medicine and Biology*, *83*, 127377. DOI: 10.1016/j.jtemb.2023.127377 PMID: 38183919

Kumar, S., Prasad, S., Yadav, K. K., Shrivastava, M., Gupta, N., Nagar, S., Bach, Q.-V., Kamyab, H., Khan, S. A., Yadav, S., & Malav, L. C. (2019). Hazardous heavy metals contamination of vegetables and food chain: Role of sustainable remediation approaches - A review. *Environmental Research*, *179*, 108792. DOI: 10.1016/j.envres.2019.108792 PMID: 31610391

Kumar, S., Singh, R., & Sharma, P. (2022). Nickel toxicity in plants: Mechanisms and remediation strategies. *Ecotoxicology and Environmental Safety*, *234*, 113370.

Kundzewicz, Z. W. (2007). Precipitation as an Indicator of Climate Change. *Climate Research*, *34*(3), 199–207.

Lakhouil, S., Boukheir, S., Bouaouine, H., Baron, E. L., Disdier, A., Vuillerme, V., & Naamane, S. (2022). Erosion Effect in the Degradation of Coated and Uncoated Glass Solar Mirrors Highlighted by Coupled Accelerated Erosion/Aging Tests. *Coatings*, *12*(8), 1087. DOI: 10.3390/coatings12081087

Lakhouil, S., Boukheir, S., Chbihi, A., & Naamane, S. (2021). Outdoor exposure approach in a harsh environment toward the study of corrosion aspects of primary solar reflectors. *Solar Energy*, *230*, 1070–1084. DOI: 10.1016/j.solener.2021.11.013

Lakhouil, S., Boukheir, S., Naamane, S., & Bouaouine, H. (2023). Mechanical resistance assessment by real outdoor study of an anti-wear coating on glass solar mirror. *AIP Conference Proceedings*, *2815*, 020007. DOI: 10.1063/5.0149405

Ling, Y., Wang, D., Peng, Y., Peng, D., & Li, Z. (2025). Cross-stressful adaptation to drought and high salinity is related to variable antioxidant defense, proline metabolism, and Dehydrin b expression in white clover. *Agronomy (Basel)*, *15*(1), 126. DOI: 10.3390/agronomy15010126

Li, S., Khoso, M. A., Xu, H., Zhang, C., Liu, Z., Wagan, S., Dinislam, K., & Liu, L. (2024). WRKY Transcription Factors (TFs) as Key Regulators of Plant Resilience to Environmental Stresses: Current Perspective. *Agronomy (Basel)*, *14*(10), 2421. DOI: 10.3390/agronomy14102421

Liu, X., Zhang, Y., & Li, Q. (2023). Effects of heavy metal pollution on soil microbial communities and plant health. *Applied Soil Ecology*, *182*, 104631.

Lu, C., Li, L., Liu, X., Chen, M., Wan, S., & Li, G. (2023). Salt stress inhibits photosynthesis and destroys chloroplast structure by downregulating chloroplast development–related genes in Robinia pseudoacacia seedlings. *Plants*, *12*(6), 1283. DOI: 10.3390/plants12061283 PMID: 36986971

Lundberg, S. M., & Lee, S. I. (2017). A unified approach to interpreting model predictions. *Advances in Neural Information Processing Systems*, 30.

Lundstad, E., Brugnara, Y., Pappert, D., Kopp, J., Samakinwa, E., Hürzeler, A., Andersson, A., Chimani, B., Cornes, R., Demarée, G. R., Filipiak, J., Gates, L., Ives, G. L., Jones, J., Jourdain, S., Kiss, A., Nicholson, S. E., Przybylak, R., Jones, P., & Brönnimann, S. (2023, Jan 19). The global historical climate database HCLIM. *Scientific Data*, *10*(1), 44. Advance online publication. DOI: 10.1038/s41597-022-01919-w PMID: 36658229

Luo, J., Zhang, S., Peng, J., Zhu, X., Lv, L., Wang, C., Li, C., Zhou, Z., & Cui, J. (2017). Effects of Soil Salinity on the Expression of Bt Toxin (Cry1Ac) and the Control Efficiency of Helicoverpa armigera in Field-Grown Transgenic Bt Cotton. *PLoS One*, *12*(1), 1–13. DOI: 10.1371/journal.pone.0170379 PMID: 28099508

Macaluso, D., Licciardo, F., & Carbone, K. (2024). Farming of Medicinal and Aromatic Plants in Italy: Structural Features and Economic Results. *Agriculture*, *14*(1), 151. DOI: 10.3390/agriculture14010151

MacQueen, J. (1967). Some methods for classification and analysis of multivariate observations. Proceedings of the Fifth Berkeley Symposium on Mathematical Statistics and Probability, Volume 1. *Statistics*, *5*, 281–298.

Mandal, S., Raju, R., Kumar, A., Kumar, P., & Sharma, P. C. (2018). Current status of research, technology response and policy needs of salt-affected soils in India-a review. *Journal of the Indian Society of Coastal Agricultural Research*, *36*, 40–53.

Mansinhos, I., Gonçalves, S., & Romano, A. (2024). How climate change-related abiotic factors affect the production of industrial valuable compounds in Lamiaceae plant species: A review. *Frontiers in Plant Science*, *15*, 1370810. DOI: 10.3389/fpls.2024.1370810 PMID: 39049861

Matal, A., Karim, M., Bannouri, O., Naamane, S., & Bouaouine, H. (2017). Setting Parameters of an Erosion Test Bench for Solar Reflectors. *International Renewable and Sustainable Energy Conference (IRSEC)*. Tangier, Morocco. DOI: 10.1109/IRSEC.2017.8477290

Matal, A., Karim, M., & Naamane, S. (2020). Development and calibration of an experimental test bench simulating solar reflectors erosion. *Solar Energy*, *201*, 724–731. DOI: 10.1016/j.solener.2020.03.028

Matczak, P., Lewandowski, J., Choryński, A., Szwed, M., & Kundzewicz, Z. W. Flood risk governance arrangements in Europe, *Proc. IAHS*, 369, 195–199, , 2015. DOI: 10.5194/piahs-369-195-2015

Ma, Y., Liu, W., Qiao, Y., Qiao, W., Yang, H., Zhong, Y., Yang, H., Wang, H., Li, Y., Dong, B., & Liu, M. (2023). Effects of soil salinity on foxtail millet osmoregulation, grain yield, and soil water utilization under varying water conditions. *Agricultural Water Management*, *284*, 108354. DOI: 10.1016/j.agwat.2023.108354

Mbuvha, R., Yaakoubi, Y., Bagiliko, J., Potes, S. H., Nammouchi, A., & Amrouche, S. (2024, Jan 1). Leveraging AI for Climate Resilience in Africa: Challenges, Opportunities, and the Need for Collaboration. RELX Group (Netherlands). DOI: 10.2139/ssrn.4815919

Merrouni, A. A., Conceição, R., Mouaky, A., Silva, H. G., & Ghennioui, A. (2020). CSP performance assessment including soiling measurements for Morocco and Portugal reveals similar electricity yield and cost in both countries. *Renewable Energy*, *162*, 1777–1792. DOI: 10.1016/j.renene.2020.10.014

Merrouni, A. A., Mezrhab, A., Ghennioui, A., & Naimi, Z. (2017). Measurement, comparison and monitoring of solar mirror's specular reflectivity using two different Reflectometers. *Energy Procedia*, *119*, 433–445. DOI: 10.1016/j.egypro.2017.07.045

Merrouni, A. A., Wolfertstetter, F., Mezrhab, A., Wilbert, S., & Pitz-Paal, R. (2015). Investigation of Soiling Effect on Different Solar Mirror Materials under Moroccan Climate. *Energy Procedia*, *69*, 1948–1957. DOI: 10.1016/j.egypro.2015.03.194

Mishra, A. K., Das, R., George Kerry, R., Biswal, B., Sinha, T., Sharma, S., Arora, P., & Kumar, M. (2023). Promising management strategies to improve crop sustainability and to amend soil salinity. *Frontiers in Environmental Science*, *10*, 962581. DOI: 10.3389/fenvs.2022.962581

Mohammadi Alagoz, S., Hadi, H., Toorchi, M., Pawłowski, T. A., Asgari Lajayer, B., Price, G. W., Farooq, M., & Astatkie, T. (2023). Morpho-physiological responses and growth indices of triticale to drought and salt stresses. *Scientific Reports*, *13*(1), 8896. DOI: 10.1038/s41598-023-36119-y PMID: 37264097

Mohammadi, M., & Mohammadi, A. (2024, Jan 1). Empowering Distributed Solutions in Renewable Energy Systems and Grid Optimization. Springer Nature, 141-155. DOI: 10.1007/978-3-031-57567-9_7

Morton, M. J., Awlia, M., Al-Tamimi, N., Saade, S., Pailles, Y., Negrão, S., & Tester, M. (2019). Salt stress under the scalpel dissecting the genetics of salt tolerance. *The Plant Journal*, *97*(1), 148–163. DOI: 10.1111/tpj.14189 PMID: 30548719

Mosgaard, M. A., Bundgaard, A. M., & Kristensen, H. S. (2022). ISO 14001 practices – A study of environmental objectives in Danish organizations. *Journal of Cleaner Production*, *331*, 129799. DOI: 10.1016/j.jclepro.2021.129799

Mouniane, Y., Vibhute, A. D., Kadmi, Y. (2025). Using Phytoremediation to Treat Industrial and Pharmaceutical Wastewater: Innovations and Future Prospects. In: Biotechnology Approaches to Industrial and Pharmaceutical Wastewater Treatment. pp. 277–298. IGI Global Scientific Publishing. DOI: 10.4018/979-8-3693-8487-9.ch009

Muccitelli, S., Pozzi, C., D'Ascanio, R., & Magaudda, S. (2023). Environmental Contract: A Collaborative Tool to Improve the Multilevel Governance of European MPAs. *Sustainability (Basel)*, *15*(10), 8174. DOI: 10.3390/su15108174

Múnera-Echeverri, A., Múnera-Echeverri, J. L., & Segura-Sánchez, F. (2024). Biopesticidal potential of nanostructured lipid carriers loaded with thyme and rosemary essential oils against common ornamental flower pests. *Colloids and Interfaces*, *8*(5), 55. DOI: 10.3390/colloids8050055

Munir, N., Hasnain, M., Roessner, U., & Abideen, Z. (2022). Strategies in improving plant salinity resistance and use of salinity resistant plants for economic sustainability. *Critical Reviews in Environmental Science and Technology*, *52*(12), 2150–2196. DOI: 10.1080/10643389.2021.1877033

Munns, R., & Tester, M. (2008). Mechanisms of salinity tolerance. *Annual Review of Plant Biology, 59*(1), 651–681. DOI: 10.1146/annurev.arplant.59.032607.092911 PMID: 18444910

Mustafa, S., Sabir, R. M., Sarwar, A., Safdar, M., Al Ansari, M. S., & Hussain, S. (2024). Precision Agriculture and Unmanned Aerial Vehicles (UAVs). In Khang, A. (Ed.), *Agriculture and Aquaculture Applications of Biosensors and Bioelectronics* (pp. 83–108). IGI Global., DOI: 10.4018/979-8-3693-2069-3.ch005

Muzammil, H., Zaman, M., Shahid, M. A., Safdar, M., Majeed, M. D., & Sabir, R. M. (2023). The Impacts of Climate Change on Monsoon Flood Situations in Pakistan. *Environmental Sciences Proceedings., 25*(1), 63. DOI: 10.3390/ECWS-7-14255

Naamane, S., Chbihi, A., Boukheir, S., Mouadine, M., Tahri, K., Bouaddi, S., & Bouhamidi, M. H. (2020). Towards the choice of concentrated solar power and photovoltaic materials for Moroccan conditions. *AIP Conference Proceedings, 150013*, 150013. Advance online publication. DOI: 10.1063/5.0029152

Naeem, M., Ali, S., Rehman, A., & Farooq, S. (2025). Ethical Concerns in Artificial Intelligence (AI): The Role of Governance Mechanism in Finance. In *Generative AI for Web Engineering Models* (pp. 193-202). IGI Global.

Naeem, M., Khan, A., Rehman, A., Farooq, S., Mehboob, A., Abdali, A. S., & Ahmad, B. (2025). Does Artificial Intelligence With Blockchain Reduce the Costs of the Financial Sector? In *Generative AI for Web Engineering Models* (pp. 147-160). IGI Global.

Naeem, M., Mehboob, A., Shaikh, M. R., & Rehman, A. (2024). Does Green Innovation Linked between Environmental Governance and Firm Performance? A System of Generalised Method of Moments in Pakistan.

Naeem, M., Siraj, M., Ali, S., Rehman, A., & Farooq, S. (2025). The Role of Artificial Intelligence in Risk Management: Practices of the Banking Sector. In *Generative AI for Web Engineering Models* (pp. 83–106). IGI Global.

Nair, N. V., & Nayak, P. K. (2023). Exploring Water Quality as a Determinant of Small-Scale Fisheries Vulnerability. *Sustainability (Basel), 15*(17), 13238. DOI: 10.3390/su151713238

Najafi, H., Shrestha, P. K., Rakovec, O., Apel, H., Vorogushyn, S., Kumar, R., Thober, S., Merz, B., & Samaniego, L. (2024, May 2). High-resolution impact-based early warning system for riverine flooding. *Nature Communications, 15*(1), 3726. Advance online publication. DOI: 10.1038/s41467-024-48065-y PMID: 38698000

National Audit Office. (2022). Environmental compliance and enforcement. https://www.nao.org.uk/wp-content/uploads/2022/05/Environmental-compliance-and-enforcement.pdf

Nearing, G., Cohen, D., Dube, V., Gauch, M., Gilon, O., Harrigan, S., Hassidim, A., Klotz, D., Kratzert, F., Metzger, A., Nevo, S., Pappenberger, F., Prudhomme, C., Shalev, G., Shenzis, S., Tekalign, T., Weitzner, D., & Matias, Y. (2024, Mar 20). Global prediction of extreme floods in ungauged watersheds. *Nature*, *627*(8004), 559–563. DOI: 10.1038/s41586-024-07145-1 PMID: 38509278

Neidlinger, H. H., & Schissel, P. (1986). Polymer glazings for silver mirrors. *Solar Energy Materials*, *14*(3-5), 327–339. DOI: 10.1016/0165-1633(86)90056-0

Nguyen, T. T., Jewik, J., Bansal, H., Sharma, P., & Grover, A. (2023, Jan 1). ClimateLearn: Benchmarking Machine Learning for Weather and Climate Modeling. Cornell University. https://doi.org/DOI: 10.48550/arXiv.2307

Nikalje, G. C., Srivastava, A. K., Pandey, G. K., & Suprasanna, P. (2018). Halophytes in biosaline agriculture: Mechanism, utilization, and value addition. *Land Degradation & Development*, *29*(4), 1081–1095. DOI: 10.1002/ldr.2819

Nimma, D., Devi, O. R., Laishram, B., Ramesh, J. V. N., Boddupalli, S., Ayyasamy, R., Tirth, V., & Arabil, A. (2025). Implications of climate change on freshwater ecosystems and their biodiversity. *Desalination and Water Treatment*, *321*, 100889. DOI: 10.1016/j.dwt.2024.100889

Nofal, O. M., & van de Lindt, J. W. (2020). Understanding flood risk in the context of community resilience modeling for the built environment: Research needs and trends. *Sustainable and Resilient Infrastructure*. Advance online publication. DOI: 10.1080/23789689.2020.1722546

Nogueira, E., Gomes, S., & Lopes, J. M. (2025). Unveiling triple bottom line's influence on business performance. *Discover Sustainability*, *6*(1), 43. DOI: 10.1007/s43621-025-00804-x

Noh, Y., Jafarinejad, S., & Anand, P. (2024). A Review on Harnessing Renewable Energy Synergies for Achieving Urban Net-Zero Energy Buildings: Technologies, Performance Evaluation, Policies, Challenges, and Future Direction. *Sustainability (Basel)*, *16*(8), 3444. DOI: 10.3390/su16083444

Noor, I., Sohail, H., Sun, J., Nawaz, M. A., Li, G., Hasanuzzaman, M., & Liu, J. (2022). Heavy metal and metalloid toxicity in horticultural plants: Tolerance mechanism and remediation strategies. *Chemosphere*, *303*, 135196. DOI: 10.1016/j.chemosphere.2022.135196 PMID: 35659937

Nordgren, A. (2022, Feb 9). Artificial intelligence and climate change: ethical issues. Emerald Publishing Limited, 21(1), 1-15. DOI: 10.1108/JICES-11-2021-0106

O'Neill, M. J. (2010). A model of environmental control and effective work. *Facilities., 28*(3/4), 118–136. DOI: 10.1108/02632771011023104

OECD (Organisation for Economic Co-operation and Development). (2020). Environmental enforcement and compliance: Best practices. OECD Publishing. https://doi.org/DOI: 10.1787/9789264280359-en

Olawade, D. B., Wada, O. Z., David-Olawade, A. C., Fapohunda, O., Ige, A. O., & Ling, J. (2024). Artificial intelligence potential for net zero sustainability: Current evidence and prospects. In *Next Sustainability* (Vol. 4, p. 100041). Elsevier BV., DOI: 10.1016/j.nxsust.2024.100041

Ologunorisa, E. T. (2001) An Assessment of Flood Risk in the Niger Delta, Nigeria. Unpublished Ph. D. Thesis, University of Port-Harcourt, Nigeria 303pp.

Omuto, C. T., Kome, G. K., Ramakhanna, S. J., Muzira, N. M., Ruley, J. A., Jayeoba, O. J., Raharimanana, V., Owusu Ansah, A., Khamis, N. A., Mathafeng, K. K., Elmobarak, A. A., Vargas, R. R., Koetlisi, A. K., Dembele, D., Diawara, M., Mbaikoubou, M., Maria, R. M., Adam Boukary, I., Malatji, A., & Nyamai, M. (2024). Trend of soil salinization in Africa and implications for agro-chemical use in semi-arid croplands. *The Science of the Total Environment, 951*, 175503. DOI: 10.1016/j.scitotenv.2024.175503 PMID: 39147045

Ortiz, A., & Sansinenea, E. (2023). Genetically Modified Plants Based on Bacillus Genes and Commercial Bacillus-Based Biopesticides for Sustainable Agriculture. *Horticulturae, 9*(9), 963. DOI: 10.3390/horticulturae9090963

Oudghiri, M., Yamani, B., Benlemlih, N., El Aammouri, S., Abid, N., Brhadda, N., Bouhassoun, S., Ziri, R., Chriqui, A., Aoujil, F. Z., El Bakkali, M., Mouniane, Y., & Ibriz, M. (2025). Effect of Heavy Metals on the Morphological and Physiological Responses of the Torro Plus Variant of *Zea mays. Journal of Environment and Earth Science, 7*(4), 165–179. DOI: 10.30564/jees.v7i4.8208

Pai, V., & Chandra, S. (2022). Exploring factors influencing organizational adoption of artificial intelligence (AI) in corporate social responsibility (CSR) initiatives. *Pacific Asia Journal of the Association for Information Systems, 14*(5), 4. DOI: 10.17705/1pais.14504

Pakistan Meteorological Department. (2020). Climate Data Processing System.

Pakistan Weather Portal. (2023, Apr 30). Exploring the complexity of El Niño and its effects on rainfall in Pakistan. Retrieved from https://pakistanweatherportal.com/2023/04/30/exploring-the-complexity-of-el-nino-and-its-effects-on-rainfall-in-pakistan/

Pandey, A. K., Zorić, L., Sun, T., Karanović, D., Fang, P., Borišev, M., Wu, X., Luković, J., & Xu, P. (2022). The Anatomical Basis of Heavy Metal Responses in Legumes and Their Impact on Plant–Rhizosphere Interactions. *Plants*, *11*(19), 2554. DOI: 10.3390/plants11192554 PMID: 36235420

Pan, J., Du, L., Wu, H., & Liu, X. (2024). Does environmental law enforcement supervision improve corporate carbon reduction performance? Evidence from environmental protection interview. *Energy Economics*, *132*, 107441. DOI: 10.1016/j.eneco.2024.107441

Pantalona, G., Tsalakanidou, F., Nikolopoulos, S., Kompatsiaris, I., Lombardo, F., Norbiato, D., Ferri, M., Kovats, L., & Haberstock, H. (2021). Decision support system for flood risk reduction policies: The case of a flood protection measure in the area of Vicenza. *Data & Policy*, *3*, e26. DOI: 10.1017/dap.2021.23

Paraskevopoulou, A. T., Kontodaimon Karantzi, A., Liakopoulos, G., Londra, P. A., & Bertsouklis, K. (2020). The Effect of Salinity on the Growth of Lavender Species. *Water (Basel)*, *12*(3), 618. DOI: 10.3390/w12030618

Pasupuleti, V., Thuraka, B., Kodete, C. S., & Malisetty, S. (2024). Enhancing Supply Chain Agility and Sustainability through Machine Learning: Optimization Techniques for Logistics and Inventory Management. *Logistics (Basel)*, *8*(3), 73. DOI: 10.3390/logistics8030073

Pató, Z., & Mandel, T. (2022, Oct 22). Energy Efficiency First in the power sector: incentivising consumers and network companies. Springer Science+Business Media, 15(8). DOI: 10.1007/s12053-022-10062-9

Paul, S. (2018, Jul 4). Managing the environment using GIS. Geospatial World. Retrieved from https://www.geospatialworld.net/article/managing-the-environment-using-gis/

Pearl, J. (2022). Causal Diagrams for Empirical Research (With Discussions). In H. Geffner, R. Dechter, & J. Y. Halpern (Eds.), Probabilistic and Causal Inference (p. 255-316). ACM. DOI: 10.1145/3501714.3501734

Pescheux, A. C., Le Baron, E., & Raccurt, O. (2019). Characterization of different Moroccan sands to explain their potential negative impacts on CSP solar mirrors. *Solar Energy*, *194*, 959–968. DOI: 10.1016/j.solener.2019.11.020

Pissolato, M. D., Martins, T. S., Fajardo, Y. C. G., Souza, G. M., Machado, E. C., & Ribeiro, R. V. (2024). Stress memory in crops: What we have learned so far. *Theoretical and Experimental Plant Physiology*, *23*(3), 535–565. DOI: 10.1007/s40626-024-00315-6

Porter, M. E., & van der Linde, C. (1995). Toward a new conception of the environment-competitiveness relationship. *The Journal of Economic Perspectives*, *9*(4), 97–118. DOI: 10.1257/jep.9.4.97

Prasanna, R. P. I. R., Jayasundara, J. M. S. B., Naradda Gamage, S. K., Ekanayake, E. M. S., Rajapakshe, P. S. K., & Abeyrathne, G. A. K. N. J. (2019). Sustainability of SMEs in the competition: A systemic review on technological challenges and SME performance. *Journal of Open Innovation*, *5*(4), 100. DOI: 10.3390/joitmc5040100

Premavathi, T. (2023, Aug 21). A Comparative Analysis of ARIMA Model and LSTM Network in Predicting COVID-19 Outcomes in India. European Organization for Nuclear Research. DOI: 10.5281/zenodo.8267919

Prudhomme, C. (2002). Hydrological Modelling of the Upper Thames Basin Using the Environment Agency's Precipitation-Runoff Model. *Hydrology and Earth System Sciences*, *6*(2), 193–212.

Psarrou, I., Oreopoulou, A., Tsimogiannis, D., & Oreopoulou, V. (2020). Extraction Kinetics of Phenolic Antioxidants from the Hydro Distillation Residues of Rosemary and Effect of Pretreatment and Extraction Parameters. *Molecules (Basel, Switzerland)*, *25*(19), 4520. DOI: 10.3390/molecules25194520 PMID: 33023142

Qin, H., Wang, Z., Sha, W., Song, S., Qin, F., & Zhang, W. (2024). Role of Plant-Growth-Promoting Rhizobacteria in Plant Machinery for Soil Heavy Metal Detoxification. *Microorganisms*, *12*(4), 700. DOI: 10.3390/microorganisms12040700 PMID: 38674644

Qu, Y., Wang, C., Wei, Y., Wu, L., & Zheng, N. (2024). Does Eco-Innovation of Emerging Market Firms Benefit from Knowledge Spillovers of MNC in a Multi-dimensional Task Environment? *MIR. Management International Review*, *64*(3), 527–565. DOI: 10.1007/s11575-024-00546-4

Raffo, A., Baiamonte, I., De Benedetti, L., Lupotto, E., Marchioni, I., Nardo, N., & Cervelli, C. (2023). Exploring volatile aroma and non-volatile bioactive compounds diversity in wild populations of rosemary (*Salvia rosmarinus* Schleid.). *Food Chemistry*, *404*, 134532. DOI: 10.1016/j.foodchem.2022.134532 PMID: 36242966

Raillani, B., Ouali, H. A. L., Amraqui, S., Moussaoui, M. A., Jeyar, M., & Mezrhab, A. (2019). The Impact of Aerosol Optical Depth (AOD) On Decreasing the Optical Efficiency of CSP Solar Power Plants-A case study: *Eastern Morocco. International Conference on Intelligent Systems and Advanced Computing Sciences (ISACS)*. Taza, Morocco. DOI: 10.1109/ISACS48493.2019.9068909

Rajput, P., Singh, A., Agrawal, S., Ghazaryan, K., Rajput, V. D., Movsesyan, H., Mandzhieva, S., Minkina, T., & Alexiou, A. (2024). Effects of environmental metal and metalloid pollutants on plants and human health: Exploring nano-remediation approach. *Stress Biology*, *4*(1), 27. DOI: 10.1007/s44154-024-00156-y PMID: 38777953

Rana, I. A., & Routray, J. K. (2016, Jul). Risk assessment of flood-prone urban areas of Pakistan. Paper presented at the International Conference on Disaster Management: From Polar Region to the Local Communities, National Institute of Development Administration (NIDA), Bangkok, Thailand.

Randall, M., Sun, F., Zhang, Y., & Jensen, M. B. (2019). Evaluating Sponge City volume capture ratio at the catchment scale using SWMM. *Journal of Environmental Management*, *246*, 745–757. DOI: 10.1016/j.jenvman.2019.05.134 PMID: 31226528

Rane, N. L., Choudhary, S. P., & Rane, J. (2024). Artificial Intelligence and machine learning in renewable and sustainable energy strategies: A critical review and future perspectives. *SSRN*, *2*(3), 80–102. DOI: 10.2139/ssrn.4838761

Rao, M. J., & Zheng, B. (2025). The Role of Polyphenols in Abiotic Stress Tolerance and Their Antioxidant Properties to Scavenge Reactive Oxygen Species and Free Radicals. *Antioxidants*, *14*(1), 74. DOI: 10.3390/antiox14010074 PMID: 39857408

Rashid, A., Biswas, P., Biswas, A., Nasim, M A A., Gupta, K. D., & George, R. (2024, Jun 22). Present and Future of AI in Renewable Energy Domain: A Comprehensive Survey. Cornell University. https://doi.org/DOI: 10.48550/arXiv.2406

Raza, A. (2024). Climate Change Impacts on Crop Productivity and Food Security: An Overview. In Kanga, S., Singh, S. K., Shevkani, K., Pathak, V., & Sajan, B. (Eds.), *Transforming Agricultural Management for a Sustainable Future. World Sustainability Series*. Springer., DOI: 10.1007/978-3-031-63430-7_8

Rehman, S., Yang, J., Zhang, J., Zhang, L., Hao, X., Song, R., Chen, S., Wang, G., & Hua, L. (2025). Salt stress in wheat: A physiological and genetic perspective. *Plant Stress*, *16*, 100832. DOI: 10.1016/j.stress.2025.100832

Renn, O. (2008) White Paper on Risk Governance: Toward an Integrative Framework//Global Risk Governance; Springer: Dordrecht, The Netherlands, pp. 3–73.

Riyazuddin, R., Nisha, N., Ejaz, B., Khan, M. I. R., Kumar, M., Ramteke, P. W., & Gupta, R. (2022). A Comprehensive Review on the Heavy Metal Toxicity and Sequestration in Plants. *Biomolecules, 12*(1), 43. DOI: 10.3390/biom12010043 PMID: 35053191

Rizvi, A., Ahmed, B., Khan, M. S., Rajput, V. D., Umar, S., Minkina, T. M., & Lee, J. (2022). Maize associated bacterial microbiome linked mitigation of heavy metal stress: A multidimensional detoxification approach. *Environmental and Experimental Botany, 200*, 104911. DOI: 10.1016/j.envexpbot.2022.104911

Roberts, O. (2025). A Qualitative Exploration on Artificial Intelligence and Renewable Energy Integration in Supply Chains.

Roberts, H., Hine, E., Taddeo, M., & Floridi, L. (2023, Jan 1). Global AI governance: barriers and pathways forward. RELX Group (Netherlands). DOI: 10.2139/ssrn.4588040

Rolnick, D., Donti, P. L., Kaack, L. H., Kochanski, K., Lacoste, A., Sankaran, K., Ross, A. S., Milojevic-Dupont, N., Jaques, N., Waldman-Brown, A., Luccioni, A., Maharaj, T., Sherwin, E. D., Mukkavilli, S. K., Körding, K., Gomes, C. P., Ng, A. Y., Hassabis, D., Platt, J., & Bengio, Y. (2022, Feb 7). Tackling Climate Change with Machine Learning. *ACM Computing Surveys, 55*(2), 1–96. DOI: 10.1145/3485128

Rosenbaum, P. R., & Rubin, D. B. (1983). The central role of the propensity score in observational studies for causal effects. *Biometrika, 70*(1), 41–55. DOI: 10.1093/biomet/70.1.41

Ruggerio, C. A. (2021). Sustainability and sustainable development: A review of principles and definitions. *The Science of the Total Environment, 786*, 147481. DOI: 10.1016/j.scitotenv.2021.147481 PMID: 33965820

Rumelhart, D. E., Hinton, G. E., & Williams, R. J. (1986). Learning representations by back-propagating errors. *Nature, 323*(6088), 533–536. DOI: 10.1038/323533a0

Safdar, M., Shahid, M. A., Zaman, M., Rasul, F., Muzammal, H., Raza, A., Sabir, R. M., & Zafar, U. (2023). Drought Monitoring with Multiple Indices and Management through Various Techniques: A Review. *Engineering Proceedings, 56*(1), 307. DOI: 10.3390/ASEC2023-16602

Seyyedi, S. R., Kowsari, E., Gheibi, M., Chinnappan, A., & Ramakrishna, S. (2024). A comprehensive review integration of digitalization and circular economy in waste management by adopting artificial intelligence approaches: Towards a simulation model. *Journal of Cleaner Production, 460*, 142584. DOI: 10.1016/j.jclepro.2024.142584

Shabbir, R., Javed, T., Afzal, I., Sabagh, A. E., Ali, A., Vicente, O., & Chen, P. (2021). Modern Biotechnologies: Innovative and Sustainable Approaches for the Improvement of Sugarcane Tolerance to Environmental Stresses. *Agronomy (Basel)*, *11*(6), 1042. DOI: 10.3390/agronomy11061042

Shah, I. A., Jhanjhi, N. Z., & Ashraf, H. (2024). Logistics With the Internet of Things: Challenges, Perspectives, and Applications. In *Navigating Cyber Threats and Cybersecurity in the Logistics Industry* (pp. 172-195). IGI Global.

Shah, I. A., Jhanjhi, N. Z., & Brohi, S. (2024). Cybersecurity issues and challenges in civil aviation security. *Cybersecurity in the Transportation Industry*, 1-23.

Shah, I. A., Jhanjhi, N., & Ujjan, R. M. A. (2024). Industry 4.0: use of digitalization in healthcare. In *Advances in Computational Intelligence for the Healthcare Industry 4.0* (pp. 174-193). IGI Global.

Shah, I. A., Jhanjhi, N., & Brohi, S. N. (2024). Use of Emerging Technologies in Healthcare 4.0. *Advances in Medical Technologies and Clinical Practice*, *4*(0), 280–296. DOI: 10.4018/979-8-3693-2333-5.ch015

Shah, I. A., Laraib, A., & Hussain, F. (2024). Logistics Industry in the Context of the Blockchain Technology. In *Navigating Cyber Threats and Cybersecurity in the Logistics Industry* (pp. 214–235). IGI Global. DOI: 10.4018/979-8-3693-3816-2.ch008

Shah, I. A., & Sial, Q. (2024). *Advances in Computational Intelligence for the Healthcare Industry 4.0*. IGI Global. DOI: 10.4018/979-8-3693-2333-5

Shahzad, A., Ullah, S., Dar, A. A., Sardar, M. F., Mehmood, T., Tufail, M. A., Shakoor, A., & Haris, M. (2021). Nexus on climate change: Agriculture and possible solution to cope future climate change stresses. *Environmental Science and Pollution Research International*, *28*(12), 14211–14232. DOI: 10.1007/s11356-021-12649-8 PMID: 33515149

Shakespear, S., Sivaji, M., & Kumar, V. (2024). Navigating Through Harsh Conditions: Coordinated Networks of Plant Adaptation to Abiotic Stress. *Journal of Plant Growth Regulation*. Advance online publication. DOI: 10.1007/s00344-023-11224-4

Shakri, T., Che-Othman, M. H., Md Isa, N., Sukiran, N. L., & Zainal, Z. (2022). Morpho-physiological and stress-related gene expression of rice varieties in response to salinity stress at early vegetative stage. *Agriculture*, *12*(5), 638. DOI: 10.3390/agriculture12050638

Shang, C., Zhou, Q., Nkoh, J. N., Liu, J., Wang, J., Hu, Z., & Hussain, Q. (2024). Integrated physiological, biochemical, and transcriptomic analyses of *Bruguiera gymnorhiza* leaves under long-term copper stress: Stomatal size, wax crystals and composition. *Ecotoxicology and Environmental Safety*, *281*, 116609. DOI: 10.1016/j.ecoenv.2024.116609 PMID: 38905937

Shokri, N., Hassani, A., Sahimi, M. (2024). Multi-scale soil salinization dynamics from global to pore scale: A review. Reviews of Geophysics, 62(4), e2023RG000804. DOI: 10.1029/2023RG000804

Silva, H. F., Silva, N. F., Oliveira, C. M., & Matos, M. J. (2021). Heavy Metals Contamination of Urban Soils—A Decade Study in the City of Lisbon, Portugal. *Soil Systems*, *5*, 27. DOI: 10.3390/soilsystems5020027

Singh, A., Rajput, V. D., Lalotra, S., Agrawal, S., Ghazaryan, K., Singh, J., Minkina, T., Rajput, P., Mandzhieva, S., & Alexiou, A. (2024). Zinc oxide nanoparticles influence on plant tolerance to salinity stress: Insights into physiological, biochemical, and molecular responses. *Environmental Geochemistry and Health*, *46*(5), 148. DOI: 10.1007/s10653-024-01921-8 PMID: 38578547

Singh, K. M., Baksi, S., Rani, S., Jha, A. B., Dubey, R. S., & Sharma, P. (2024). NanoBoost: Maximizing crop resilience and yield via nanopriming under salt stress. *Environmental and Experimental Botany*, *226*, 105937. DOI: 10.1016/j.envexpbot.2024.105937

Soni, S., Jha, A. B., Dubey, R. S., & Sharma, P. (2024). Nanowonders in agriculture: Unveiling the potential of nanoparticles to boost crop resilience to salinity stress. *The Science of the Total Environment*, *925*, 171433. DOI: 10.1016/j.scitotenv.2024.171433 PMID: 38458469

Srinivas, H., & Nakagawa, K. (2015). Urbanization and Informalization of Waste Management in Asian Developing Countries: Challenges and Opportunities in the Twenty-First Century. *Habitat International*, *47*, 22–31.

Stahl, B C., Andreou, A G., Brey, P., Hatzakis, T., Кириченко, A M., Macnish, K., Shaelou, S L., Patel, A., Ryan, M., & Wright, D. (2020, Dec 16). Artificial intelligence for human flourishing – Beyond principles for machine learning. Elsevier BV, 124, 374-388. DOI: 10.1016/j.jbusres.2020.11.030

Stavins, R. N. (2003). Market-based environmental policies: What can we learn from U.S. experience (and related research)? Resources for the Future. https://doi.org/DOI: 10.2139/ssrn.421720

Suarez, P., Anderson, W., Mahal, V., & Lakshmanan, T. R. (2005). Impacts of flooding and climate change on urban transportation: A systemwide per-formance assessment of the Boston Metro Area. *Transportation Research Part D, Transport and Environment, 10*(3), 231–244. DOI: 10.1016/j.trd.2005.04.007

Su, B., Huang, H., & Zhang, N. (2015). Dynamic risk assessment method for urban waterlogging based on scenario simulation. *J. Tsinghua Univ. Sci. Technol., 55*, 684–690.

Surówka, E., Grigore, M. N., Ślesak, I., Dubert, F., & Miszalski, Z. (2025). Halophytes in the Bioeconomy: New Opportunity and Challenge for Agriculture. In Arora, J., Grigore, M. N., & Joshi, A. (Eds.), *Sustainable Utilisation and Bioengineering of Halophytes. Sustainable Development and Biodiversity* (Vol. 38). Springer. DOI: 10.1007/978-981-97-9819-3_16

Sutter, F., Wette, J., Wiesinger, F., Fernández-García, A., Ziegler, S., & Dasbach, R. (2018). Lifetime prediction of aluminum solar mirrors by correlating accelerated aging and outdoor exposure experiments. *Solar Energy, 174*, 149–163. DOI: 10.1016/j.solener.2018.09.006

Taddeo, M., Tsamados, A., Cowls, J., & Floridi, L. (2021). Artificial intelligence and the climate emergency: Opportunities, challenges, and recommendations. In One Earth, *4*(6), p. 776). Elsevier BV. DOI: 10.1016/j.oneear.2021.05.018

Tang, H., Du, L., Xia, C., & Luo, J. (2024). Bridging gaps and seeding futures: A synthesis of soil salinization and the role of plant-soil interactions under climate change. *iScience, 27*(9), 110804. DOI: 10.1016/j.isci.2024.110804 PMID: 39286506

Tarroum, M., Ben Romdhane, W., Ali, A. A. M., Al-Qurainy, F., Al-Doss, A., Fki, L., & Hassairi, A. (2021). Harnessing the Rhizosphere of the Halophyte Grass Aeluropus littoralis for Halophilic Plant-Growth-Promoting Fungi and Evaluation of Their Biostimulant Activities. *Plants, 10*(4), 784. DOI: 10.3390/plants10040784 PMID: 33923476

Tazerji, S. S., Nardini, R., Safdar, M., Shehata, A. A., & Duarte, P. M. (2022). An Overview of Anthropogenic Actions as Drivers for Emerging and Re-Emerging Zoonotic Diseases. *Pathogens (Basel, Switzerland), 11*(11), 1376. DOI: 10.3390/pathogens11111376 PMID: 36422627

Tessema, N., Yadeta, D., Kebede, A., & Ayele, G. T. (2023). Soil and Irrigation Water Salinity, and Its Consequences for Agriculture in Ethiopia: A Systematic. *Agriculture, 13*(1), 109. DOI: 10.3390/agriculture13010109

Tofah, M. L., Mseddi, K., Al-Abbasi, O. K., Ben Yazid, A., Khechine, A., Gdoura, R., & Khannous, L. (2023). A New Lavender (*Lavandula multifida* L.) Ecotype from Arid Tunisia, with Differential Essential Oil Composition and Higher Antimicrobial Potential. *Life (Chicago, Ill.)*, *13*(1), 103. DOI: 10.3390/life13010103 PMID: 36676052

Tok, Y C., & Chattopadhyay, S. (2023, Mar 15). Identifying threats, cybercrime and digital forensic opportunities in Smart City Infrastructure via threat modeling. Elsevier BV, 45, 301540-301540. DOI: 10.1016/j.fsidi.2023.301540

Tursunalieva, A., Alexander, D. L. J., Dunne, R., Li, J., Riera, L., & Zhao, Y. (2024). Making Sense of Machine Learning: A Review of Interpretation Techniques and Their Applications. *Applied Sciences (Basel, Switzerland)*, *14*(2), 496. DOI: 10.3390/app14020496

Uddin, M. M., Zakeel, M. C. M., Zavahir, J. S., Marikar, F. M. M. T., & Jahan, I. (2021). Heavy Metal Accumulation in Rice and Aquatic Plants Used as Human Food: A General Review. *Toxics*, *9*(12), 360. DOI: 10.3390/toxics9120360 PMID: 34941794

Ullah, K., & Zhang, J. (2020). GIS-based flood hazard mapping using relative frequency ratio method: A case study of Panjkora River Basin, eastern Hindu Kush, Pakistan. *PLoS One*, *15*(3), e0229153. DOI: 10.1371/journal.pone.0229153 PMID: 32210424

UNCHS. (1981) Settlement Planning for Disasters, Nairobi, World Meteorological Organization (W.M.O) (1975) Drought and Agriculture.

UNEP (United Nations Environment Programme). (2022). *Strengthening environmental law compliance and enforcement*. UNEP Publishing.

Van der Maaten, L., & Hinton, G. (2008). Visualizing data using t-SNE. *Journal of Machine Learning Research*, *9*, 2579–2605.

Vlahos, N., Berillis, P., Levizou, E., Patsea, E., Panteli, N., Demertzioglou, M., Morfesis, K., Voudouri, G., Krigas, N., Kormas, K., Antonopoulou, E., & Mente, E. (2023). Investigating Salinity Effects in Brackish Aquaponics Systems: Evidencing the Co-Cultivation of the Halophyte *Crithmum maritimum* with the Euryhaline *Sparus aurata*. *Applied Sciences (Basel, Switzerland)*, *13*(6), 3385. DOI: 10.3390/app13063385

Wang, D., Bai, L., Huang, X., Yan, W., & Li, S. (2022). Size-dependent acute toxicity and oxidative damage caused by cobalt-based framework (ZIF-67) to *Photobacterium phosphoreum*. *The Science of the Total Environment*, *851*, 158317. DOI: 10.1016/j.scitotenv.2022.158317 PMID: 36037900

Wang, X., Liu, X., Su, Y., & Shen, H. (2025). Rice Responses to Abiotic Stress: Key Proteins and Molecular Mechanisms. *International Journal of Molecular Sciences*, *26*(3), 896. DOI: 10.3390/ijms26030896 PMID: 39940666

Wang, Y. L., Lee, Y. H., Chou, C. L., Chang, Y. S., Liu, W. C., & Chiu, H. W. (2024). Oxidative stress and potential effects of metal nanoparticles: A review of biocompatibility and toxicity concerns. *Environmental Pollution*, *346*, 123617. DOI: 10.1016/j.envpol.2024.123617 PMID: 38395133

Wang, Z., Wang, T., Zhang, X., Wang, J., Yang, Y., Sun, Y., Guo, X., Wu, Q., Nepovimova, E., Watson, A. E., & Kuca, K. (2024). Biodiversity conservation in the context of climate change: Facing challenges and management strategies. *The Science of the Total Environment*, *937*, 173377. DOI: 10.1016/j.scitotenv.2024.173377 PMID: 38796025

Waqar, M., Bhatti, I., & Khan, A. H. (2024). AI-powered automation: Revolutionizing industrial processes and enhancing operational efficiency. *Revista de Inteligencia Artificial en Medicina*, *15*(1), 1151–1175.

Waqas, W., Yuan, Y., Ali, S., Zhang, M., Shafiq, M., Ali, W., Chen, Y., Xiang, Z., Chen, R., Ikhwanuddin, M., & Ma, H. (2024). Toxic effects of heavy metals on crustaceans and associated health risks in humans: A review. *Environmental Chemistry Letters*, *22*(3), 1391–1411. DOI: 10.1007/s10311-024-01717-3

Wette, J., Sutter, F., Fernández-García, A., Ziegler, S., & Dasbach, R. (2016). Comparison of Degradation on Aluminum Reflectors for Solar Collectors due to Outdoor Exposure and Accelerated Aging. *Energies*, *9*(11), 916. DOI: 10.3390/en9110916

Wiesinger, F., Sutter, F., Fernández-García, A., Reinhold, J., & Pitz-Paal, R. (2016). Sand erosion on solar reflectors: Accelerated simulation and comparison with field data. *Solar Energy Materials and Solar Cells*, *145*, 303–313. DOI: 10.1016/j.solmat.2015.10.036

Wiesinger, F., Sutter, F., Fernández-García, A., Wette, J., & Hanrieder, N. (2021). Sandstorm erosion on solar reflectors: A field study on height and orientation dependence. *Energy*, *217*, 119351. DOI: 10.1016/j.energy.2020.119351

Wiesinger, F., Sutter, F., Fernández-García, A., Wette, J., Wolfertstetter, F., Hanrieder, N., Schmücker, M., & Pitz-Paal, R. (2020). Sandstorm erosion on solar reflectors: Highly realistic modeling of artificial aging tests based on advanced site assessment. *Applied Energy*, *268*, 114925. DOI: 10.1016/j.apenergy.2020.114925

Wiesinger, F., Sutter, F., Wolfertstetter, F., Hanrieder, N., Fernández-García, A., Pitz-Paal, R., & Schmücker, M. (2018). Assessment of the erosion risk of sandstorms on solar energy technology at two sites in Morocco. *Solar Energy*, *162*, 217–228. DOI: 10.1016/j.solener.2018.01.004

Wiesinger, F., Sutter, F., Wolfertstetter, F., Hanrieder, N., Wette, J., Fernández-García, A., & Pitz-Paal, R. (2017). Sandstorm erosion simulation on solar mirrors and comparison with field data. *AIP Conference Proceedings*, *1850*, 130014. DOI: 10.1063/1.4984508

Wu, X., Wang, Z., Guo, S., Liao, W., Zeng, Z., & Chen, X. (2017). Scenario-based projections of future urban inundation within a coupled hydrodynamic model framework: A case study in Dongguan City, China. *Journal of Hydrology (Amsterdam)*, *547*, 428–442. DOI: 10.1016/j.jhydrol.2017.02.020

Xin, Z., Zhang, H., & Shi, J. (2022, Sep 1). Application Based on Artificial Intelligence in Substation Operation and Maintenance Management. *Computational Intelligence and Neuroscience*, *2022*, 1–12. DOI: 10.1155/2022/7509532

Yan, J., & Wang, X. (2022). Unsupervised and semi-supervised learning : The next frontier in machine learning for plant systems biology. *The Plant Journal*, *111*(6), 1527–1538. DOI: 10.1111/tpj.15905 PMID: 35821601

Ye, Y., Medina-Velo, I. A., Cota-Ruiz, K., Moreno-Olivas, F., & Gardea-Torresdey, J. L. (2019). Can abiotic stresses in plants be alleviated by manganese nanoparticles or compounds? *Ecotoxicology and Environmental Safety*, *184*, 109671. DOI: 10.1016/j.ecoenv.2019.109671 PMID: 31539809

Zaim, S., Bourrach, E., & El Alami, H. (2021). Heavy metal pollution in agricultural soils near industrial areas in Morocco. *Environmental Monitoring and Assessment*, *193*(4), 208. PMID: 33755795

Zamani, H., Díaz, F., Dehghani, M., Metzler, D., & Bendersky, M. (2022, Jul 6). Retrieval-Enhanced Machine Learning. https://doi.org/10.114

Zeng, Z., Lai, C., Wang, Z., Wu, X., Huang, G., & Hu, Q. (2020). Rapid simulation of urban rainstorm and flood based on WCA2D and SWMM models. *Adv. Water Sci.*, *31*, 29–38.

Zhang, S., Zhang, M., Qiao, Y., Li, X., & Li, S. (2022). Does improvement of environmental information transparency boost firms' green innovation? Evidence from the air quality monitoring and disclosure program in China. *Journal of Cleaner Production*, *357*, 131921. DOI: 10.1016/j.jclepro.2022.131921

Zhang, W. P., Surigaoge, S., Yang, H., Yu, R. P., Wu, J. P., Xing, Y., Chen, Y., & Li, L. (2024). Diversified cropping systems with complementary root growth strategies improve crop adaptation to and remediation of hostile soils. *Plant and Soil*, *502*, 7–30. DOI: 10.1007/s11104-023-06464-y

Zhang, X., Ibrahim, Z., Khaskheli, M. B., Raza, H., Zhou, F., & Shamsi, I. H. (2024). Integrative Approaches to Abiotic Stress Management in Crops: Combining Bioinformatics Educational Tools and Artificial Intelligence Applications. *Sustainability (Basel)*, *16*(17), 7651. DOI: 10.3390/su16177651

Zhao, W., Chen, J., & Feng, L. (2023). Heavy metal-induced oxidative stress and its implications for sustainable crop production. *Plant Physiology and Biochemistry*, *190*, 85–97.

Zheng, J., Zhang, C., Li, Y., You, Q., Liu, J., Wang, L., & Wu, Y. (2019). Research on prevention and control of waterlogging in subway stations based on InfoWorks ICM model. *J. Beijing Norm. Univ. Nat. Sci.*, *55*, 648–655.

Zhu, Y. X., Gong, H. J., & Yin, J. L. (2019). Role of silicon in mediating salt tolerance in plants: A review. *Plants*, *8*(6), 147. DOI: 10.3390/plants8060147 PMID: 31159197

Zidan, K., Mandi, L., Hejjaj, A., Ouazzani, N., & Assabbane, A. (2024). Soil fertility and agro-physiological responses of maize (*Zea mays*) irrigated by treated domestic wastewater by hybrid multi-soil-layering technology. *Journal of Environmental Management*, *351*, 119802. DOI: 10.1016/j.jenvman.2023.119802 PMID: 38134502

Ziogas, V., Tanou, G., Morianou, G., & Kourgialas, N. (2021). Drought and salinity in citriculture: Optimal practices to alleviate salinity and water stress. *Agronomy (Basel)*, *11*(7), 1283. DOI: 10.3390/agronomy11071283

Zong, Z., & Guan, Y. (2024). AI-driven intelligent data analytics and predictive analysis in Industry 4.0: Transforming knowledge, innovation, and efficiency. *Journal of the Knowledge Economy*, 1–40. DOI: 10.1007/s13132-024-02001-z

About the Contributors

Mohamed Bouzelmad is a research scholar in the Laboratory of Physics, Energy and Information Processing, Polydisciplinary Faculty of Ouarzazate, University Ibn Zohr, Agadir-Morocco. His research areas are Solar Energy, Heat transfer, Engineering, and Energy.

Manoj Kumar is a dedicated software engineering graduate and silver medalist from Indus University. With a strong academic background and a passion for innovative technologies, I have consistently demonstrated excellence in both academics and professional endeavors. Currently, I am working as a Senior Software Engineer at a reputed software house, where I focus on developing robust software solutions and contributing to cutting-edge research in the field of software engineering. My interests include system architecture, data analytics, and emerging technologies that drive digital transformation.

El Bakali Mohamed is an Associate Professor at the Higher Institute of Nursing Professions and Health Techniques (ISPITS) in Kénitra, Morocco. He holds a Ph.D. in Epidemiology and Biostatistics and has made significant scholarly contributions in maternal and child health, public health, and health biology. A specialist in Data Science, he has supervised numerous research projects across undergraduate, master's, and doctoral levels. His current teaching responsibilities encompass advanced courses in Epidemiology, Biostatistics, and Research Methodology, serving diverse health disciplines. Actively engaged in multidisciplinary research collaborations, Dr. El-Bakali is dedicated to enhancing health education quality and advocating evidence-based healthcare practices.

Index

L

M

N

P

R

S

Z

www.ingramcontent.com/pod-product-compliance
Lightning Source LLC
LaVergne TN
LVHW081331300926
849161LV00008B/1941